INFORMATION TECHNOLOGY, PLANT PATHOLOGY AND BIODIVERSITY

Information Technology, Plant Pathology and Biodiversity

Edited for the British Society for Plant Pathology and the Systematics Association by

Paul Bridge
International Mycological Institute, Egham, Surrey, UK

Peter Jeffries
Research School of Biosciences, University of Kent, Canterbury, UK

David R. Morse
Computing Laboratory, University of Kent, Canterbury, UK

and

Peter R. Scott
Information Institute, CAB INTERNATIONAL, Wallingford, UK

CAB INTERNATIONAL
in association with the British Society for Plant Pathology and the Systematics Association

New India Publishing Agency
New Delhi - 110088, (INDIA)

First Indian Print: 2005

New India Publishing Agency
101, Vikas Surya Plaza, CU Block, L.S.C. Mkt.,
Pitam Pura, New Delhi - 110 088, (INDIA)
Phone: 011-27341717, Fax: 011-27341616
E-mail: spjain_niph@rediffmail.com
Web: www.bookfactoryindia.com

ISBN: 81-89422-08-1

Printed at :
Salasar Imaging Systems
Lawrence Road, New Delhi

Contents

Part Five – Computer-based Species Identification

Part Six – Applications of Computer-based Species Identification

Part Seven – Passing on Knowledge in Education and Training

Contributors

R. Allkin, 'Plantas do Nordeste', Centre for Economic Botany, Royal Botanic Gardens, Kew, Richmond, Surrey TW9 3AE, UK

J.E. Anderson, BIOSIS, 2100 Arch Street, Philadelphia, PA 19103–1399, USA

R. Ausher, Ministry of Agriculture and Rural Development, Extension Service, Department of Crop Protection, PO Box 7054, Tel-Aviv 61070, Israel

F.A. Bisby, Biodiversity Informatics Research Group, School of Biological Sciences, University of Southampton, Southampton SO16 7PX, UK. Present address: Department of Botany, School of Plant Sciences, PO Box 221, University of Reading, Reading RG6 6AS, UK

S. Blackmore, The Natural History Museum, Cromwell Road, London SW7 5BD, UK

P. Blaise, Institute of Plant Sciences, Section Phytomedicine/Pathology, Swiss Federal Institute of Technology, Universitätstr. 2, CH 8092 Zürich, Switzerland

L. Boddy, School of Pure and Applied Biology, University of Wales, PO Box 915, Cardiff CF1 3TL, UK

P. Bridge, International Mycological Institute, Bakeham Lane, Egham, Surrey TW20 9TY, UK

T.N. Bryant, Medical Statistics and Computing, University of Southampton, Southampton General Hospital, Southampton SO16 6YD, UK

R. Campbell, Blackwell Science Ltd, Osney Mead, Oxford OX2 0EL, UK

E.D. Chesmore, Environmental Electronics Research Group, Department of Electronic Engineering, University of Hull, Hull HU6 7RX, UK

P. Cochrane, BT Laboratories, Martlesham Heath, Ipswich, Suffolk IP5 7RE, UK

M.J. Dallwitz, CSIRO Division of Entomology, GPO Box 1700, Canberra, ACT 2601, Australia

J. Diederich, Department of Mathematics, University of California, Davis, CA 95616, USA

M. Edwards, Computing Laboratory, University of Kent, Canterbury, Kent CT2 7NF, UK and School of Sciences, University of Buckingham, Buckingham MK18 1EG, UK. Present address: 33 George Street, Berkhamsted, Herts HP4 2EG, UK

O.P. Femminella, Environmental Electronics Research Group, Department of Electronic Engineering, University of Hull, Hull HU6 7RX, UK

R. Fortuner, 4 rue des Jardins, 17130 Montendre, France

M.J. Jeger, Department of Phytopathology, Wageningen Agricultural University, POB 8025, 6700EE Wageningen, The Netherlands

P. Jones, Agricultural and Biological Department, University of Florida, PO Box 110570, Rogers Hall, Gainesville, FL 32611–0570, USA

S.B. Jones, CAB INTERNATIONAL, Wallingford, Oxon OX10 8DE, UK

J. Lebbe, Laboratoire Organisation et Evolution des Systèmes, Université Pierre et Marie Curie, 4 Place Jussieu, 75252 Paris Cedex 05, France

W.H. Los, Zoological Museum Amsterdam, University of Amsterdam, Mauritskade 61, 1092 AD Amsterdam, The Netherlands

A. Marshall, ITpro AG, R-1008.5.11. CH.4002 Basel, Switzerland

S.S. M'Boob, FAO Regional Office for Africa, PO Box 1628, Accra, Ghana. Present address: FAO Representative, Box 2, Dar es Salaam, Tanzania

A. McLean-Inglis, Blackwell Science Ltd, Osney Mead, Oxford OX2 0EL, UK

J. Milton, Department of Mathematics, University of California, Davis, CA 95616, USA

A. Morgan, School of Pure and Applied Biology, University of Wales, PO Box 915, Cardiff CF1 3TL, UK

C.W. Morris, Department of Computer Studies, University of Glamorgan, Pontypridd CF37 1DL, UK

D.R. Morse, Computing Laboratory, University of Kent at Canterbury, Canterbury, Kent CT2 7NF, UK

J.D. Mumford, Centre for Environmental Technology, Imperial College of Science, Technology and Medicine, Silwood Park, Ascot, Berks SL5 7PY, UK

N.S. Murali, Danish Institute for Plant and Soil Science, Department of Plant Pathology and Pest Management, Lottenborgvej 2, DK-2800 Lyngby, Denmark

M. Murguía, Asociación de Biólogos Amigos de la Computación, A.C. 28 de Agosto No. 32 Col. Escandón, 11870 México D.F., Mexico

S. Nagarajan, Directorate of Wheat Research, Indian Council of Agricultural Research, Post Box 158, Kunjpura Road, Karnal 132 001, Haryana, India

A.C. Newton, Scottish Crop Research Institute, Invergowrie, Dundee DD2 5DA, UK

G. Norton, Cooperative Research Centre for Tropical Pest Management, Gehrmann Laboratories, University of Queensland, Brisbane, Qld 4072, Australia

T.A. Paine, CSIRO, Division of Entomology, GPO Box 1700, Canberra, ACT 2601, Australia

R.J. Pankhurst, Royal Botanic Gardens Edinburgh, 20A Inverleith Row, Edinburgh EH3 5LR, UK

R.W. Payne, Statistics Department, IACR-Rothamsted, Harpenden, Herts AL5 2JQ, UK

S.A. Rudgard, CAB INTERNATIONAL, Wallingford, Oxon OX10 8DE, UK

G.R. Sandlant, International Institute of Entomology, 56 Queen's Gate, London SW7 5JR, UK

P.H. Schalk, ETI, University of Amsterdam, Mauritskade 61, 1092 AD Amsterdam, The Netherlands

C.Y.L. Schotman, CAB INTERNATIONAL, Wallingford, Oxon OX10 8DE, UK

G.L. Schumann, Department of Microbiology, University of Massachusetts, Fernald Hall, Amherst, MA 01003, USA

P.R. Scott, Information Institute, CAB INTERNATIONAL, Wallingford, Oxon OX10 8DE, UK

B.J.M. Secher, Danish Institute for Plant and Soil Science, Department of Plant Pathology and Pest Management, Lottenborgvej 2, DK 2800 Lyngby, Denmark

Z.R. Shen, Department of Plant Protection, College of Plant Science and Technology, China Agricultural University, 2 Yuanmingyuan Xilu, Beijing 100094, China

I.M. Smith, European and Mediterranean Plant Protection Organization, 1 rue Le Nôtre, 75016 Paris, France

P.H.A. Sneath, Department of Microbiology and Immunology, University of Leicester, Leicester LE1 9HN, UK

T.M. Stewart, Department of Plant Science, Massey University, Palmerston North, New Zealand

M.D. Swarbrick, Environmental Electronics Research Group, Department of Electronic Engineering, University of Hull, Hull HU6 7RX, UK

A. Sweetmore, CAB INTERNATIONAL, Wallingford, Oxon OX10 8DE, UK. Present address: 15A Benson Lane, Crowmarsh Gifford, Oxon OX10 8ED, UK

G.M. Tardivel, Computing Laboratory, University of Kent at Canterbury, Canterbury, Kent CT2 7NF, UK

R. Vignes, Laboratoire Organisation et Évolution des Systèmes, Université Pierre et Marie Curie, 4 Place Jussieu, 75252 Paris Cedex 05, France

J.L. Villaseñor, Instituto de Biologia, UNAM Departamento de Botánica, Apartado Postal 70–367, 04510 México D.F., Mexico

I.M. White, International Institute of Entomology, 56 Queen's Gate, London SW7 5JR, UK

C. Winter, BT Laboratories, Martlesham Heath, Ipswich, Suffolk IP5 7RE, UK

B.C. Zhang, CAB INTERNATIONAL, Wallingford, Oxon OX10 8DE, UK. Present address: 219 Beatrice Street, Toronto, Ontario, M6G 3E9 Canada

E.J. Zurcher, CSIRO, Division of Entomology, GPO Box 1700, Canberra, ACT 2601, Australia

Preface

1996 was the year in which the British Society for Plant Pathology (BSPP) took several initiatives in handling information electronically, including:

- Development of the BSPPWeb World Wide Web site, to keep Members and others informed of the Society's activities, and to provide electronic links with the global community of plant pathologists on the Web.
- Launch of the Web journal *Molecular Plant Pathology On-Line.*
- Parallel publication of the printed journal *Plant Pathology* in electronic form.
- Organization of the Presidential Conference on *Unlocking the Future – Information Technology in Plant Pathology.*

1996 was also the 21st anniversary of the publication of the Systematics Association volume *Biological Identification with Computers* (ed. R. Pankhurst), which had been produced from an earlier Symposium in Cambridge. The Systematics Association decided to mark the occasion by organizing an anniversary meeting on Computer-based Species Identification.

BSPP and the Systematics Association recognized the opportunity to join forces and develop a combined programme for a Conference on the twin themes:

- *Unlocking the Future –Information Technology in Plant Pathology*
- *Computer-based Species Information*

This Conference, held at the University of Kent at Canterbury, 16–19 December 1996, provided the resource from which this book was developed.

The Organizing Committee of the Conference became the editors of the book. We greatly appreciate the cooperation of speakers in making their presentations available for publication to a tight schedule. Recognizing what Peter Scott's Presidential Address calls 'the incredible pace of change', we decided it was essential to work fast to provide a snapshot picture of our subject at the beginning of 1997. In doing this we recognize, with excitement, that the picture will date rapidly. We aim therefore to provide a baseline to which future development can be referred.

The Conference included an open forum at which two public figures, a biologist and an information technologist, took a look at 'The Road Ahead' for the interface between their disciplines. The book closes with their contributions.

The programme of the Conference remains accessible from the BSPPWeb site. This is linked to the abstracts, and to a vast array of further links to related information on the World Wide Web. This remarkable resource has an ongoing dynamic of its own, and we suggest that readers regard it as complementary to this book. It can be accessed from

http://www.bspp.org.uk/meeting/dec96con.htm

We thank the many people who supported the organization of the Conference and the compilation of this book, especially Christine Davies for much needed secretarial support.

Paul Bridge, Systematics Association
Peter Jeffries, University of Kent at Canterbury
David Morse, University of Kent at Canterbury
Peter Scott, British Society for Plant Pathology

April 1997
CAB INTERNATIONAL, Wallingford OX10 8DE
Fax: +44 (0)1491 833508/E-mail: p.scott@cabi.org

The Incredible Pace of Change: Information Technology in Support of Plant Pathology

P.R. Scott

Information Institute, CAB INTERNATIONAL, Wallingford, Oxon OX10 8DE, UK
Fax: +44 (0)1491 833508/E-mail: p.scott@cabi.org

The Incredible Pace of Change

Technophile or technophobe, we are all exposed to the incredible pace of technological change. The year before last, I gave talks from acetates written in coloured pens and laid on an overhead projector. Last year, I prepared the acetates in advance on a 486 PC and printed them out to take with me to the meeting. For the BSPP Conference in December 1996, of which this book is a synthesis, I took my Pentium laptop to the meeting, connected it to the digital projector and showed actual pages from the World Wide Web. Next year this will be commonplace, and will not be the high-wire act that it may have seemed in 1996!

The pace of change is accelerating. This is fuelled by Moore's Law (Gates, 1996) which states that the power of computers doubles every 18 months. And it is driven by demand from users who come to enjoy and then expect higher speeds, higher power, and the friendlier face that advancing technology offers.

As scientists and plant pathologists, we are missing something if we do not recognize that we have on our hands a truly revolutionary era in the history of information management and communications. The powerful combination of personal computing with digital telecommunications has only started to show its potential, for example in the adolescent fervour of the World Wide Web.

This chapter is based on the 1996 Presidential Address to the British Society for Plant Pathology. It was presented at the Society's Conference in Canterbury, UK, 16–19 December 1996, entitled Unlocking the Future: Information Technology in Plant Pathology. An extended version is published in Plant Pathology (1997) 46, 615–635.

What has just started is, in my opinion, a revolution in human communications comparable in significance with the invention of printing.

Information Technology in Support of Plant Pathology

Handling facts to produce information

The information mountain in plant pathology

One of the tasks of CAB INTERNATIONAL (CABI), like other abstracting services, is to acquire the worldwide literature of the disciplines it covers. In CABI's case, to cover agriculture and related disciplines, this means more than 12,000 periodicals and 5000 books, in more than 50 languages and from more than 100 countries. They arrive every day at CABI Headquarters, typically in two or three mailbags. The CABI Information Institute (CABI, 1997a) has 90 professional Information Scientists whose task is to scan the content of this mountain of material, select what is relevant to the disciplines we cover, prepare English abstracts, and index the resulting records for CABI's bibliographic databases. Every year nearly 200,000 records are added, about 10,000 of them in plant pathology.

Mastering the mountain would indeed be a daunting task if we were limited to the technology of printing. Fortunately, information technology (IT) renders the task hardly daunting at all. The database content that is presented in print as, say, *Review of Plant Pathology* can also be presented on CD-ROM (CABI, 1997d), via the World Wide Web (CABI, 1997b), or in other electronic media. In these formats, it is the work of a few seconds to select all the abstracts on, say, *Gaeumannomyces graminis* relating to biological control in Australia (CABI, 1997b). Furthermore, the result can immediately be viewed, stored, printed, e-mailed, word-processed, or the original document ordered. For many pathologists, this has transformed the task of bibliographic research from pain to pleasure.

The use of database management technology for selecting and organizing what is wanted from a mountain of bibliographic data was one of the first and remains one of the most powerful of the applications of IT to plant pathology. It is a remarkable tool for handling facts to produce information.

Data management

There are so many applications of IT to general data management in plant pathology that I shall merely list some headings here:

- data capture.
- data monitoring.
- data storage.
- data manipulation.

- data analysis.
- stock control.
- record keeping.

The use of IT in most of these is now taken for granted. However, the incredible pace of change has been prominent. At the start of my career, plant pathologists handled most of these tasks manually. Each of us can think of particular applications of IT that have been important to us. Here are some of mine.

Names of pathogens. For my colleagues who index the CABI bibliographic databases, nomenclature would be a nightmare without IT. Management of synonymy through nomenclatural databases is essential. An example in plant pathology is the publication by the American Phytopathological Society (APS), in printed and electronic formats, of Lists of Common Names of Plant Diseases, such as Wiese *et al.* (1994). Through the APS World Wide Web site (APS, 1997) it is the work of a moment to look up 'take-all' and find that it has been used to refer to:

Gaeumannomyces graminis (Sacc.) Arx & D. Olivier var. *tritici* J. Walker
G. graminis (Sacc.) Arx & D. Olivier
G. graminis (Sacc.) Arx & D. Olivier var. *avenae* (E.M. Turner) Dennis

Pathogen cultures. There are numerous databases of culture collections, briefly reviewed by Scott (1991). An example is the Genetic Resource Collection of CABI's International Mycological Institute (IMI), accessible through IMI (1997). Here, a search for '*Gaeumannomyces graminis*' produces a list of a number of cultures familiar to me including:

[IMI-224172] J. Walker – P.R. Scott (PO76/56) – M. Holden, Rothamsted (G1) – IMI, 1978.

Molecular data. This is probably the area in which the greatest use is now made of IT in support of plant pathology. The recent pace of change in molecular genetics has been as remarkable as that in IT, and parallels it in many ways. It is now a simple matter to search one of the many molecular sequence databases, such as the EMBL Nucleotide Sequence Database (EBI, 1996).

Interpreting information to produce knowledge

Taxonomic information systems

If nomenclatural databases are an application of IT in handling facts to produce information, then taxonomic information systems are the corresponding application in interpreting information to produce knowledge. Taxonomic information systems are well represented in this volume. The DELTA format (DEscription Language for TAxonomy) (Dallwitz *et al.*, Chapter 19, this volume), for example, is a whole system for encoding taxonomic descriptions and providing a standard for data exchange and interpretation.

Viruses of Plants (Brunt *et al.*, 1996) is an example of a printed book derived directly from DELTA-format data in the VIDE (Virus Identification Data Exchange) database. The book can readily be updated because it is directly linked to the database. Furthermore, the whole database is now mounted on the World Wide Web with interactive access, as *Plant Viruses Online* (Brunt *et al.*, 1996 onwards).

Molecular information systems

A glance at resources like those provided by the Center for Advanced Research in Biotechnology, University of Maryland (Pedersen, 1997) provides a graphic insight into the global array of linked information systems for recording, interpreting and modelling molecular information: 'These Internet based services illustrate how the exponential growth in electronic information can be made available to the layman using WWW browsers.'

An example from the Protein Data Bank of the Brookhaven National Laboratory provides a single glimpse of a remarkable world that is developing at an incredible pace. The Data Bank (Brookhaven National Laboratory, 1997) holds an archive of experimentally determined three-dimensional structures of biological macromolecules. One of these is a cutinase molecule from the plant pathogen *Fusarium solani* f.sp. *pisi*, whose structure can be displayed using a virtual reality viewer. On the PC monitor, the molecule can be enlarged, rotated, and viewed from any side.

Geographic information systems

There is still much scope for plant pathologists to exploit the power of Geographic Information Systems (GIS). In a GIS, digital information is presented in the form of a map on which layers of information are superimposed. A simple example is provided by the *Crop Protection Compendium* (CABI, 1997c), which allows the geographic distribution of a pathogen to be mapped and overlaid with the distribution of another organism (for example, a host plant or a parasite). Figure 1.1 shows this for *Puccinia polysora* and its maize host, plotted on a climatic map of the world.

At the Cooperative Research Centre for Tropical Pest Management (CRCTPM), Brisbane, a system for matching the climatic requirements of a species against a global database of climatic parameters has been developed, called CLIMEX (Sutherst and Maywald, 1985). An application is presented in Risk Analysis below. Other comparable systems include the Australian National University's BIOCLIM (Kohlmann *et al.*, 1988).

Multimedia

The *Crop Protection Compendium*, mentioned above, is a new example of the application of multimedia technology to disease, pest and weed management.

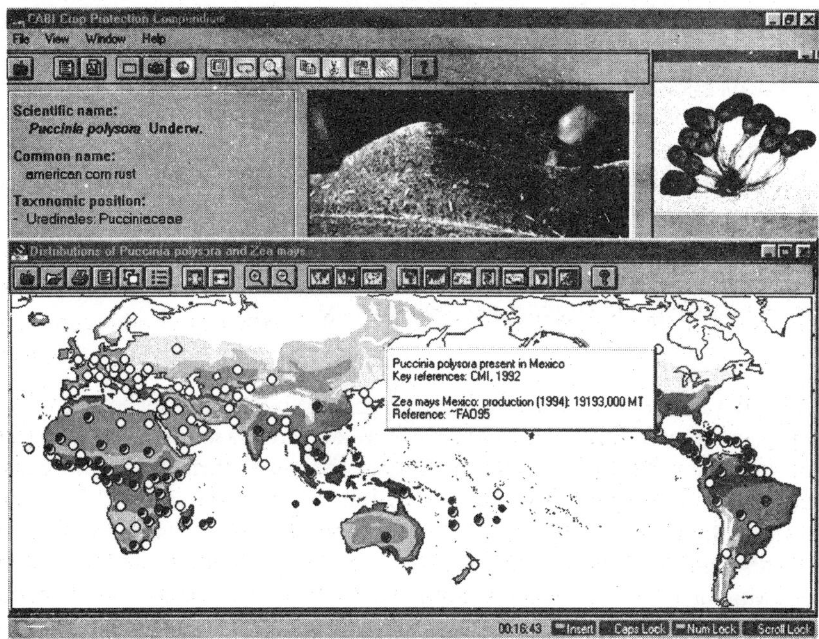

Fig. 1.1. Global distribution of *Puccinia polysora* (small dark dots) and of maize (large pale dots), plotted on a climatic map. The cursor has been pointed to the record for Mexico to show the data that support the symbols. From the *Crop Protection Compendium* (CABI, 1997c). Other windows from this multimedia application are also visible.

CABI's concept of an Electronic Compendium is an integrated information system combining text, images, a GIS, taxonomic data, bibliographic records, diagnostic keys and glossary (Zhang *et al.*, 1995; Sweetmore *et al.*, Chapter 11, this volume). These are packaged to allow flexible navigation throughout the system.

Module 1 (CABI, 1997c) provides information on approximately 1000 disease, pest and weed species and their natural enemies, with a specific focus on South-east Asia. It is published on CD-ROM but, like most electronic information resources, is independent of medium and will be developed for the World Wide Web. The core of the system is its relational database architecture, which allows retrieval of specific information through narrowing down from many disease or pest species to few, based on attributes like country, crop, part of plant, and symptom.

There is considerable further scope for development of multimedia applications in plant pathology. Combinations of text and images have been encouraged by World Wide Web technology. Among many of these the following three each exemplify a different aspect:

- illustrated identification and information sheets on global crop pests, by the Cornell International Institute for Food, Agriculture and Development (Rueda and Shelton, 1996)
- illustrations of plant virus particles, by Rothamsted Experimental Station (Antoniw, 1994)
- a multimedia system on cotton and its pests in francophone Sub-Saharan Africa, by the Centre de Coopération Internationale en Recherche Agronomique pour le Développement (CIRAD), Montpellier (Girardot, 1994).

Using knowledge to support decision making

Diagnostic tools

The opportunities presented by personal computing for the development of a new generation of diagnostic tools have been seized on by several groups. In this context, Jacques Lebbe and Régine Vignes (Chapter 4, this volume) have discussed the concepts used in identification of an unknown species. Ian White and Graham Sandlant (Chapter 24, this volume) have compared the costs and benefits of multi-entry and dichotomous keys. Lynne Boddy, Colin Morris and Alex Morgan (Chapter 21, this volume) have ventured into the world of neural networks as a basis for diagnostic systems. Trevor Bryant (Chapter 28, this volume) explains probabilistic identification systems.

As examples of actual diagnostic systems, the following list provides a selection of what is being developed:

- BIKEY (Academy of Sciences, St Petersburg): an illustrated multi-entry key forming part of the DIALOBIS system (Lobanov *et al.*, 1996).
- CABIKEY (CABI, UK): an illustrated multi-entry key (White and Sandlant, Chapter 24, this volume).
- BugMatch (CRCTPM, Queensland): multimedia information package on IPM including a graphical key (CRCTPM, 1996b).
- GENCOMEX (ABAC, Mexico): a multi-entry key driven from a single screen (Murguía and Villaseñor, Chapter 27, this volume).
- IdentifyIt (ETI, Amsterdam): an illustrated multi-entry key forming part of the LINNAEUS II system (ETI, 1996).
- INTKEY (CSIRO, Canberra): a multi-entry key forming part of the DELTA system (Dallwitz *et al.*, Chapter 19, this volume).
- LucID (CRCTPM, Brisbane): an illustrated multi-entry key and builder (CRCTPM, 1996a).
- MALHERB (INRA, Versailles): an illustrated dichotomous/polychotomous key forming part of the HYPP system (INRA, 1996).
- PANKEY (Royal Botanic Garden, Edinburgh): a multi-entry key (Pankhurst, Chapter 26, this volume).
- Pictorial Key (ETI, Amsterdam): an illustrated dichotomous key forming part of the LINNAEUS II system (ETI, 1996).

- TAXAKEY (CABI, UK): an illustrated dichotomous key (White and Sandlant, Chapter 24, this volume).

Using knowledge to make predictions

Risk analysis

The CLIMEX system (see Geographic Information Systems above) has been used to predict the potential distribution of a species, for example to map the areas of eucalypt forest in Australia that are at risk from attack by the Asian gypsy moth (CRCTPM, 1997). The prediction is based on the known climatic requirements of the species elsewhere. These are used by CLIMEX to compute a potential distribution map in Australia, and then superimposed on a map of eucalypt distribution. This is an example of the direct application of computerized decision support to risk analysis. The context is usually plant quarantine, and specifically Pest Risk Analysis (PRA).

Passing on knowledge in education and training

New media

Books, journals, lectures, seminars, broadcasts – the traditional media through which knowledge is passed on have stood the test of time. By comparison, the new media that IT offers are untried, and in their youth can hardly be expected yet to have had their rough edges smoothed. But experiments abound and Pierce Jones (Chapter 31, this volume) cites a couple of examples in our field. The University of Florida is pioneering the use of CD-ROM and now the Internet as media for transferring knowledge from the university to extension staff and growers. The FAIRS (Florida Agricultural Information Retrieval System) discs and Web pages are impressive (FAIRS, 1997). What they lack in portability compared with their printed equivalents they make up for in presentation and interactivity.

Interactive training

Opportunities for applying IT and the new media to make training a newly interactive process are only starting to be seized. Gail Schumann (Chapter 30, this volume) describes the benefits of IT to student teaching, and also the drawbacks and the new instructional challenges that need to be addressed. Terry Stewart (Chapter 33, this volume) describes the novel resource, DIAGNOSIS, that adopts the style of adventure game software to challenge students with agricultural scenarios in which they must diagnose problems in crop protection. It aims to provide a simulation of reality: tools are available to help with the diagnostic task (such as a spade, a microscope, a conversation with a farmer), some of the tools having costs associated with their use that have to be charged to a finite budget. There are certain to be many more such experiments before long.

Storing and disseminating information

The World Wide Web

Figure 1.2 shows part of the Home Page of the British Society for Plant Pathology at its World Wide Web site (BSPP, 1997a). Superficially, it appears to be a collection of text and graphics that might have been taken from the contents section of a brochure. In reality, it is a point on a global information network of almost unimaginable extent. On this portion of a page, 12 items are underlined (and are coloured in the original) to show that they are links to other sites on the World Wide Web. A mouse click on any of these links causes the content of the linked site to be displayed, no matter where that site is located.

Clicking, say, 'Unlocking the future: information technology in plant pathology' presents the programme of the Conference on which this book is based. The programme has more than 100 underlined links, mostly to speakers' institutions. Each of these contains many further links to sites related to their activities. A click on any of the titles of a conference paper connects with its entry on the Web page containing abstracts of the Conference. This has more than 100 further links, to Web pages referred to in the abstracts or providing e-mail contact with the authors. In turn, each of these contains many more links, and so on.

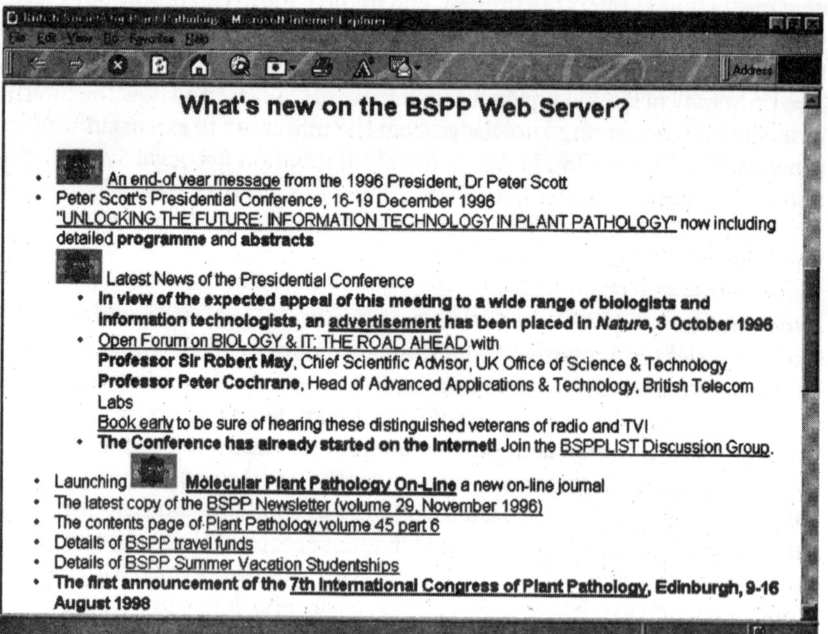

Fig. 1.2. Part of the Home Page of the British Society for Plant Pathology's World Wide Web site (BSPP, 1997a).

And a glance back at Fig. 1.2 shows that there are 11 other primary links on this portion of a page, each of which is linked again, and again ... The BSPP Home Page is thus a truly remarkable resource of information related to plant pathology, immediately available from any computer in the world that has access to the World Wide Web by connection to the Internet. It includes a link to the *Plant Pathology Internet Guide Book* (Kraska, 1997), which is itself a Web site whose specific purpose is to provide links to other pathology-related Web sites!

Electronic publishing

Some of the most immediately dramatic developments in the application of IT to plant pathology will be felt through radical changes in the world of publishing. Publishing need no longer be the preserve of institutions with the resources to manage the printing, binding, storage and transport of journals and books. Since authors create the value of published material, and since the cost of its dissemination and storage will fall within the reach of organizations representing them, they have the potential to play the leadership role in the new world of publishing. A publisher's view of developments and prospects is given by Campbell and McClean-Inglis (Chapter 38, this volume).

Making conventional journals electronic. The first move in this scenario of change is being made by publishers. All the major publishers are already offering electronic mimics of their printed journals. For example subscribers to the BSPP's *Plant Pathology* can find an exact electronic copy, page for page, via the publisher's Web site (Blackwell Science, 1997). In the short term, production costs are increased because both printed and electronic media are being serviced. At some time in the future, the printed journal is likely to be phased out, leaving the electronic version with much reduced costs for both production and distribution.

Consolidation, indexing and abstracting. While readers of, say, *Plant Pathology* can thus readily find their journal on the World Wide Web, those with a general interest in the discipline of plant pathology need a service that consolidates what is available across the world's publishers and languages. An example is PEST CABWeb (CABI, 1997b), which uses the World Wide Web to allow subscribers to an abstracting journal such as *Review of Plant Pathology* to search its contents interactively (see 'The information mountain in plant pathology' above). The results of a search are presented with English abstracts and bibliographic citations, so that the full articles of interest can be selected and ordered. In the future, there will be great scope for linking Web databases that provide this service of consolidation directly with the full articles on the Web.

Real electronic journals. Electronic mimics of printed journals represent a transitional phase in the new technology's impact on publishing. They are being followed by real electronic journals, which need not be constrained by the conventions of the printed page, and which can add value such as:

- speed, because the printing, binding and distribution phases are lost.
- economy, because the costs of these phases are avoided.
- graphics, which have negligible cost even in colour.
- new media, such as moving images, models, simulations.
- networking, through Web links to related publications.
- cross-media links, for example between citations of molecular sequences and their actual records in sequence databases.
- dynamic archiving, through the progressive addition of links to create connections forward in time.

BSPP's *Molecular Plant Pathology On-Line* is a first step in this direction (Fig. 1.3; BSPP, 1997b). It uses electronic communications to receive, referee, edit and publish its contents. It has an entirely electronic format, published on the Web, and is being offered free. It has mouse-click links between text, figures, tables and references, and welcomes colour pictures.

Electronic communication as publication. The Internet, the World Wide Web and electronic mail offer entirely new methods of communication and archiving,

Fig. 1.3. Part of the contents page of the Web journal *Molecular Plant Pathology On-Line* (BSPP, 1997b).

which amount to publishing. For example, the American Phytopathological Society has conducted an on-line symposium on the currently vexed question of quarantine issues for Karnal bunt of wheat (*Tilletia indica*), and has archived the result on its Web site (APS, 1997).

Plant pathologists have not yet experimented with the concept of the electronic preprint – a draft publication posted to a Web site with an open invitation to readers to respond with comments, which are also posted to the site. The author may then modify the draft. This approach to a kind of open refereeing is well established in certain disciplines, notably high-energy physics (Ginsparg, 1994). It challenges tradition, but is very likely to be introduced more widely by enthusiasts since the cost of doing so no longer constrains experiments of this sort.

Opportunities for developing countries

Fears have been expressed that the new technology will widen the gap in information access between developed and developing countries. The reverse is probably the case. The infrastructure required by the new IT is significant, but is spreading rapidly throughout the world, even in Africa. IT is versatile and, where it is available, greatly reduces the need for investment in the traditional fabric of libraries. For example, a well designed new library in a developing country can maximize the use of CD-ROM and Internet resources, and rationalize the extent of shelving. The result can even leap-frog over a traditional library in the industrial world, with its ponderous archival practices. Opportunities for the Internet in developing rural communities have been specifically reviewed by Richardson (1996), and the use of IT in transferring agricultural information in developing countries has been reviewed by Zijp (1994).

Conclusions

We have just glimpsed the opportunities that IT presents in support of plant pathology, in terms of:

- handling facts to produce information.
- interpreting information to produce knowledge.
- using knowledge to support decision making.
- using knowledge to make predictions.
- passing on knowledge in education and training.
- storing and disseminating information.

The extent of the opportunity is truly revolutionary. Our reaction to the revolution, even if it includes some elements of fear, bewilderment or amusement, should emphatically be one of excitement. We have only just seen the beginning of this!

References

Antoniw, J. (1994) *Electron micrographs of plant viruses*. Rothamsted Experimental Station, Harpenden, UK. World Wide Web page at http://www.res.bbsrc.ac.uk/cdm/plantpath/virusems/

APS (1997) *APSnet. Plant Pathology On-Line*. American Phytopathological Society, St Paul, USA. World Wide Web page at http://www.scisoc.org/

Blackwell Science (1997) *Blackwell Science*. Blackwell Science, Oxford, UK. World Wide Web page at http://www.blackwell-science.com/

Brookhaven National Laboratory (1997) *Protein Data Bank*. Brookhaven National Laboratory, USA. World Wide Web page at http://www.pdb.bnl.gov/index.html

Brunt, A.A., Crabtree, K., Dallwitz, M.J., Gibbs, A.J., Watson, L. and Zurcher, E.J. (1996) *Viruses of Plants*. CAB International, Wallingford, UK.

Brunt A.A., Crabtree, K., Dallwitz, M.J., Gibbs, A.J., Watson, L. and Zurcher, E.J. (1996 onwards) *Plant Viruses Online: Descriptions and Lists from the VIDE Database*. Version: 16th January 1997. Australian National University, Canberra, Australia. World Wide Web page at http://biology.anu.edu.au/Groups/Mes/vide/

BSPP (1997a) *British Society for Plant Pathology*. British Society for Plant Pathology, UK. World Wide Web page at http://www.bspp.org.uk

BSPP (1997b) *Molecular Plant Pathology On-Line*. British Society for Plant Pathology, UK. World Wide Web journal at http://www.bspp.org.uk/mppol/

CABI (1997a) CAB International Information Institute. CAB International, Wallingford, UK. World Wide Web page at http://www.cabi.org/infolib/infolib.htm

CABI (1997b) PEST CABWeb: WebSPIRS Search. CAB International, Wallingford, UK. World Wide Web page at http://pest.cabweb.org/cgi-dos/webspirs.bat

CABI (1997c) *Crop Protection Compendium: Module 1*. Multimedia CD-ROM. CAB International, Wallingford, UK.

CABI (1997d) PlantPathCD 1973–1996. Bibliographic database on CD-ROM. CAB International, Wallingford, UK.

CRCTPM (1996a) *LucID: Identification Tool for Teachers, Taxonomists and Ecologists*. Cooperative Research Centre for Tropical Pest Management, Brisbane, Australia. World Wide Web page at http://www.ctpm.uq.edu.au/software/lucid.html

CRCTPM (1996b) *BugMatch*. Cooperative Research Centre for Tropical Pest Management, Brisbane, Australia. World Wide Web page at http://www.ctpm.uq.edu.au/software/bugmatchcotton.html

CRCTPM (1997) *The Role of Geographic Information Systems*. Cooperative Research Centre for Tropical Pest Management, Brisbane, Australia. World Wide Web page at http://www.modeling.ctpm.uq.edu.au/sware/pra/spat4.htm

EBI (1996) *European Bioinformatics Institute: The EMBL Nucleotide Sequence Database*. European Bioinformatics Institute, Hinxton, UK. World Wide Web page at http://www.ebi.ac.uk/ebi_docs/embl_db/ebi/topembl.html

ETI (1996) *Welcome to Linnaeus II from ETI – version 1.2*. Expert Center for Taxonomic Identification, Amsterdam, Netherlands. World Wide Web page at http://wwweti.eti.bio.uva.nl/demo/folnvgtr/hnvgtr.html

FAIRS (1997) *Florida Agricultural Information Retrieval System*. University of Florida, Gainesville, USA. World Wide Web site at http://hammock.ifas.ufl.edu/.

Gates, W. (1996) *The Road Ahead*. Viking, London, UK.

Ginsparg, P. (1994) First steps towards electronic research communication. *Computers in Physics* 8, 390.

Girardot, B. (1994) *CIRAD-CA: COTON-DOC*. Centre de Coopération Internationale en Recherche Agronomique pour le Développement, Montpellier, France. World Wide Web page at http://www.cirad.fr/logiciels/coton-doc/home.html

IMI (1997) *The International Mycological Institute Culture Collections*. International Mycological Institute, Egham, UK. World Wide Web page at http://www.cabi.org/institut/imi/grc.htm.

INRA (1996) HYPP: Hypermédia pour la Protection des Plantes. Institut National pour la Recherche Agronomique, Versailles, Francc. World Wide Web page at http://www.inra.fr/user/productions/publications/dpenv/o-hypp.htm

Kohlmann, B., Nix, H., Shaw, D.D. (1988) Environmental predictions and distributional limits of chromosomal taxa in the Australian grasshopper *Caledia captiva* (F.). *Oecologia* 75, 483–493.

Kraska, T. (1997) *Plant Pathology Internet Guide Book*. University of Hannover, Hannover, Germany. World Wide Web page at http://www.ifgb.uni-hannover.de

Lobanov, A., Dianov, M., Ryss, A. and Schilow, W. (1996) The pictorial computerized identification system BIKEY as a part of the DIALOBIS biological encyclopaedias on CD-ROM. In: *Unlocking the future: information technology in plant pathology*. British Society for Plant Pathology, Presidential Conference, December 1996, Canterbury UK. Abstracts in World Wide Web page at http://www.bspp.org.uk/meetings/abst0096.htm

Pedersen, J.T. (1997) *Bringing Molecular Modelling and Databases to the People*. Center for Advanced Research in Biotechnology, University of Maryland, USA. World Wide Web page at http://iris4.carb.nist.gov/www/databases.html

Richardson, D. (1996) The Internet and rural development: recommendations for strategy and activity. FAO, Rome, Italy. Food and Agriculture Organization of the United Nations, Rome, Italy. World Wide Web page at http://www.fao.org/waicent/faoinfo/sustdev/cddirect/cddo/contents.htm

Rueda, A. and Shelton, A.M. (1996) *Global Crop Pests: Identification and Information*. Cornell University, Ithaca, USA. World Wide Web page at http://www.nysaes.cornell.edu/ent/hortcrops/index.html

Scott, P.R. (1991) The universal issue: information transfer. In: Hawksworth, D.L. (ed.) *The Biodiversity of Microorganisms and Invertebrates: its Role in Sustainable Agriculture*. CAB International, Wallingford, UK, pp. 245–265.

Sutherst, R.W. and Maywald, G.F. (1985) A computerised system for matching climates in ecology. *Agriculture, Ecosystems and Environment* 13, 281–299.

Wiese, M.V., Murray, T.D. and Forster, R.L. (1994) *Diseases of Wheat* (Triticum *spp. L.*). American Phytopathological Society, St Paul, Minnesota, USA.

Zhang, B.C., Scott, P.R., Schotman, C.Y.L. (1995) Information and management for agriculture and natural resources. In: *Proceedings, 1st Asian Information Meeting, Hong Kong, 27–30 September 1995*. Learned Information Ltd, Oxford, UK, pp. 211–217.

Zijp, W. (1994) Improving the transfer and use of agricultural information: a guide to Information Technology. *World Bank Discussion Papers* 247, 105 pp.

Development of Computer-based Systems in Systematics

2

P.H.A. Sneath

Department of Microbiology and Immunology, University of Leicester, Leicester LE1 9HN, UK
Fax: +44 (0)116 2525030

Introduction

This contribution is a brief overview of the development of computer-based systems for systematics. Historical aspects are covered by Sneath and Sokal (1973), Hull (1988) and Vernon (1988), so that it is mainly the major conceptual and technical advances that are mentioned here. Today most systematics is to some extent numerical. Computers and numerical taxonomic programs are now standard resources in museums and systematic laboratories (as was predicted by Ehrlich, 1961), and they are increasingly available to general users of systematic information. The trend toward numerical treatment can be seen in microbiology in the proportion of papers that contain some form of quantitative relationship between organisms, which has risen from about 5% in 1960 to over 70% in the last few years (Sneath, 1995a). The only exception to this trend has been Hennigian cladistics, which in inception is barely numerical, but which in practice is increasingly so (Hull, 1988; Sneath, 1995a). This trend is intimately connected to the availability of computers. Before these became widely available it took hours to calculate correlation coefficients even with the aid of mechanical calculators, and matrix arithmetic (as used in ordination and discriminant analysis) was only feasible for the smallest problems. Now there are many computer packages for systematics, starting with the NT-SYS system of James Rohlf and his colleagues (Rohlf *et al.*, 1971) and the CLUSTAN package of David Wishart (briefly described, together with other early computer programs,

This chapter is based on the Keynote Address to the conference of the Systematics Association in Canterbury, UK, 16–19 December 1996, entitled *Computer-Based Species Identification.*

 Information Technology, Plant Pathology and Biodiversity (eds P. Bridge, P. Jeffries, D.R. Morse and P.R. Scott)

in Cole, 1969). Packages such as PHYLIP of Joseph Felsenstein and PAUP of David Swofford and his colleagues for reconstructing phylogeny were developed a little later. Many more packages have since been developed for diverse applications; most are reviewed by Sackin (1987) and Pankhurst (1991). Some early reminiscences on computers are given by Sneath (1984).

Numerical methods were seldom applied to systematics before computers were available, for two reasons. The first was the labour of computation mentioned above. A few techniques had been developed which were related to discrimination or identification (though this was not very clearly separated from classification). These included Pearson's Coefficient of Racial Likeness and Heinke's multivariate analyses of races of herring. Fisher's Discriminant Functions and Mahalanobis' multivariate distances were explicit procedures for identification, and were the conceptual foundation of much later work. There were few ideas on quantification of taxonomic relationships; though Adanson in the 19th century proposed a logic approximating to equal weighting of characters, he did not develop a clear numerical analysis. These early essays are reviewed by Sokal and Sneath (1963) and Sneath and Sokal (1973). Systematic relationships are highly multivariate, and require the analysis of numerous different characters. Furthermore, systematics needs to marshal large amounts of information, and thus makes heavy demands on data-processing, even for straightforward applications such as check-lists, nomenclators and museum accessions. Good data-processing based on information theory was therefore needed.

The second reason that little was achieved before the availability of computers was that there was little incentive to provide a logical framework for systematics before any practical applications were in sight. Most early workers viewed systematics as a complex art – so complex that it would not be feasible to reduce it to a series of set procedures, let alone to make it numerical. This logical basis was provided by numerical taxonomy between the mid-1950s and mid-1960s. It rested in part on concepts of information content and the general and polythetic nature of biological classification, which goes back to the 19th century philosophers of science such as Whewell and Mill. This was brought to the attention of taxonomists by a key paper of Gilmour (1937). Numerical taxonomy defined a logical progression, which is briefly described in the following section.

Numerical Taxonomy

It is not possible to attack the problem if one must first require the solution to the problem. Therefore one must, at the start, set aside considerations of phylogeny, diagnosis, discrimination between groups, and the like, because these can only be determined from later steps. This precludes *a priori* character weighting in the usual subjective sense (objective weighting can depend on information content of characters). The initial step is thus to consider organisms (at various levels, individuals, species, etc., or generally, Operational Taxonomic Units, OTUs) and their characters, and to determine the correspondences between the characters, i.e. homology in the broad sense. This has led to much misunderstanding, and it has only been the availability of molecular sequences that has clarified the distinction between correspondences in characters (such as chemical correspondence) and evolutionary homologies (which are, of course, deduced from a later stage in the process). This stage leads to a table of organisms and their character states. From this one can calculate relationships (more generally, similarities or resemblances) between organisms, usually as an OTU by OTU similarity matrix. The next stage is to find taxonomic structure. Two main approaches are used, cluster analysis or its close ally, tree construction, and ordination which gives a representation in a space of a few dimensions. Many of the early concepts were discussed in a symposium of the Systematics Association (Heywood and McNeill, 1964). The meeting on numerical taxonomy in 1968 (reported in Cole, 1969) and that on identification in 1973 (reported in Pankhurst, 1975a) were notable in that computing underlay all the contributions. The result of these steps is to permit numerical circumscription of taxa, and, to facilitate the construction of systematic databases and numerical methods of identification.

Though computers were used for academic work in the mid-1950s, there were few advances in programming for systematics until high-level languages became available in the early 1960s. Technical advances in this area due to computers include the routine determination of homologous sites in molecular sequences (by algorithms that match them with allowance for gaps and insertions), the use of many similarity coefficients, and effective techniques of cluster analysis. Programs for trees are discussed later under molecular systematics. Notable here is the work of Williams and Dale (1965) and Lance and Williams (1966), Lance *et al.* (1968), Needleman and Wunsch (1970) and Queen and Korn (1980). Ordination techniques have been extended by computer-intensive methods, in particular non-metric multidimensional scaling (Kruskal, 1964) and principal co-ordinate analysis (Gower, 1966). In the last ten years there has been a great increase in the use of microcomputers in systematics, with very effective interactive graphics. Excellent search procedures are now available on the global internet. Electronic mail is now the preferred way of transmitting programs and data. Meetings on computing for systematics (Fortuner, 1993), and text books on this (Pankhurst, 1991) are now appearing. Similar developments are seen in related fields such as ecology (Orloci, 1978) and genetics (Weir, 1990).

Databases in Systematics

The preparation of descriptions of taxa by computer methods is now a very active field. Some of this work is associated with methods for numerical identification (discussed later). This is particularly so in microbiology, where a main reason for a numerical taxonomy is to define taxa, both known taxa and new ones, in such a way that diagnostic tables are prepared automatically. Other work is connected with broader aims, as shown by a Systematics Association symposium volume edited by Allkin and Bisby (1984). In a few areas, taxonomic data can now be captured by automatic methods. In microbiology, this has been possible for some time for biochemical reactions and cell constituents (Goodfellow *et al.*, 1985). Morphometrics is now entering this phase; early attempts at scanning of images (e.g. Rohlf and Archie, 1984) and analysis of shape (Gower, 1975; Bookstein, 1978) are now leading to major advances (Rohlf and Marcus, 1993). Automation of molecular sequences is well advanced.

In the broad field of systematics, much attention has been given to standardization of taxonomic descriptions, often while preparing specific databases. Pankhurst (1972, 1978b) proposed computer methods for data capture and printing taxonomic descriptions. The DELTA data format of Dallwitz (1980) has become an international standard. Early examples of systematic databases include those of Perring and Walters (1962) on computerized mapping, and of Gómez-Pompa and Nevling (1973) on a regional flora. Other landmarks include the work of Morris (1974) and the Vicieae database (Adey *et al.*, 1984), and there are now many more. Australian workers have been particularly active, e.g. Watson and Dallwitz (1988, 1994) and Hyland and Wiffin (1994). These developments are leading to new ways of preparing taxonomic monographs, regional floras and inventories of biological diversity. Bibliographic and nomenclatural databases are being established, and automatic translation to other languages is now practicable if the terminology is standardized (for example, to Greek, Watson *et al.*, 1988).

Such databases offer many advantages, in making available much data that would otherwise remain unpublished. Computer languages to exploit them have greatly improved, and offer numerous facilities such as colour graphics.

Numerical Identification

Numerical identification has grown greatly from its origins in discriminant analysis. The 1996 Conference at which this chapter was presented as a Keynote Address fittingly commemorates the twenty-first anniversary of the publication in 1975 of a seminal symposium of the Systematics Association (Pankhurst, 1975a). The field has diversified into many sophisticated methods, of which a historical résumé is given by Pankhurst (1991). Early work was reviewed by Sneath and Sokal (1973). It owes much to two groups of workers, Richard Pankhurst and his colleagues and the late Stephen Lapage and his colleagues. It is significant that many of the advances have come from active practitioners of diagnostic taxonomy, perhaps because the difficulties are seldom apparent until the techniques are applied to difficult data sets.

Among the earliest suggestions on numerical identification were those by Maccacaro (1958) on diagnostic keys and by Gyllenberg (1963) on information needed to define taxa. The first on-line identification (Boughey *et al.*, 1968) and the first working batch-processing system (Lapage *et al.*, 1970) marked the introduction of computers to the field. Methods to generate diagnostic keys by computer (see below) soon followed. The contribution of Morse (1974) has been influential, both for identification and databases. These developments were consolidated in an important book (Pankhurst, 1978a), and reviews of statistical aspects soon appeared (Payne and Preece, 1980; Willcox *et al.*, 1980).

Numerical methods for identification can be roughly grouped into three, diagnostic keys, polyclaves and distance models. The last two overlap, as they are variants of matching methods.

One of the earliest papers on numerical properties of keys is that of Maccacaro (1958), who investigated the probability that a diagnosis would be correct. A great impetus came from methods to produce a key by computers (Morse, 1968, 1971; Hall, 1970; Pankhurst, 1970a,b; Dallwitz, 1974 and a number of other papers). Examples of such keys soon appeared (e.g. Watson and Milne, 1972; Barnett and Pankhurst, 1974). But despite early attempts (Möller, 1962), it is not easy to incorporate probability statements into keys.

Polyclaves, developed for computers from the earlier 'peek-a-boo' punched card systems, and sometimes referred to as multiple-entry keys, received only a little attention at first (Goodall, 1968; Morse, 1971). The early on-line program of Boughey *et al.* (1968) used this principle. They are now attracting more interest because of their flexibility and capacity to handle missing characters. They can also take some account of within-taxon variability (Pankhurst and Aitchison, 1975; Duncan and Meacham, 1986; Wilson and Partridge, 1986). Allied work is that of Rypka *et al.* (1967) and Pankhurst (1983) on character sets that are needed to distinguish taxa.

Taxonomic distance methods (direct descendants of discriminant analysis, but much simplified) have become extremely important in microbiology, partly because they can accommodate much variability within taxa (which renders the use of keys difficult), and also because they can be readily associated with the probability of correct identification. There is a tendency to restrict the term probabilistic identification to those methods which use Bayesian inference. But it should be noted that all these methods are, in principle, probabilistic if statistical probabilities can be attached to distances. Such probabilities were, of course, incorporated in the original non-Bayesian discriminant analysis.

Helge Gyllenberg made two important contributions here (Sneath, 1995b). First, he proposed the automatic production of taxon descriptions from numerical taxonomies, with the aim of providing sufficient information to give reliable identification (Gyllenberg, 1963). Second, he suggested a limit, or envelope, to a formal taxonomic description, such that an unknown strain that fell outside the envelope would not be considered identified with that taxon (Gyllenberg, 1964, 1965). More generally, he proposed that distances could imply probabilities of correct identification. Both these concepts seem obvious today, but they were far from being accepted at the time. The concept of an envelope was referred to by Sneath and Sokal (1973) as a taxon-radius model, but of course the envelope need not represent a simple radius about a taxon centre. This is analogous to the peculiarity index of Hall (1965, 1968) and the variability limits of Morse (1974) in setting bounds to acceptable identities.

Gyllenberg (1965) gave the first clear description of a distance model for bacteria, though Beers and Lockhart (1962) had suggested that distances could be used for identification. His paper also enunciated clearly that an unknown may not be identifiable at all, either from lack of characters or from lack of taxa in the system. Dybowski and Franklin (1968) proposed a Bayesian model, but its acceptance was held back by lack of good databases and rigorous test methods. Work on these, however, soon led to the first computer identification service for bacteria, developed by Lapage *et al.* (1970).

The success of computer methods for identifying bacteria depended critically on databases of high quality, and on excellent test-reproducibility, first studied by Lapage *et al.* (1970). It has led to work on the quality of taxon descriptions in microbiology by our group at Leicester University (Sackin, 1987; Priest and Williams, 1993). This includes effective ways to assess test-reproducibility, overlap between taxa and the presence of misclassified individuals whose inclusion in a taxon degrades the quality of the taxon description. Distance methods are probably the most efficient in terms of the numbers of characters required to distinguish a given number of taxa (Sneath and Chater, 1978). A reliable diagnostic system can be produced from a high-quality numerical taxonomy of almost any group of bacteria, and without this a reliable system usually cannot be constructed. Industrial firms now produce automated instruments for numerical identification from biochemical tests, and these instruments contain software based on the developments mentioned above. An excellent review is that of Schindler (1984, unfortunately not translated to a widely spoken language). The power of computing has even revived the use of discriminant analysis (Sielaff *et al.*, 1982). A variety of chemical analytic methods are now used (Goodfellow *et al.*, 1985). This is a far cry from the comment of a distinguished expert on an admittedly difficult group of bacteria over twenty years ago (Trejo, 1970): 'Anyone who has been faced with the problem of objectively identifying an unknown streptomycete strain can appreciate the sense of futility thus engendered'.

Swift advances are also occurring outside microbiology, and these too are associated with the production of high-quality databases (Pankhurst, 1975b). Most earlier taxonomic monographs contained numerous gaps in the information that only became apparent when it was needed in tabular format. Australian botanists have made outstanding databases, combined with software for interactive identification (Watson and Dallwitz, 1983, 1994; Hyland and Whiffin, 1994). Examples are now appearing in zoology and other disciplines (e.g. Fortuner and Wong, 1984 on nematodes; von Hayek, 1990, on beetles). Many more recent examples are reviewed elsewhere in this volume, e.g. Dallwitz *et al.*, Chapter 19. Various expert systems, which commonly work in a similar way to polyclaves, are being introduced, and these have been reviewed by Pankhurst (1991). Neural networks are also being explored (Chun *et al.*, 1993; Boddy *et al.*, Chapter 21, this volume). More difficult is the development of satisfactory systems that can learn from new unknowns and correct themselves; an early attempt is that of Niemelä and Gyllenberg (1975). A major challenge will be to apply nucleic acid probes to identification in a wide range of organisms.

Molecular Systematics

The advent of molecular sequences has been one of the major advances in systematics. Huge molecular databases are being accumulated, together with sophisticated computer methods for alignment, for searches and for making trees. This has always been associated with computers, from the early databases of Dayhoff *et al.* (1965) and the first molecular phylogeny to attract wide notice (Fitch and Margoliash, 1967). Yet much of this has been rather isolated from other areas of systematics. Most work is directed to reconstructing phylogeny, with little consideration of why, except that it is always interesting to know evolutionary origins and it is felt to represent some form of objective truth. This attitude is partly due to difficulties with interpreting character variation, and uncertainty over the proper aims of taxonomy. In bacteriology, recent molecular studies have defined many unsuspected clades of bacteria, and the view is sometimes expressed by bacteriologists that these clades, when more is known, will turn out to be more phenetic than existing phenetic groups – an interesting philosophical point.

Methods of tree-construction have only come into their own with molecular sequences. Their development has been reviewed by Felsenstein (1982), Sneath (1995a) and Edwards (1996). They are all highly computer-intensive. Exhaustive searches of all tree topologies is still impossible for more than a moderate number of OTUs. One area has resisted computing, that of DNA–DNA pairing, despite its theoretical attractions of expressing the whole genome (with some major reservations). This is partly because test-reproducibility is not very good, but mainly because computers cannot store the information in the DNA until complete genomes can be easily sequenced. Yet one may expect some breakthroughs here in the near future.

In some areas, molecular sequence studies are being applied outside straightforward phylogenetics. The most noteworthy is in population genetics, which closely interfaces with parts of systematics. Two recent meetings considered this field (Baumberg *et al.*, 1995; Harvey *et al.*, 1996), with discussions of questions as varied as the origins of viruses, extinction rates, and lateral gene transfer. All these questions rely for their answers on intensive computation.

Conclusions

One may express the hope that the education of systematists will include adequate understanding of both statistics and computation. Such advances were pioneered by Furlow *et al.* (1971) and are becoming accepted in postgraduate study. There is some danger that the widely available computer packages will be viewed as 'black boxes' and used uncritically. One may hope also that systematists will experiment widely with new algorithms now that these are so readily transferable, as there is always a lag before they become incorporated into program packages. Yet the advances that have been made in the last forty years are truly remarkable, and the close links between information science and systematics imply extensive usage of computers. Perhaps we should take some time to review the aims of systematics; computers are marvellous servants, but systematists should decide what they should do for us.

References

Adey, M.E., Allkin, R., Bisby, F.A., White, R.J. and MacFarlane, T.D. (1984) The Vicieae database: an experimental taxonomic monograph. In: Allkin, R. and Bisby, F.A. (eds) *Database in Systematics*. Academic Press, London, pp. 175–188.

Allkin, R. and Bisby, F.A. (eds) (1984) *Database in Systematics*. Academic Press, London, 329 pp.

Barnett, J.A. and Pankhurst, R.J. (1974) *A New Key to the Yeasts*. North Holland, Amsterdam, 273 pp.

Baumberg, S., Young, J.P.W., Wellington, E.M.H. and Saunders, J.R. (eds) (1995) *Population Genetics of Bacteria*. Cambridge University Press, Cambridge, 348 pp.

Beers, R.J. and Lockhart, W.R. (1962) Experimental methods in computer taxonomy. *Journal of General Microbiology* 28, 633–640.

Bookstein, F.L. (1978) *The Measurement of Biological Shape and Shape Change*. Springer-Verlag, New York, 384pp.

Boughey, A.S., Bridges, K.W. and Ikeda, A.G. (1968) An automated biological identification key. *Museum of Systematic Biology, University of California, Irvine, Research Series* No. 2, 1–36.

Chun, J., Atalan, E., Ward, A.C. and Goodfellow, M. (1993) Artificial neural network analysis of pyrolysis mass spectrometric data in the identification of *Streptomyces* strains. *FEMS Microbiology Letters* 107, 321–326.

Cole, A.J. (ed.) (1969) *Numerical Taxonomy*. Academic Press, London, 324 pp.

Dallwitz, M.J. (1974) A flexible computer program for generating identification keys. *Systematic Zoology* 23, 50–57.

Dallwitz, M.J. (1980) *User's Guide to the DELTA System: a General System for Encoding Taxonomic Descriptions*. CSIRO Division of Entomology, Canberra.

Dayhoff, M.O., Eck, R.V., Chang, M. and Sochard, M.R. (1965) *Atlas of Protein Sequence and Structure*. National Biomedical Research Foundation, Silver Spring, Maryland, 97 pp.

Duncan, T. and Meacham, C.A. (1986) Multiple-entry keys for the identification of angiosperm families using a microcomputer. *Taxon* 35, 492–494.

Dybowski, W. and Franklin, D.A. (1968) Conditional probability and the identification of bacteria: a pilot study. *Journal of General Microbiology* 54, 215–224.

Edwards, A.W.F. (1996) The origin and early development of the method of minimum evolution for the reconstruction of phylogenetic trees. *Systematic Biology* 45, 79–91.

Ehrlich, P.R. (1961) Systematics in 1970: some unpopular predictions. *Systematic Zoology* 10, 157–158.

Felsentein, J. (1982) Numerical methods for inferring evolutionary trees. *Quarterly Review of Biology* 57, 379–404.

Fitch, W.M. and Margoliash, E. (1967) Construction of phylogenetic trees. *Science* 155, 279–284.

Fortuner, R. (ed.) (1993) *Advances in Computer Methods for Systematic Biology*. Johns Hopkins University Press, Baltimore, 584 pp.

Fortuner, R. and Wong, Y. (1984) Review of the genus *Helicotylenchus* Steiner, 1945. 1. A computer program for the identification of the species. *Revue de Nématologie* 7, 385–392.

Furlow, J.J., Morse, L.E. and Beaman, J.H. (1971) Computers in biological systematics, a new University course. *Taxon* 20, 283–290.

Gilmour, J.S.L. (1937) A taxonomic problem. *Nature* 139, 1040–1042.

Gómez-Pompa, A. and Nevling, L.I., Jr (1973) The use of electronic data processing methods in the Flora of Veracruz program. *Contributions of the Gray Herbarium* 203, 49–64.

Goodall, D.W. (1968) Identification by computer. *BioScience* 18, 485–488.

Goodfellow, M., Jones, D. and Priest, F.G. (eds) (1985) *Computer-assisted Bacterial Systematics*. Academic Press, London, 443 pp.

Gower, J.C. (1966) Some distance properties of latent root and vector methods used in multivariate analysis. *Biometrika* 53, 325–338.

Gower, J.C. (1975) Generalized Procrustes analysis. *Psychometrika* 40, 33–51.

Gyllenberg, H.G. (1963) A general method for deriving determinative schemes for random collections of microbial isolates. *Annales Academiae Scientiarum Fennicae Series IV A. Biologica* No. 69, 1–23.

Gyllenberg, H.G. (1964) An approach to numerical descriptions of microbial populations. *Annales Academiae Scientiarum Fennicae Series IV A. Biologica* No. 81, 1–23.

Gyllenberg, H.G. (1965) A model for computer identification of microorganisms. *Journal of General Microbiology* 39, 401–405.

Hall, A.V. (1965) The Peculiarity Index, a new function for use in numerical taxonomy. *Nature* 206, 952.

Hall, A.V. (1968) Methods for showing distinctness and aiding identification of critical groups in taxonomy and ecology. *Nature* 218, 203–204.

Hall, A.V. (1970) A computer-based system for forming identification keys. *Taxon* 19, 12–18.

Harvey, P.H., Leigh Brown, A.J., Maynard Smith, J. and Nee, S. (eds) (1996) *New Uses for New Phylogenies*. Oxford University Press, Oxford, 349 pp.

Heywood, V.H. and McNeill, J. (eds) (1964) *Phenetic and Phylogenetic Classification*. Systematic Association, London, 164 pp.

Hull, D.L. (1988) *Science as a Process*. University of Chicago Press, Chicago, 586 pp.

Hyland, B.P.M. and Whiffin, T. (1994) *Australian Forest Trees: an Interactive Identification System*. CSIRO Information Services, East Melbourne, Victoria, Australia.

Kruskal, J.B. (1964) Multidimensional scaling by optimizing goodness of fit to a non-metric hypothesis. *Psychometrika* 29, 1–27.

Lance, G.W., Milne, P.W. and Williams, W.T. (1968) Mixed-data classificatory programs. III. Diagnostic systems. *Australian Computer Journal* 1, 178–181.

Lance, G.N. and Williams, W.T. (1966) A generalized sorting strategy for computer classification. *Nature* 212, 218.

Lapage, S.P., Bascomb, S., Willcox, W.R. and Curtis, M.A. (1970) Computer identification of bacteria. In: Baillie, A. and Gilbert, R.J. (eds) *Automation, Mechanization and Data Handling in Microbiology*. Academic Press, London, pp. 1–22.

Maccacaro, G.A. (1958) La misura delle informazione contenuta nei criteri di classificazione. *Annali di Microbiologia ed Enzimologia* 8, 231–239.

Möller, F. (1962) Quantitative methods in the systematics of Actinomycetales. IV. The theory and application of a probabilistic identification key. *Giornale di Microbiologia* 10, 29–47.

Morris, J.W. (1974) Progress in the computerisation of herbarium procedures. *Bothalia* 11, 349–354.

Morse, L.E. (1968) Construction of identification keys by computer. *American Journal of Botany* 55, 737.

Morse, L.E. (1971) Specimen identification and key construction with time-sharing computers. *Taxon* 20, 269–282.

Morse, L.E. (1974) Computer programs for specimen identification, key construction and description printing using taxonomic data matrices. *Publications of the Museum of Michigan State University, Biological Series* 5 (1), 1–128.

Needleman, S.B. and Wunsch, C.D. (1970) A general method applicable to the search for similarities in the amino acid sequence of two proteins. *Journal of Molecular Biology* 48, 443–453.

Niemelä, T.K. and Gyllenberg, H.G. (1975) Simulation of a computer-aided self-correcting classification method. In: Pankhurst, R.J. (ed.) *Biological Identification with Computers*. Academic Press, London, pp. 137–151.

Orloci, L. (1978) *Multivariate Analysis in Vegetation Research*, 2nd edn. W. Junk, The Hague, 450 pp.

Pankhurst, R.J. (1970a) Key generation by computer. *Nature* 227, 1269–1270.

Pankhurst, R.J. (1970b) A computer program for generating diagnostic keys. *Computer Journal* 12, 145–151.

Pankhurst, R.J. (1972) A method for data capture. *Taxon* 21, 549–558.

Pankhurst, R.J. (ed.) (1975a) *Biological Identification with Computers*. Academic Press, London, 333 pp.

Pankhurst, R.J. (1975b) Identification methods and the quality of taxonomic descriptions. In: Pankhurst, R.J. (ed.) *Biological Identification with Computers*. Academic Press, London, pp. 237–247.

Pankhurst, R.J. (1978a) *Biological Identification. The Principles and Practice of Identification Methods in Biology*. Edward Arnold, London, 104 pp.

Pankhurst, R.J. (1978b) The printing of taxonomic descriptions by computers. *Taxon* 27, 35–38.

Pankhurst, R.J. (1983) An improved algorithm for finding diagnostic taxonomic descriptions. *Mathematical Biosciences* 65, 209–218.

Pankhurst, R.J. (1991) *Practical Taxonomic Computing*. Cambridge University Press, Cambridge, 202 pp.

Pankhurst, R.J. and Aitchison, R.R. (1975) A computer program to construct polyclaves. In: Pankhurst, R.J. (ed.) *Biological Identification with Computers*. Academic Press, London, pp.73–78.

Payne, R.W. and Preece, D.A. (1980) Identification keys and diagnostic tables: a review. *Journal of the Royal Statistical Society A* 143, 253–282.

Perring, F.H. and Walters, S.M. (1962) *Atlas of the British Flora*. Botanical Society of the British Isles and Thomas Nelson, London, 432 pp.

Priest, F.G. and Williams, S.T. (1993) Computer-assisted identification. In: Goodfellow, M. and O'Donnell, A.G. (eds) *Handbook of New Bacterial Systematics*. Academic Press, London, pp. 361–381.

Queen, C.L. and Korn, L.J. (1980) Computer analysis of nucleic acids and proteins. *Methods in Enzymology* 65, 595–609.

Rohlf, F.J. and Archie, J. (1984) A comparison of Fourier methods for the description of wing shape in mosquitos (Diptera: Culicidae). *Systematic Zoology* 33, 302–317.

Rohlf, F.J. and Marcus, L.F. (1993) A revolution in morphometrics. *Trends in Ecology and Evolution* 8, 129–132.

Rohlf, F.J., Kishpaugh, J. and Kirk, D. (1971) *NT-SYS. Numerical Taxonomy System of Multivariate Statistical Programs*. Technical Report, State University at Stony Brook, New York.

Rypka, E.W., Clapper, W.E., Bowen, I.G. and Babb, R. (1967) A model for the identification of bacteria. *Journal of General Microbiology* 46, 407–424.

Sackin, M.J. (1987) Computer programs for classification and identification. *Methods in Microbiology* 19, 459–494.

Schindler, J. (1984) *Automatiká Diagnostika Bactérií*. Avicenum, Prague, 204 pp.

Seilaff, B.H., Matsen, J.M. and McKie, J.E. (1982) Novel approach to bacterial identification that uses the Autobac system. *Journal of Clinical Microbiology* 15, 1103–1110.

Sneath, P.H.A. (1984) Early experience with computers. *Binary, the Society for General Microbiology Computer Club Newsletter* No.1, 5–7.

Sneath, P.H.A. (1995a) Thirty years of numerical taxonomy. *Systematic Biology* 44, 281–298.

Sneath, P.H.A. (1995b) The history and future potential of numerical concepts in systematics: the contributions of H.G. Gyllenberg. *Binary* 7, 32–36.

Sneath, P.H.A. and Chater, A.O. (1978) Information content of keys for identification. In: Street, H.E. (ed.) *Essays in Plant Taxonomy*. Academic Press, London, pp. 76–95.

Sneath, P.H.A. and Sokal, R.R. (1973) *Numerical Taxonomy*. W.H. Freeman, San Francisco, 573 pp.

Sokal, R.R. and Sneath, P.H.A. (1963) *Principles of Numerical Taxonomy*. W.H. Freeman, San Francisco, 359 pp.

Trejo, W.H. (1970) An evaluation of some concepts and criteria used in the speciation of streptomycetes. *Transactions of the New York Academy of Sciences* 32, 989–997.

Vernon, K. (1988) The founding of numerical taxonomy. *British Journal of the History of Science* 21, 143–159.

von Hayek, C.M.F. (1990) A reclassification of the *Melanotus* group of genera (Coleoptera: Elateridae). *Bulletin of the Natural History Museum, Entomology* 59, 37–115.

Watson, L. and Dallwitz, M.J. (1983) *The Genera of Leguminosae-Caesalpinioideae, Anatomy, Morphology, Classfication and Keys*. Australian National University, Canberra.

Watson, L. and Dallwitz, M.J. (1988) *Grass Genera of the World.* Research School of Biological Sciences, Australian National University, Canberra.

Watson, L. and Dallwitz, M.J. (1994) *The Families of Flowering Plants.* CSIRO Information Services, East Melbourne, Victoria, Australia.

Watson, L., Damanakis, M. and Dallwitz, M.J. (1988) *Ta Gene ton Agrostodon tes Helladas.* University of Crete, Iraklion.

Watson, L. and Milne, P. (1972) A flexible system for automatic generation of special purpose dichotomous keys, and its application to Australian grass genera. *Australian Journal of Botany* 20, 331–352.

Weir, B.S. (1990) *Genetic Data Analysis.* Sinauer, Sunderland, Massachusetts, 377 pp.

Willcox, W.R., Lapage, S.P. and Holmes, B. (1980) A review of numerical methods in bacterial identification. *Antonie van Leeuwenhoek* 46, 233–299.

Williams, W.T. and Dale, M.B. (1965) Fundamental problems in numerical taxonomy. *Advances in Botanical Research* 2, 35–68.

Wilson, J.B. and Partridge, T.R. (1986) Interactive plant identification. *Taxon* 35, 1–12.

3

Handling the Information Explosion: the Challenge of Data Management

J.E. Anderson

BIOSIS, 2100 Arch Street, Philadelphia, PA 19103-1399, USA
Fax: 215-587-2074/E-mail: jeanderson@mail.biosis.org.

Introduction

Rise up, my fellow life-scientists, and throw off the bonds of techno-tyranny! Information technology is a servant – at most a colleague – not a ruler or tyrant.

Most scientists have suffered the plight of working with a knowledgeable colleague in a foreign language. Untoward time and energy is expended on communication, thereby reducing effort and knowledge available for the issue at hand. For such a collaboration to grow, both parties must work toward learning the language of the other, otherwise the labour imbalance may become detrimental, even causing termination. Regarding an information/knowledge data base as a servant or colleague gives rise to a healthy assessment of effort expenditure. Why should computer data banks not come as much to us (learning the language of science) as we to them.

What then, can life science expect of its colleague, information technology? Life scientists can expect, and should demand:

- to communicate with information technology in a manner more akin to that of humans – 'natural language'.
- to reach the knowledge, in real time, beyond the abstract of a publication, beyond the publication, to the data/ information/ knowledge from which the publication was written, and even to the author/ scientist.
- to communicate via subject matter concepts with data technology, not merely the scientific baby-talk of cooccurrence of a few terms which may or may not capture the concept.

 Information Technology, Plant Pathology and Biodiversity (eds P. Bridge, P. Jeffries, D.R. Morse and P.R. Scott)

The last of these three points is the subject of this chapter. It is related to, and even dependent upon, the first two.

Harnessing the Power of Information Technology

Information technology is a servant or a colleague, but most of us have suffered while working with a knowledgeable colleague in a foreign language. Time, effort and knowledge available for the issue at hand is reduced by the energy expended on communication. For such a collaboration to grow, both parties must work toward learning the language of the other, otherwise the labour imbalance may become detrimental. Regarding an information/knowledge database as a servant or colleague gives rise to a healthy assessment of effort expenditure. Why should computer data banks not come as much to us (learning the language of science) as we to them (we have all learned some of the language of technology through exposure to the Universal Resource Locator on the World Wide Web of the programme of the conference at which this chapter was first presented! – http://www.bspp.org.uk/dec96con.htm).

What does it matter, beyond frustration? It matters because the product of science – knowledge – is being lost in intellectual noise akin to that of an international border flea market – as opposed to the organization and decorum of a knowledge department store or library. Disregarding the possible charm of the market, why is it so noisy? It's noisy because scientist buyers and sellers in the intellectual din have to shout and gesture translations to the delivery lads who are rushing madly about, looking frantically in each stall, bumping into each other, queuing at the intersections, disappearing for electronic 'days at a time'. All the while each ignores the other; fishmonger next to clothier, palaeontologist next to pathologist. Does this sound like the Internet? And that's only the tip of the iceberg.

Organizing the Content of Large Information Resources

We all go to any library without giving thought that someone has organized the books and magazines in some logical fashion. And they are organized, not by size, not by binding colour, not by author's first name, but by content – the Dewey Decimal System, say. And someone has provided a way to find things – the card catalogue, now usually a personal computer. Where is the Dewey Decimal System of today's science information on the Internet?

Well, what's the problem? The fact is that library organization systems only became necessary as libraries outgrew private collections in single rooms where one could browse the titles, perhaps sending a servant up one of those quaint ladders on wheels – the YAHOO of yesteryear. Just as servants, even on roller skates, can't browse huge random book stacks, the knowledge of science today must be in some way organized for access. The retrieval and classification systems being developed for large information resources like the World Wide Web (YAHOO, LYCOS and many others) are an early attempt to impose some organization retrospectively. They are at a transitional stage where the raw power of IT has run well ahead of the development of intelligent tools to make the best use of it. An undergraduate student LYCOS retrieval of 12,000 citations from a search of a virus name, akin to a lorry load of books from a liveried servant, is ridiculous!

The problem, then, is twofold. The information market is disorganized – the fish stalls are sprinkled among the jewellery and furs; and the delivery boys don't speak the language. A noun or two, along with a lot of gesturing, is no way to study or share the effect of plant disease on world food supplies. In part because of the continuing information explosion, science must transfer more of the burden of communication with information technology to information technology. If we, as the ultimate life form, don't train and delegate more responsibility to our lackeys (runners to computer knowledge banks), we'll not have time and energy to do what only scientists can do: science.

Retrieval Systems for the Information Explosion

Consider simultaneously the runners and the market; and then, in fairness, our responsibility to straighten up the scientific flea market of science knowledge. When the market was in the village square and the database was a few hundred thousand records, it was entirely sufficient, if quaint, to grunt one word to the runner (a single term query). The market was not so big that, even if the runner only had a faint notion of the quest, one could not paw through the resulting basket, retain the Delicious apple variety, and discard the 'horse' apples, both the Osage oranges and the road droppings (Table 3.1). Distinguishing information about apple scab as a disease, from apple scab as a fungal organism (Ascomycetes), from apple scab as part of injured fruit healing, was annoying but do-able. Given time, all those books about 'Arizona' as a pleasant place to visit are discernable from publications about the bacterium *Arizona* under study by the scientist (Table 3.2). '*Abas*' as in 'Chrysophyta' (phytoplankton) is miles from '*Abas*' in 'Sauria' (lizards). To make matters worse, the runners are absolutely insensitive to both case and font, happily ignoring all printing conventions about what is a subset of which. As an aside, however, the net shopping results were and are a measure of knowledge density, or paucity, in the information market.

But the village and market grew and grew, doubling and doubling, again and again. The knowledge density did not necessarily change, only the size of the shopping area and therefore the results. Since the runners now came back with much more than we were willing to sort through, we taught them the notion of dealing with two words at a time, and to bring it back only if it was

Table 3.1. To retrieve information on apples, and then apples as fodder for horses.

Key Token	Key Term Type	Variant Token	Variant Term Type	Modifier Token	Modifier Term Type	Classifier Token	Classifier Term Type
Racehorse	Organisms: common name					Equidae	Taxonomic
Horse apple	Organisms: common name	Osage orange	Organisms: common name			Moraceae	Taxonomic
Horse apple	Miscellaneous descriptors	Horse dung	Miscellaneous descriptors	Fertilizer	Detail		
Apple	Organisms: common name			Fodder	Detail	Rosaceae	Taxonomic
Apple-cv.-Delicious (Delicious apple)	Organisms: common cultivar	*Malus domestica*	Organism: genus	Food	Detail	Rosaceae	Taxonomic

Old query: 'apple' would retrieve apples as well as Osage oranges and dung.
New query: apple *modified* by food = Organism.
Old query: 'apple and horse' would retrieve Osage oranges which are also referred to as 'horse apples' as well as papers referring to the vernacular use of the phrase 'horse apples' (i.e. horse dung).
New query: apple modified by fodder = Organism, horse = Organism: common name.

Table 3.2. To retrieve information specifically discussing *Arizona* as harmful food bacteria.

Key Token	Key Term Type	Variant Token	Variant Term Type	Modifier Token	Modifier Term Type	Classifier Token	Classifier Term Type
Arizona	Organisms: genus			Food contaminant		Enterobacteriaceae	Taxonomic
Arizona	Geopolitical locations					USA	Geographic

Old query: Arizona.
Problem: Information regarding the geopolitical area in the USA.
New query: Arizona *modified by* contaminant = Organism.

'fish and fresh'. There were still occasional problems: the results of the phrase 'horse and apple', while not complicated by table fare, still confused Osage oranges with stable fertilizer sources and equine food source information. Nor is 'hard red wheat scab' as a pathogen the same as 'hard red wheat scab' as a disease epidemic distribution or even as a protective covering (Table 3.3). At least 'Rhodophyta *Abbottella*' (red algae) can be separated from 'Gastropoda *Abbottella*' (snails) by simply inserting 'and'. And then came 'or' and 'not' and finally, when those were not enough to cope, 'eor', 'co-occurrence', 'proximity' and 'juxtapose'. All of these 'relational operators' for our nouns, and clever combinations of our terms of science could approximate the knowledge and language of science. But we still avoided teaching, or demanding that our runners learn, the language of science.

Instead, we hired translators who learned both some of the science and the clever operators. These specialists made careers (as librarians and information specialists) of approximating our information wants and needs with sophisticated combinations of the clever operators and then translating these to our runners for us. But still, when we want 'information about ... ', we settle for 'has mention of, some place, for some reason ... ', and then have to spend time culling the unwanted. We want information about the organism 'Connecticut virus' and we waste time culling information about an influenza outbreak in the northeast United States caused by an unrelated virus (Table 3.4).

But now, the unthinkable: a shopping mega-mall has come to the village. It is no more organized, just bigger, and is adding hugely to the confusion. Its name is Internet, it is only the vanguard of more to come, and it is proletarian! Once again, we are dealing directly with our runners, but now they're returning with barrows, not baskets; train loads not lorry loads. 12,000,000 LYCOS citations for a genus species search, indeed!

Information Objects and New Indexing Concepts

But help is at hand in the form of a few simple concepts:

1. Information objects. Stop and think. Nuggets of information are what we seek, not papers or Web sites or databases. Those are just unit containers or sources. If one craves liquorice jelly beans, one would hate to overlook pound packages that might contain a few. Those black beans in mixed lots are just as tasty, if they can be found. A relevant information object in a single sentence or record in an otherwise unrelated paper or data base may be no less useful than that in a whole paper or computer bank on the topic.

2. Objects tagged by terms each with a fixed list of types: An invisible main 'tag' on each object (Key Term) with optional tags (e.g., Modifier, Classifier, Variant) could characterize the object, thereby resolving ambiguity by creating the context, as in Table 3.5.

Table 3.3. To retrieve information specifically discussing hard red wheat scab as a pathogen.

Key Token	Key Term Type	Variant Token	Variant Term Type	Modifier Token	Modifier Term Type	Classifier Token	Classifier Term Type
Hard red wheat scab	Organisms: common name	*Gibberella zeae*	Organisms: genus	Pathogen	Role	Ascomycetes	Taxonomic
Hard red wheat scab	Diseases			Fungal disease	Disease affiliation		

Old query: hard *adjacent* red *adjacent* wheat *adjacent* scab.
Problem: would also retrieve papers where hard red wheat scab was discussed as a disease with no mention of the disease organism. In addition, the query would retrieve papers where the disease organism is mentioned but in a molecular biology context and not as a pathogen.
New query: hard red wheat scab *modified by* pathogen = Organism: common name.

Table 3.4. To retrieve information on an influenza virus outbreak in Connecticut.

Key Token	Key Term Type	Variant Token	Variant Term Type	Modifier Token	Modifier Term Type	Classifier Token	Classifier Term Type
Connecticut virus	Organisms: common name			Pathogen		Rhabdoviridae	Taxonomic
Connecticut	Geopolitical locations					USA	Geographic
Influenza virus	Organisms: common name			Pathogen		Orthomyxoviridae	Taxonomic
Influenza	Diseases			Respiratory disease; viral disease	Disease affiliation		

Old query: Influenza *and* Virus *and* Connecticut.
Problem: query would retrieve papers on the Connecticut virus as well as papers in which these terms cooccur at random.
New query: Influenza = Disease, Influenza virus = Organism, Connecticut < USA = Geopolitical location.

The above mentioned examples of ambiguity and confusion can be resolved as information objects which provide an improved level of knowledge approximation (over that of relational operators). A hypothetical database addressed with tools that resolve these ambiguities could produce results like those in Tables 3.1–3.4.

Since the natural language index concept 'a eats b' or 'b hosts a' has so much more meaning and context than 'a and b are in here somewhere, for some reason', these sharper tools are less likely to work across disciplines. Whereas 'and' has the same meaning in both chemistry and biology, 'eats' has a dubious relationship in chemistry, which will need to develop its own relationships, perhaps 'a is a reagent with b'.

While logical relational operators approximate the knowledge content of an object in an effort to provide relevance, this new indexing is much less of an approximation, which improves relevance. Conversely, recent informal studies at BIOSIS show a meagre (approximately 10%) relevancy overlap at the top of the rank-ordered search results of YAHOO, LYCOS, etc. That's because these runners don't speak the language of science and the village mart is in disarray.

BIOSIS is applying these 'New Indexing' principles to its 1997 bibliographic data. The problem is that none of the runners can capitalize on the power– yet. BIOSIS staff will spend much of next year educating scientists and runners on the possibilities. By the time the world is up to speed, a whole year's worth of data will be available. But this is only the beginning. The concept should easily extend to life science knowledge of any sort: bibliographic, database, Web site, metadata, etc. The BIOSIS 'world view' of knowledge, its source, and availability looks like a plant root system, potentially reaching all the way to the author/researcher/scientist, Fig. 3.1.

Table 3.5. Examples of terms and types for tagging objects.

Terms	Types
Key Term	Organisms, Diseases, Methods, Geopolitical, Persons,
Modifier Term	Role, Gender, Developmental Stage, New taxa indicators, ...
Classifier Term	Taxonomic, Geopolitical.
Variant Term	Common names, Synonyms, Slang, ...

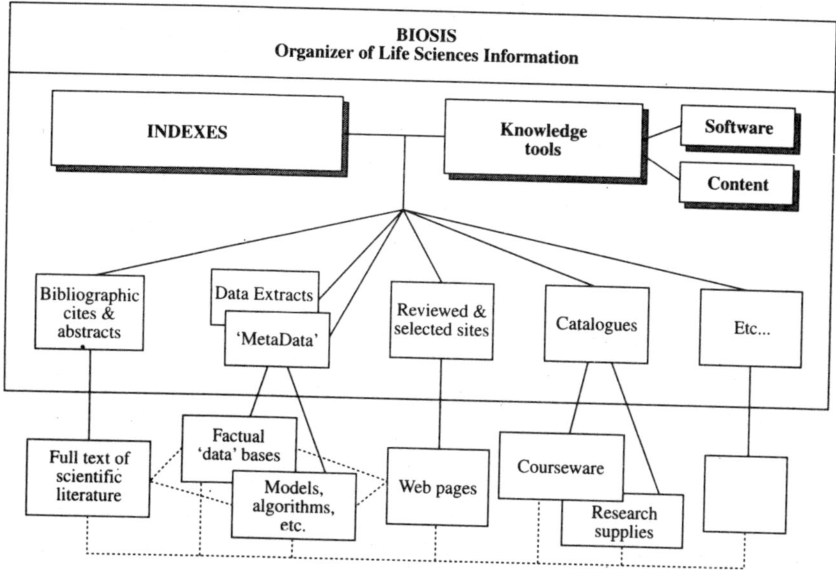

Fig. 3.1. Conceptual diagram of the BIOSIS 'world view' of knowledge that will become accessible through the application of 'new indexing' tools.

Modelling Taxonomic Descriptions for Identification

J. Lebbe and R. Vignes

Laboratoire Organisation et Évolution des Systèmes, Université Pierre et Marie Curie, 4 Place Jussieu, 75252 Paris Cedex 05, France
Fax: +3301 4427 5963/E-mail: lebbe @ ccr.jussieu.fr

Introduction

Taxonomic names are the main access keys to biological information. So the identification of life forms is crucial for all human activities related to biology. This stands to reason for both biological and medical studies but it is also true for industrial domains. Even the naturalist knowledge of the general public is based on the identification of living plants and animals, or fossils.

The tasks of systematists are precisely to inventory, to study and to structure biodiversity. Because dissemination of this knowledge relies on identification, it is the systematists' responsibility to provide reliable and efficient identification tools.

The identification of a biological specimen is its assignment to a taxon previously known and placed in a classification. Assigning an object to a concept is a universal process, for instance: diagnosing a disease or an engine failure, recognizing a word, a shape, or a pattern, identifying phytosociological associations, molecules or rocks etc. Because of the general scope of identification, works related to this subject are very numerous and also concern various domains outside biology. So research on biological identification is naturally placed in a perspective larger than systematics.

Even today, the printed key is the most frequently used identification method in biology. But since the sixties, identification is a topic touched by the computer revolution. Many computer programs were and are developed. One traditionally distinguishes monothetic methods, where taxa can possibly be discarded by only one character, as in keys (single-access) or most computer-aided identification tools (free or multi-access) and polythetic methods where several

 Information Technology, Plant Pathology and Biodiversity (eds P. Bridge, P. Jeffries, D.R. Morse and P.R. Scott)

characters are used simultaneously to evaluate a matching score between the specimen to be identified and the taxa (Pankhurst, 1978). The methods are also classified as probabilistic and non-probabilistic.

As is suggested by these classifications, most of the methodological work in biological identification is oriented towards process conception. In this chapter we discuss the importance of taxonomic knowledge representation in obtaining meaningful identification tools. Diedrich *et al.* (Chapter 5, this volume) discuss some issues in the structure of taxonomic databases, and the sections on *Computer-based Species Identification* address the application of taxonomic information to identification from many perspectives.

Representations of Concepts for Identification

Independently of the internal procedure, the principle of all identification methods, with or without a computer, is to compare a representation of an individual object, that is the specimen, with the representations of concepts that are already known, that is the taxa. Representing individual objects is often more straightforward than concepts. We can distinguish four types of concept representations. The comparison between these types is important because taxonomic description, which is the main concept representation used in biological identification, belongs to one of these types and because there is a strong interdependence between the representations and the procedure comparing the representations.

Four types of concept representations (Table 4.1)

A concept always refers to a set of instances and its representation has to derive from their descriptions. This can be done in many different ways. For example, one particular instance, real or virtual, may be chosen or constructed as paragon of the concept. In this case, the representation of the concept is the description of its prototypical instance. This type of concept representation was conceived in cognitive science to explain the properties of our mental representations (Rosch, 1975). This also has to be related to frame-based knowledge representation in artificial intelligence (Minsky, 1975). The main drawback of this type is its inability to represent variability limits explicitly.

In statistics and data analysis, a concept is usually represented by a set of instance descriptions, without particularizing one as prototype. For example, if we want to discriminate two species with an artificial neural network, we have to present to the network the descriptions of a sample of specimens. From this initial representation of the species the network will derive a more concise internal representation. More generally, from this type of representation, any kind of other representation could be computed.

A concept, here a taxon, is at once one element (a class of a taxonomic

Table 4.1. The four types of concept representations.

	Example of representation	Advantages	Drawbacks
Prototypical example	Adult lion has a mane and weighs 200 kg	Conciseness	Does not represent variability
Set of instances	Lion 1 has a mane and weighs 183 kg, Lion 2 has a mane and weighs 217 kg, Lion 3 does not have a mane and weighs 146 kg	Base of all representations	Needs a lot of space
Characterization	If a mammal has a mane and weighs more than 100 kg it is an adult lion	Extreme conciseness	May record very little information
Statistical summary	Adult lions may or may not have a mane and weigh between 150 and 250 kg	Conciseness Represents variability May record a lot of information	Over-generalization (see below)

classification) and a set of instances (set of specimens) (Cracraft, 1983). This duality is met again in the different representations of concepts. If the first two types of concept representation are directly related to the representation of instances, the two next consider concepts as individual entities.

In many expert systems the representation of a concept is equivalent to a characterization of the concept, i.e. a set of attributes which allows the concept to be discriminated from others. This representation is equivalent to a rule. The basic reasoning is: if the attributes of a specimen are in the set recorded for the concept, the specimen belongs to it. This type of concept representation is common in systematics. For example, the different attributes of a path in an identification key form a partial characterization of the taxon identified at the end of the path. The disjunction of all the paths leading to a taxon is a full characterization. The diagnosis of a taxon is another example of such characterization. The main advantage of this type of representation is its possible extreme conciseness: we record only the information needed for discrimination, but due to this conciseness a characterization requires very limited knowledge.

The last possible type of concept representation is a statistical summary of the multidimensional variability within the concept. Such a concept description is more exhaustive than a characterization because it records the full polymorphism of the instances of the concept even for the attributes that cannot provide complete discrimination. Taxonomic descriptions belong to this type of concept representation.

Taxonomic descriptions

Taxonomic descriptions are generally expressed in a pseudo-natural language, coding description elements, one for each descriptor. Each element formulates a statistical summary of the distribution of values due to biological polymorphism. Theoretically, any type of descriptive statistic could be used (range, histogram, central tendency, dispersion measure, etc.) but ranges are the most common. In the case of a qualitative descriptor, the range is a set of values (example: the petal colour is red, pink or rose) and in the case of a quantitative descriptor, the range is an interval (example: the petal size is between 12 and 22 mm). For example, in the description of the genus *Trifolium* H. Coste notes two possible shapes for the calyx and various numbers of grains: 'Calyx tubular or bell-shaped, with 5 teeth equal or unequal; corolla almost always marcescent. Fruit straight or almost straight, 1 to 2 seeds round, rarely 3 to 6 ...' (Coste, 1900–1906).

It is the double nature of a taxon (a concept and also a set of instances) that accounts for information such as 'teeth equal or unequal'. If this part of the description does not give information for one specimen of *Trifolium*, it is informative for the genus *Trifolium* as a whole because it rejects the fact that the calyx teeth are always equal or always unequal.

Usually one speaks about probabilistic and non-probabilistic methods for biological identification. But, independently of the identification method, a taxonomic description is always a statistical summary of the biological polymorphism. Even if a description element has only one value, the real meaning of this value is a distribution, the described concept being a class of specimen (Jardine and Sibson, 1971). But is it the mean value, the most frequent value, or the only possible value? The answer to this question requires a semantic model of the description.

The main drawback of taxonomic descriptions is over-generalization. In such a description each description element is recorded separately and can be interpreted as a conjunction of attributes: 'Calyx tubular or bell-shaped and with 5 teeth and teeth equal or unequal; ...'. Even if the text expresses that all the combinations are possible (for instance: Calyx tubular and teeth equal, Calyx tubular and teeth unequal, Calyx bell-shaped and teeth equal, Calyx bell-shaped and teeth unequal) no one is sure, even H. Coste, that these combinations really appear in nature. The over-generalization clearly results from the conjunctive bias.

To build an identification tool (key, computer-aided identification system, etc.) systematists can use four sources of knowledge: personal expertise, observation of specimens and special types, texts, and sometimes previously computerized knowledge. All these sources are useful but the textual sources are critical for the future of biological identification. A big project like *Systematics Agenda 2000* will succeed only if we are able to make use of the 10^6 monographs, flora or fauna that are available.

In a monograph, a flora or a fauna, we find diagnoses, taxonomic descriptions and identification keys. All these texts give information about the different taxa, but they are not equivalent. Among the textual sources, taxonomic descriptions are the most important because they are the most informative. Both in diagnoses and keys, a limited number of descriptors are used, different for each subset of the taxa. So, these knowledge sources are restricted to a specific identification context: if some important attributes are lacking, diagnoses and keys are useless for identification.

In conclusion, to computerize the collected knowledge about biodiversity meaningfully we need to know precisely how to interpret taxonomic descriptions. Because there are many different ways of creating a statistical summary of a distribution of values, the interpretation is complex. In all the monographs, flora, and fauna written by systematists, the meaning of taxonomic descriptions is probably not always the same. So, we believe that a large knowledge re-engineering of taxonomic descriptions is the key issue in developing the use of computer-aided identification in biology.

Knowledge representation is an important domain in computer science and particularly in artificial intelligence. In order to derive scientific benefit from the different works on identification and knowledge representation, we think that the first step is to clarify by a model what the accumulated knowledge about taxa for identification represents. It is critical because the meaning and the properties of the representations have to be taken into account within the identification procedure.

Modelling Taxonomic Descriptions

To determine the meaning of taxonomic descriptions, we have to know what is the question that can be answered by these descriptions. Moreover, this question has to be precise enough to allow for the definition of a procedure aiming at constructing the taxonomic descriptions from a set of specimen descriptions. This does not imply that the descriptions have to be constructed automatically with a computer, but only that the procedure defined is a realistic model of how the descriptions are or were constructed.

Different construction procedures can have the same type of output. So in a computer format, like DELTA, it is not enough to have a semantic representation of the taxonomic knowledge. For instance, if we record three colours for the flowers of a species, is it the exhaustive list of colours observed at least once, or at least in 10% of specimens etc.? If we record a single median value for a numerical descriptor, is it the observed median in a sample or the estimated one?

To differentiate the significance of different characters is important because, according to the interpretation, the identification algorithm may not always be the same. Moreover, it is dangerous for the quality of results if the meaning of a description is not completely defined. For example, if some algorithms interpret the previous three colours as an equiprobable distribution instead of a simple list of possible values without direct probabilistic interpretation, the result will probably be meaningless.

Most of the biological identification systems are based on taxonomic descriptions. But they use neither the same computer formats to express the taxonomic descriptions nor the same methods to compare the representation of the specimen to be identified and the representations of the concepts. Are these only superficial differences or the consequences of different taxonomic description interpretations? How to decide which one to use?

Very few abstract evaluations of the algorithms proposed for biological identification are published. In this domain most papers concern the description of the program function or empirical evaluation of the result. But a detailed description of the algorithm properties should be very useful at least to make the user confident that for his or her data set the program will be able to run well.

In such a description, we need to define first the global criterion to be optimized by the identification algorithm. This could be minimizing error rate or identification cost, etc. or even a complex criterion concerning user interaction. There could also be more theoretical criteria, such as order independence (the result may have to stay the same with input order modification), monotony (a result may have not to be falsified after a further step), etc. Without such precise criteria in mind we cannot assess if the algorithm succeeds or not. Because the value of the criterion does not depend only on the algorithm but also critically on the input, we need to characterize in formal terms what could be these inputs in a model of taxonomic descriptions. Only then will it be possible to specify an algorithm and to give the proof that it satisfies the criterion when applied to taxonomic descriptions fulfilling the model.

Many models have useful characteristics. We propose some comments about three simple ones. In each case the taxonomic descriptions require answers to different questions, precisely enough to suggest how to compare the representation of a specimen and the representations of the taxa.

Model 1: recording the possibility space

In this first model, a taxonomic description is the answer to the question: for the taxon, what is the exhaustive list of states for each description element? That is: what is the exhaustive list of colours, the exhaustive size range, etc.

With this model, a conceptually simple identification procedure exists based on abductive reasoning. If the specimen does not present a state possible for the taxon, it cannot belong to this taxon. When only one taxon remains in the list of possible taxa, the identification of the specimen is made. This is the main reasoning used in INTKEY (Dallwitz, 1993 and Chapter 19, this volume), ONLINE (Pankhurst, 1991 and Chapter 26, this volume) and XPER (Lebbe *et al.*, 1989).

The simplicity of this model is attractive but unfortunately it requires a complete knowledge of the polymorphism to give exact results. But how to know the true possibility space of a taxon? How this could be inferred from the specimen description? Another drawback of this model is over-generalization: a taxon that should be discarded could be kept in the list of possible taxa, making the identification process longer than necessary.

Model 2: recording the sampled space

In this case we ask the following question: What are the known states for each description element? If we record only this information in a taxonomic description we are in a second model. We obtain the same set of states or a range as above but the meaning is different. The difference is parallel to population description *versus* sample description in statistics. Knowing if our description belongs to model 1 or model 2 is important in practice because in a simple computer format we cannot distinguish these two different representations.

The main drawback of this model is that the result of the description phase depends clearly on sampling pressure. At one extreme, for the numerous taxa known only from one specimen no polymorphism will be recorded even if we know that it does exist.

If all the polymorphism is not recorded then a specimen might not be compatible with the description of its own species. A numerical matching method is a possible identification procedure. The specimen is identified to the taxon including the combination of states which is the most similar to its description. Many different similarity measures can be used. But how to justify the use of one of these measures or another? Moreover, how can the decision procedure be endorsed and, more precisely, how much dissimilarity should require that a taxon be definitively discarded?

Model 3: recording conditional probabilities

A third model can be derived from the answer to the question: What is the estimated probability distribution for each description element? For discrete descriptors we obtain a set of states with a probability for each one, and for continuous descriptors, distribution parameters could be recorded.

The Bayesian method, which is used in several identification systems, uses this model and is able to minimize the identification error rate. But, unfortunately, it is not a realistic model for most published taxonomic descriptions. The conditional probabilities are almost always unknown and impossible to deduce with certainty from monographs. At worst, if some description elements are lacking, this model will represent the absence of information with equiprobability, introducing confusion with the case where equiprobability is explicitly known.

Moreover the Bayesian method, like most of the matching methods, is not monotone. The best decision could become the worst after adding new description elements and could even become the new best one if the description is again extended.

More complex models?

As we have seen, each proposed model has advantages and drawbacks. We do not want to defend one particular model. Our goal is to claim: (i) the necessity to specify the model used in an identification system; and (ii) the necessity to have a model compatible with textual taxonomic descriptions. To be meaningfully compared, the taxonomic descriptions have to rely on the same interpretation, which is not always explicit. The true nature of a taxonomic description must be defined enough to allow for a definition of an automatic procedure aiming at constructing taxonomic descriptions from a set of specimen descriptions.

Research is still needed on the biological practices related to the construction of taxonomic descriptions and on knowledge representation. For example, more complex models have to be defined so as to rely on a realistic procedure for constructing taxonomic descriptions, minimizing over-generalization, taking account of variation and uncertainty, and compatible with textual taxonomic descriptions. Based on such a model it should be possible to enhance identification algorithms, ensure they are monotone, and optimize a clear set of criteria.

Conclusions

Computer identification systems are becoming more and more numerous. But the classical key remains the main identification method. If we want to develop a computer-aided systematics and convince potential users that it is useful we have to reject 'black-box' software: the methods have to be published. They have to be endorsed and these endorsements have to rely on a semantic model of taxonomic descriptions.

Computer-aided systematics will be an important part of the future of systematics (Lebbe, 1995). Its development forces us to prepare a new generation of systematists. The teaching must be methodological, with a biological part and a computer science part. It must be generic, with problems and solutions not related to only one specific taxonomic group. Modelling taxonomic descriptions has a central importance in the development of this new domain.

References

Coste, H. (1900–1906) *Flore Descriptive et Illustrée de la France*. Librairie Scientifique et Technique Albert Blanchard.

Cracraft, J. (1983) The significance of phylogenetic classifications for systematic and evolutionary biology. In: Felsenstein, J. (ed.) *Numerical Taxonomy*. Springer-Verlag, Berlin, Germany, pp. 1–21.

Dallwitz, M. (1993) DELTA and INTKEY. In: Fortuner, R. (ed.) *Advances in Computer Methods for Systematic Biology*. The Johns Hopkins University Press, Baltimore, USA.

Jardine, N. and Sibson, R. (1971) *Mathematical Taxonomy*. John Wiley & Sons Ltd, London, 286 pp.

Lebbe, J. (1995) Systématique et informatique, Systématique et Biodiversité, *Biosystema 13*, Société Française de Systématique, Paris, 71–79.

Lebbe, J., Vignes, R. and Dedet, J.P. (1989) Computer aided identification of insect vectors. *Parasitology Today* 5, 301–304.

Minsky, M. (1975) A framework for representing knowledge. In: Winston, P. (ed.) *The Psychology of Computer Vision*. McGraw-Hill, New York, pp. 211–281.

Pankhurst, R.J. (1978) *Biological identification*. Edward Arnold, London, pp. 55–67.

Pankhurst, R.J. (1991) *Practical Taxonomic Computing*. Cambridge University Press, pp. 11–43.

Rosch, E. (1975) Cognitive representations of semantic categories. *Journal of Experimental Psychology* 104, 192–233.

A General Structure for Biological Databases

5

J. Diederich[1], R. Fortuner[2] and J. Milton[1]

[1]Department of Mathematics, University of California, Davis, CA 95616, USA; [2]4 rue des Jardins, 17130 Montendre, France (Correspondant du Muséum National d'Histoire Naturelle, Paris) Fax: +33 546 703659/E-mail: fortuner@wanadoo.fr

Introduction

The Project NEMISYS (Nematode Identification System) was launched in 1987 to create an identification system that would not be restricted to a particular identification method but would have the potential to use and support all identification approaches. The need for a multi-method approach to identification was argued by Fortuner (1989, 1993). A tool-based expert workstation which can meet this requirement was described by Diederich and Milton (1989, 1993a), and a NEMISYS prototype was developed using several identification tools as examples (Diederich and Milton, 1993b). NEMISYS later became GENISYS (General Identification System) when it became apparent that the concepts we developed applied outside nematology.

One of the most critical lessons learned in NEMISYS was the importance of laying a solid foundation for data and knowledge representation. Methods employed by others in developing databases for identification systems seemed inadequate in addressing the difficulties we experienced with nematological data. In recent years, the construction of a database has subsumed the entire project and given it a different orientation.

From the very beginning, the standard data decomposition of entity/property/value used by database designers (and often used for biological databases, e.g. Lebbe, 1991; Lebbe and Vignes, Chapter 4, this volume) was used for the definition of a list of characters. However, biological characters do not easily and comfortably fit into this form. Early attempts at defining a list of characters revealed that this alone would not solve the problem of a lack of homogeneity in the data. It was necessary to define new concepts such as basic properties,

 Information Technology, Plant Pathology and Biodiversity (eds P. Bridge, P. Jeffries, D.R. Morse and P.R. Scott)

name extensions, general/specific states, and state-based relationships (Diederich, 1997) and to propose guidelines to ensure that these concepts would be used correctly (Diederich *et al.*, 1997) in creating a list of characters that could be consistently and uniformly expressed. These concepts are not only important for creating morpho-anatomical databases in nematology, but they can also play a central role in other areas.

Here, we assume that it is accepted that it is better to have access to several different approaches to identification than to be restricted to a single one, and that it is better to have a database that can be used for multiple purposes. Ideally, the data in any database is used through a database management system (DBMS). However, there are limitations on what can be done in existing commercial DBMSs to support biological data and their applications. This is not surprising because DBMSs are typically business oriented and generally ill-suited to the complexity of biological data. What is needed is a Bio-DBMS; a DBMS that is created specifically for biological data and applications. In fact, part of our project will ultimately involve the definition of the architecture of a Bio-DBMS, and what we present here are some preliminary requirements for it. Naturally, a Bio-DBMS would be built on top of a commercial DBMS since it would not be practical to create a Bio-DBMS from scratch.

In this context we will: (i) describe some of the problems with traditional characters and demonstrate the need to develop and enforce a general discipline in building biological databases; (ii) give some examples of information, other than simple data, that will be needed in any application; (iii) give examples of non-morphological data that can be captured using our methodology; and (iv) discuss some key requirements for a Bio-DBMS.

Definitions

Here, the word 'character' will be used in its widest sense to mean either an abstract concept that can be used to describe and differentiate a species or other taxa, or a characteristic in an actual specimen and that can be used to identify this specimen. A character will be called 'traditional' if it is in the form found in traditional keys, e.g. organ-X cylindrical, or if it is composed of a characteristic that can be used to differentiate two taxa, e.g. shape of organ-X, and a set of values taken by this characteristic in the taxa or specimens considered, e.g. cylindrical, ovoid, spherical, etc. In many computer identification applications, characters are: (i) given in their traditional form; (ii) numerically coded and stored in a data matrix; or (iii) coded using the DELTA code and stored in text files (Dallwitz, 1980; Webster, 1988).

Problems with Traditional Characters

Selection of a subset of characters

In all the cases we have observed so far, identification aids use a subset of characters selected by an 'expert' (in the knowledge engineering sense of the term) out of all possible characters as being 'good identification criteria'. The states taken by these selected criteria in different taxa are also defined by the expert. We will not discuss the methods of character and state selection but emphasize the fact that there is selection. On the one hand, this character selection and pre-processing by the expert is compulsory for the creation of the matrix used by taxonomic programs or the text files used by DELTA applications (Dallwitz *et al.*, Chapter 19, this volume). On the other hand, it eliminates from the application database all the characters not selected by the expert and makes them unavailable to: (i) another expert who does not agree with the first one about the list of 'good identification criteria'; (ii) an expert using a different identification approach; and (iii) an expert who wants to use the unselected characters outside of identification, e.g. for taxonomic studies. This results in much duplication of effort, as each expert has to build a database from scratch.

Use of complex characters

Adding to this basic difficulty is the fact that many traditional characters, and many computer-coded characters, are complex entities that group several simpler concepts. For example, the character 'disk with a central zone densely covered with flat, heart-shaped, epidermal denticles' found in a database on lampreys (Cruette, Paris, 1991, personal communication) indicates that a central zone is present, that it is densely covered with epidermal denticles, and that these denticles have a particular outline (heart-shaped) and cross-section (flat). This type of character is well suited to traditional or computer keys (either an unknown has such a disk or not), but it is not suited to other approaches. For example, it would be difficult to compute a coefficient of similarity between a species with the character above and one with a central zone sparsely covered with flat, heart-shaped, epidermal denticles.

Classes for quantitative characters

A final problem is that many coding methods transform quantitative values into classes before storage. Often, lengths are recorded, not as the actual value in a specimen, a population or a taxon, but as a class value. This again is well suited to keys (if a taxon belongs to the same class as the unknown it will be retained; it will be eliminated if it does not), but not to other approaches, such as computation of the NEMAID similarity coefficient (Fortuner and Wong, 1985), which uses actual values.

The NEMISYS *approach*

Description

The solution adopted for NEMISYS and now GENISYS was to store as many characters as possible, in particular all the characters found in published descriptions of taxa. Actual numerical values and qualitative states of the characters, as they were recorded in the literature, were also stored. Complex characters were decomposed into elementary characters using the entity/property/value decomposition. Any selection of characters for particular uses or applications would be done *a posteriori*, from this general pool of characters.

It was quickly discovered that this was not enough to ensure a homogeneous representation of the characters. This problem was partly solved by enforcing strict guidelines for the decomposition of characters. Using morphological characters only, we defined an entity as the parts composing an organism, starting with the organism down to organs, tissues, cells, etc., all the way down to molecules. Doing this, we found that most of the properties used in the original nematode list of characters come from a very short list of basic properties which are given in Table 5.1 (Diederich, 1997). To these properties are attached the traditional states or values, including synonyms, general states, and fuzzy states for measurements and quantities.

Typical examples of the main types of characters (i.e. real numbers, integers, and qualitative states) with actual values from the description of a nematode (*Helicotylenchus dihystera*) are:

Body	length	725 μm
Lateral field lines	number	4
Tail	shape	dorsally curved

Note that these are individual values, but the database includes fields for range, mean and standard deviation for an individual population and for a composite

Table 5.1. Basic properties for morpho-anatomical data. The list of properties is broken down into four groups relating to: appearance, location, dimensions and quantity of the morpho-anatomical structure described by the properties.

Appearance	Dimension	Placement/Location	Quantity
posture	length	position relative to*	presence
shape	height	distance to*	quantity
kind	width	orientation	number
texture	diameter	angle	
arrangement	depth		
symmetry	ratio of*		
	size		

*Relational properties.

description of the corresponding species. The complex lamprey character given above would be decomposed into three parts linked together in a hierarchy of parts (disk has a central zone; central zone has denticles), and each part would be described by basic properties and the appropriate states. For example, the denticles would have the following properties and states:

Denticles	presence	present
	arrangement	dense
	width	flat
	shape	heart

The guiding principle here is that it is easier to construct complex entities from simple ones rather than the other way around. In addition, there is a greater chance to record all the important information.

Number of characters

Typically, systematic databases include a relatively small set of characters selected for a particular purpose, e.g. 30 or 50 characters selected for identification (i.e. characters which are easy to observe, clearly differentiate existing species, not variable, etc.). Thiele (1993) used 108 characters for his analysis of *Banksia*. Presumably, these characters were selected because of their value in representing systematic relationships. Compared to these very small numbers, our current nematode (order Tylenchida) list of characters includes 272 biological structures described by over 1000 characters (1 character = 1 structure, 1 property). The potential for growth is staggering as these 272 structures could be described by 20 properties each, which represents 5440 potential characters, and this number would be far greater if states were included in the count. Instead of including only a subset of characters selected for a particular purpose, a GENISYS database aims at storing every possible character to serve as a general pool of characters available for any purpose.

Requirements

Using GENISYS characters in DELTA applications

The data stored in the form described can be used by applications specifically created for it, such as the tools defined within the NEMISYS and GENISYS projects (Diederich and Milton, 1993b). However, storing data for use by *ad hoc* applications only would be an insufficient justification for such an effort. A critical factor in the general acceptance of a new method or approach in data management is its compatibility with existing databases and applications. With morphological data, particularly in botany, much of the data has been recorded in DELTA databases (Dallwitz *et al.*, Chapter 19, this volume), and it is essential that GENISYS-style data can be used with existing DELTA applications.

DELTA applications such as ONLINE (Pankhurst, 1991 and Chapter 26, this volume), INTKEY (Dallwitz, 1993), etc., are generic programs that can be used with any data, provided they are presented as DELTA codes in specially constructed files. Obviously, these applications cannot use GENISYS data directly. Fortunately, characters stored in the GENISYS format can be converted to the complex characters used in DELTA applications. If an expert wants to use the data stored in a GENISYS database with a DELTA identification program, a mapping from characters in the former to those in the latter is required. This can be done by creating the three text files required by DELTA applications (Webster, 1988), i.e. the CHARS file, a list of characters and character states selected by the expert, the ITEMS file, with the species specific values, and the character specifications file (SPECS). The CHARS file will need to be created in the usual way. For example, starting with the GENISYS lamprey data above, the expert will decide that, out of, say, 20 characters considered to be suitable for identification of lampreys, character number 9 refers to the disk, its central zone and denticles with the following codes, corresponding to states seen in known taxa:

#9. Disk/
1. with a central zone densely covered with flat, heart-shaped, epidermal denticles/
2. with a central zone sparsely covered with flat, heart-shaped, epidermal denticles/
3. with a central zone densely covered with thick, barrel-shaped, epidermal denticles/

Then, if a particular species is recorded in a GENISYS database with Denticles / presence = present; arrangement = dense; width = flat; and shape = heart, it will be presented to a DELTA application in the guise of an entry in the ITEMS file such as '... 9,1 ...'

It should be kept in mind that this is only an intermediate solution, as it is not the seamless mediation between database and application that could be provided by a Bio-DBMS. This mediation between the simple characters of GENISYS and the complex characters used in applications in the DELTA format is just one aspect of the set of requirements for large multipurpose biological databases. Here we are simply stating 'what' needs to be done, not 'how' it should be accomplished within a Bio-DBMS. An extension of the concept of a 'view', an existing mechanism within a commercial DBMS, may provide one way to do it. A view can be loosely defined as a virtual table, defined on top of a database, using the DBMS language. For the mapping of characters, using views of the simple characters would be a reasonable approach. Note that commercial DBMS have limited view mechanisms when applied to schemas, i.e. to the table definitions, so an extension of the view concept would be required in a Bio-DBMS.

Examples from systematics

The creation of a general database will be a major effort in time and money, and it needs to be used by as many people and in as many ways as possible. In particular, a general morphological database, including all the known characters of the included taxa, could be used as a source of systematic, as well as identification, characters. In fact, these two categories of characters differ in the criteria applied to select them for a particular application, but the characters would all come from the same general pool. Obviously, a GENISYS database with all existing characters can be used as such a pool.

A view mechanism, similar to the one evoked for DELTA applications, could be used to feed the proper data into other taxonomic applications. In cladistics, PAUP is one of the most used programs. As an example, here are a few characters from Thiele (1993), coded for the PAUP program:

1. Habit: erect (0); repent (1)
2. Lignotuber: absent (0); present (1)

...

50. Style color: red, yellow, golden or purplish-black (0); always yellow (1)
52. Pollen-presenter shape fusiform (0); acicular (1); linear (2); ovoid (32); conical (4); awl-shaped (5)

...

95. Adult leaf blade length (with a note saying that morphometric data is recorded as sample size, log10-transformed mean, and coded state).

The first number corresponds to a column number in a data matrix. The numbers between parentheses are the coded states entered in the rows of the matrix representing the taxa studied. In a GENISYS database, the same characters would be present as follows, where a '–' separates distinct states in the list:

Whole organism	posture	erect–repent
Lignotuber	presence	present–absent
Style	color	red–yellow–golden–purplish-black
Pollen-presenter	shape	fusiform–acicular–linear–ovoid–conical–awl-shaped
Leaf-Blade (adult)	length	actual value (sample size, mean)

Here, a 'PAUP View' would present GENISYS data in the guise of a data matrix, which could be used by any other program that uses a similar matrix of characters. The view would have to take into account the fact that Thiele lumped into a single state (50:0) what would be recorded as four separate states in the GENISYS database.

Of course, Thiele's data could have come straight from a DELTA database, if one had been created for identification within the same taxon. However, it is very likely that the set of characters selected for any identification database would have been different from the set of cladistic characters chosen by Thiele.

(Again, this is because the criteria for character selection are not the same in cladistics and identification.) Even a slight difference would have defeated the transfer of the data from one database to another. For example, if an existing DELTA identification database has:

#50. Style color/
   1. red/
   2. purplish-black/
   3. yellow, golden or always yellow/

Thiele would have been obliged to split the DELTA character state (3) between the two states of his character 50. There is no way he could have done that from the DELTA database and he would have had to go back to the original data. Note that we are not saying that DELTA forces the user to use these particular characters. Quite the opposite, the point is that DELTA leaves the user free to use any character. Freedom is fine, but then you are stuck with something that might be very difficult to use in a different application. In the GENISYS approach, each colour would be stored as a separate state, with yellow as a general state. A general state is a global expression, such as yellow, which may be divided into more specific states, such as golden (Diederich, 1996).

More generally, any coded data must be specifically defined by an expert who chooses the characters and defines the coded values, whereas our system records all the characters found in the descriptions, each in its simplest possible form and without any coding. Any selection is done later by selecting particular characters and states out of the complete pool of data.

Other Bio-DBMS requirements

The complexity of support needed within a Bio-DBMS can be appreciated in terms of the various relationships that must be defined for biological data. This support goes well beyond the task of storing the data and mapping between characters for the different purposes discussed above and it should facilitate efficient and intelligent use of the data. By efficient we mean that the Bio-DBMS should allow applications to use the data with ease, thus minimizing effort. By intelligent use we mean that the data should be properly retrieved and manipulated for the intended purpose. This does not carry the connotation that a Bio-DBMS will provide expert system capabilities, but will retrieve the intended data relative to the context of the request. Relationships play a key role in meeting this objective.

State-based character relationships

Using a view mechanism to define complex characters from simple characters is one requirement for a Bio-DBMS. Biologists also need a way to define relationships between data or between taxa. We briefly mention these as they have been presented and discussed elsewhere (Diederich and Milton, 1993b;

Pankhurst, 1993; Diederich, 1996). An example of relationships is given by dependent and summary characters. A limited concept of a dependent character is given by Pankhurst (1993): 'if character 1 (stem presence) is equal to state 1 (absent) then characters 2 through 4, which are various properties of the stem, are impossible.' We use a wider concept (Diederich and Milton, 1993b) that encompasses properties other than presence of an organ: a dependent character is applicable only if another character has a particular state. For example, the property diameter can be used only if the organ to which it refers has 'shape equal round'. Another concept is that of a summary character. Each state of a summary character implies that a number of other characters have particular states. For example, if the reproductive system of a nematode is described as amphidelphic, it implies that the number of genital branches is 2 and that each branch is directed toward opposite extremities of the worm. State-based relationships need to be defined for improved operation of identification applications, but they are not taxon dependent.

Taxon-dependent character relationships

Homologies and convergences are relations that exist between data for particular taxa. For example, the coiling of the lumen of the anterior part of the oesophagus that occurs in unrelated nematode families (Criconematidae, Belonolaimidae, Dolichodoridae) is a convergence. This fact will need to be recorded as a taxon-dependent relationship at the family level.

Given the dependency on taxa, these relationships are a form of data to be represented in the database, in addition to pure relationships built upon the list of characters. Even then, additional support for homologies and convergence is required. Whatever form of recording is used for such relationships, the standardized decomposition of characters will simplify this operation since the homologous or converging character will always be stored in the same manner, even when it refers to taxa recorded in separate databases.

Properties of characters: metadata

Characters themselves have properties. These are data about data, or metadata for short. Some are well-known such as the type of a character, e.g. real number (lengths and widths), integer, unordered multistate (nominal), and ordered multistate (ordinal), all of which, except ordinal, are generally supported by existing DBMSs. At the beginning of the NEMISYS project, we defined other kinds of metadata (Diederich *et al.*, 1989; Diederich and Milton, 1991) that are useful in identification since they characterize the qualities of the character for purposes of identification, including conspicuity, ambiguity, and variability of characters, besides their type (entered as range and scale). These metadata are taxon dependent.

In a proper database, fields must be provided for metadata with only one type of metadata per field. This would not be possible with DELTA, where all metadata and relationships have to be stored in the comment field. One major problem with this would concern the use of the information in *ad hoc* queries, such as in building indexes for the queries.

Metadata make it possible to develop the concept of endorsement, which was first presented as a 'pie in the sky' wish (Fortuner, 1993), but has recently moved into the realm of the possible (Diederich and Fortuner, 1996). Using metadata and fuzzy logic, it is possible for the system to judge the reliability of the data entered by the user.

In GENISYS databases, frequency (the percentage of specimens having a particular character state in a population) is entered as metadata when it is given in the description. It can be used for probabilistic applications (Horvitz, 1993), but it has many other uses. For example, state 1 of character 50 of Thiele (1993) given above is 'always yellow'. This means that this state would be present in 100% of the specimens considered. Obviously, the frequency metadata of GENISYS can be used to recreate Thiele's character state.

Capture of Non-morphological Data

Physiological data

So far, we have been talking of morphological data, but identification and systematics also use other kinds of characters. In the list of characters in Thiele (1993), character # 4 is:

4. Terminal buds of flowering stems transform into inflorescence axes: directly (0); after resting period (1).

Character 4 is a physiological character and it should be decomposed and stored as such. Physiology is the study of the working of organs, tissues, cells and molecules, and these are biological structures that already exist in the morphological database. This should make it possible to use a decomposition where the physiological entities would be connected to the existing morphological entities:

Flowering stems
  Terminal bulbs
    Resting period
      presence
        present / absent

where one recognizes Entity (at three levels of decomposition, two morpho-anatomical levels and one physiological level), basic property, and states.

The presence in the entity hierarchy of the same biological structures as in the morphological tables will provide an easy way to link the various kinds of data, and the same view methods will apply.

Other types of data

Similarly, biochemical data can be so decomposed with the entity being specific molecules, which constitute the bottom level of the morphological entities. In fact, the biological activity of each molecule could be described as physiological data, and ecological data could represent the relationship between an organism and its environment, depending on the molecules it emits into, or receives from this environment, emission and reception being made by particular morphological organs.

Extrapolating from this, it is possible to imagine a general database incorporating all kinds of data – morphological, molecular, physiological, ecological, etc. – which would represent all our knowledge about a particular organism. The organism, its organic structure and functions, and its relations with its environment would be captured and arranged in this huge database, and a network of views, relationships, and metadata could be defined between the various types of data to give life to this electronic monster.

The cost in time and effort for such a project would be staggering, and it takes us far from the subject of this meeting, which is identification. However, it does not cost much to keep such possibilities in mind when we start a limited project. We believe that any database should be created with a possibility of being extended to a wider use and that the data decomposition we propose is naturally suited to such an expansion. We also believe that such a database could attain its full potential only if a Bio-DBMS can be created to support a wide range of uses.

References

Dallwitz, M.J. (1980) A general system for coding taxonomic descriptions. *Taxon* 29, 41–46.

Dallwitz, M.J. (1993) DELTA and INTKEY. In: Fortuner, R. (ed.) *Advances in Computer Methods for Systematic Biology – Artificial Intelligence, Databases, Computer Vision*. The Johns Hopkins University Press, Baltimore and London, pp. 287–296.

Diederich, J. (1997) Basic properties for biological databases: character development and support. *Journal of Mathematical and Computer Modelling* 25(10), 109–127.

Diederich, J. and Fortuner, R. (1996) Endorsement of observations in identification. *IEEE International Conference on Fuzzy Systems, Sept. 8–11, 1996, New Orleans, Louisiana, USA*, pp. 175–179.

Diederich, J. and Milton, J. (1989) NEMISYS: an expert system for nematode identification. In: Fortuner, R. (ed.) *Nematode Identification and Expert-system Technology*. Plenum Publishing Corp., New York, pp. 45–63.

Diederich, J. and Milton, J. (1991) Creating domain specific metadata for scientific data and knowledge bases. *IEEE Transactions Knowledge Data Engineering* 3, 421–434.

Diederich, J. and Milton, J. (1993a) Expert workstations: a tool based approach. In: Fortuner, R. (ed.) *Advances in Computer Methods for Systematic Biology – Artificial*

Intelligence, Databases, Computer Vision. The Johns Hopkins University Press, Baltimore and London, pp. 103–123.

Diederich, J. and Milton, J. (1993b) NEMISYS: a computer perspective. In: Fortuner, R. (ed.) *Advances in Computer Methods for Systematic Biology – Artificial Intelligence, Databases, Computer Vision*. The Johns Hopkins University Press, Baltimore and London, pp. 165–179.

Diederich, J., Fortuner, R. and Milton, J. (1989) Building a knowledge base for plant-parasitic nematodes: description and specification of metadata. In: Fortuner, R. (ed.) *Nematode Identification and Expert-system Technology*. Plenum Publishing Corp., New York, pp. 65–76.

Diederich, J., Fortuner, R. and Milton, J. (1997) Construction and integration of large character sets for nematode morpho-anatomical data. *Fundamental and Applied Nematology* 20(5), 409–424.

Fortuner, R. (1989) A new description of the process of identification of plant-parasitic nematode genera. In: Fortuner, R. (ed.) *Nematode Identification and Expert-system Technology*. Plenum Publishing Corp., New York, pp. 35–44

Fortuner, R. (1993) The NEMISYS solution to problems in nematode identification. In: Fortuner, R. (ed.) *Advances in Computer Methods for Systematic Biology – Artificial Intelligence, Databases, Computer Vision*. The Johns Hopkins University Press, Baltimore and London, pp. 137–163.

Fortuner, R. and Wong, Y. (1985) Review of the genus *Helicotylenchus* Steiner, 1945. 1. A computer program for identification of the species. *Revue de Nématologie* 7, 385–392.

Horvitz, E.J. (1993) Automated reasoning for biology and medicine. In: Fortuner, R. (ed.), *Advances in Computer Methods for Systematic Biology – Artificial Intelligence, Databases, Computer Vision*. The Johns Hopkins University Press, Baltimore and London, pp. 3–27.

Lebbe, J. (1991) Représentation des concepts en biologie et en médecine. Introduction à l'analyse des connaissances et à l'identification assistée par ordinateur. Thèse de doctorat; Université Pierre et Marie Curie, Paris, xii + 282 + xxiv pp.

Pankhurst, R.J. (1991) *Practical Taxonomic Computing*. Cambridge University Press, 202pp.

Pankhurst, R.J. (1993) Principles and problems of identification. In: Fortuner, R. (ed.) *Advances in Computer Methods for Systematic Biology – Artificial Intelligence, Databases, Computer Vision*. The Johns Hopkins University Press, Baltimore and London, pp. 125–136.

Thiele, K. (1993) The holy grail of the perfect character: the cladistic treatment of morphometric data. *Cladistics* 9, 275–304.

Webster, R.D. (1988) *A Beginner's Guide to DELTA* (2nd edn). Agricultural Research Service, USDA, Beltsville, MD, 52 pp.

6

Putting Names to Things and Keeping Track: the Species 2000 Programme for a Coordinated Catalogue of Life

F.A. Bisby*

Biodiversity Informatics Research Group, School of Biological Sciences, University of Southampton, Southampton SO16 7PX, UK

Introduction

Keeping track of the different organisms involved is a task of major importance in plant pathology, as in many areas of applied biology. On some occasions, organisms that are new to science may be encountered, and the scientist becomes involved with the creation of new elements in the taxonomy: the description and naming of a new species, and its positioning in relation to the existing classification. But usually the encounter is with organisms that are already known, at least to the extent that they do have a place in the species diversity information system that we call the taxonomy (Bisby, 1984a,b). We can treat the applied biologists' interaction with this information system as having three elements: (i) accessing the reference system (keeping track of organisms and associated knowledge); (ii) identification; and (iii) patterns of diversity.

Identification of a specimen to an existing named taxon provides the entry point in many practical studies (Pankhurst, 1978), and is the subject of attention throughout this volume. Using the taxonomy to place the organism in the larger pattern of diversity may also be of value in evolutionary and ecological studies (Bisby, 1995). But most important of all, and the focus of this paper, are the simple acts of keeping track, both in the sense of keeping track of the organisms encountered in a study, and of accessing the world body of knowledge about these organisms.

* See Contributors List for present address. Fax: +44-118-975-3676/
E-mail: f.a.bisby@reading.ac.uk

Putting Names to Things

The first step is to identify and confirm the identity of the organism. As discussed elsewhere in this volume, there are many ways of doing this, in each case culminating in the supposition that the organism belongs to a particular species or other taxon with a given scientific name. What is not often stated is that doing this may be time-consuming and difficult, with real possibilities for error. This is particularly the case in plant pathology or plant protection, where the variety of organisms from quite different groups may go beyond the expertise of any one specialist. An important precaution is to follow each identification, arrived at for example through use of a key, by a 'confirmation' step involving cross-checks with descriptions based on other features, with an illustration or with a distribution map.

An important second step is to preserve a voucher specimen or culture. This serves the double function of allowing subsequent re-examination of the identification should any doubts arise, and of acting as a permanent voucher for other data associated with the record, e.g. that the disease really was observed on the host at the time and place as recorded.

The third step is to locate the species in the world information system. That is to check its accepted name, the spelling, its citation and its position in the classification. In many instances the best identification tools are local or specialist ones, created by local experts, or by others working in the same specialist field. But this very fact also means that the authors of identification systems may have used different names, classifications, or species definitions from those used by others. The differences may arise through conscious choice of particular classifications (for instance, if two arguable segregates of a species are clearly distinct in the area of study, the author may prefer a separate treatment), but may also frequently be the result of unintentional occurrences such as different traditions in different countries, lack of awareness of recent taxonomic changes widely accepted by experts on the group, or just the different dates at which identification products were created. For all these reasons there is much to be said for using the most local or immediate identification tool at the first instance, and following this with a cross-check to the wider taxonomic literature or, where available, the new generation of global species databases that maintain a responsible and up-to-date world reference system. What started out as an easy identification of a winter host plant for a fungal disease ('It is *Sarothamnus scoparius* because it clearly matches the picture in the local handbook') may thus transform itself into a firmer statement that can position the record in relation to other information worldwide ('It is *Cytisus scoparius* in the flowering plant family *Leguminosae*'). By using the most widely accepted name and classification, communication is optimized: workers in other parts of the world will notice the work, and comparability will be achieved with work carried out elsewhere.

Keeping Track: Connecting by Name

Cross referencing to the world body of knowledge is done most simply by using the name of an organism. In a library this is done by choosing relevant catalogues (books, say, about aphids, pests, true bugs, or sub-order Homoptera) and searching their indexes under the scientific and/or common name of the insect in question. But notice that the enquirer who does not know that an aphid is a bug or homopteran is at a disadvantage, either having to look this up first, or to miss some good sources. Furthermore, the library itself may not have any detailed taxonomic catalogue of the books other than by title, author, date and general subject index, and not all of the books in question will have a taxonomic index. Printed and electronic products such as *Zoological Record, Kew Record* and *The Index of Fungi* are useful for pinpointing sources for particular species, whereas more general bibliographic indexes contain only small numbers of taxa at the species level.

With the appearance of species diversity information systems on the Internet, cross-referencing becomes much easier. Web pages are indexed by search engines such as ALTAVISTA (1997), using not only the equivalent of title/author/date, but also those species names that actually occur on the fixed Web pages. Additionally many of the species diversity systems provide a name index facility for entering their database from a Web page with the name of a species. See, for instance ILDIS, LegumeWeb (1997), SINGER (1997), W3TROPICOS (1997). Frequently there is a species name box into which the user types the name. A standardized CGI interface applies this to the indexed database and returns a Web page constructed of data about that species.

Keeping Track: Automated Synonymic Indexing and Species Name Hyperlinks

Automated synonymic indexing and species name hyperlinks are two further levels of sophistication now beginning to appear in species diversity systems. Synonymic indexing is the process of locating data about a species starting from a name which turns out to be a synonym. This can be automated so that whenever users try to access species data from a synonym, they are warned of this, and transferred directly to records of the species under its accepted name. Try entering *Abrus cyaneus* to the ILDIS LegumeWeb interface: it alerts users that this is a synonym of *Abrus precatorius*, and returns a page of geographical, taxonomic and descriptive data for that species, as in Fig. 6.1.

Species name hyperlinks provide automatic linkage between data on the same species in different databases on the World Wide Web. For example, buttons provided at the bottom of each species page in LegumeWeb give a live link to data on the same species in other databases, such as the W3TROPICOS database

International Legume Database & Information Service

LEGUMEWEB

Report generated by LegumeWeb from the ILDIS World Database of Legumes

List of names matching "Abrus cyaneus"

Status	**Name**
Synonym	*Abrus cyaneus* R. Viguier

There is one name matching your enquiry. *Abrus precatorius* is the name accepted by ILDIS for the taxon located using your search string. The name you gave is a synonym.

Data for *Abrus precatorius*

Accepted name	*Abrus precatorius* L. [210, 453, 1056, 1430, 1790]
Synonyms	*Abrus abrus* (L.) Wright [269, 1430] *Abrus cyaneus* R. Viguier [269] *Abrus maculatus* Noronha [269] *Abrus minor* Desv. [269] *Abrus pauciflorus* Desv. [269] *Abrus precatorius* L. var. *novo-guineensis* Miq. [261] *Abrus squamulosus* E. Meyer [269] *Abrus tunguensis* Lima [269] *Glycine abrus* L. [269, 1290, 1408, 1409, 1414, 1430, 1587]
Vernacular names	Black-eyed Susan [2153]; Crab's Eyes [261, 1056, 1430, 1790, 1902, 2136]; Graines D'Eglise [2136]; Gwen L,glise [1790]; Jequirity Bean [1902]; Jiquirit¡ [1231]; John Crow Bead [453]; Jumbee Bead [1430]; Jumbie Bead [2136]; Jumbie Bean [1790]; Liane Reglisse [1941]; Licorice [1790]; Ojo de Cangrejo [1430]; Peon¡a [1409, 1645]; Peronila [2157]; Peron¡a [1430]; Pois Rouge [2157]; Precatory Bean [1902]; Quinquet [2157]; Red Bean Vine [1056]; Rosary Pea [1902]; R,glisse [2157]; R,glisse d'Am,rique [2157]; Tento [1231]; Wild Licorice [1430]; Wild Liquorice [1056]; Wire Wiss [1790]
Geographical records	Africa: Angola-ISO (N) [216]; Benin (N) [212]; Botswana (N) [8]; Cameroon (N) [8]; Cape Verde (N) [268]; Chad (N) [64]; Equatorial Guinea-ISO (N) [212]; Ethiopia (N) [24]; Gabon (N) [16]; Ghana (N) [212]; Guinea (N) [212]; Kenya (N) [210]; Liberia (N) [212]; Malawi (N) [8]; Mali (N) [212, 561]; Mozambique (N) [8]; Namibia-ISO (N) [232]; Niger (N) [78]; Nigeria (N) [212]; Rwanda (N) [220]; Sao Tome & Principe (N) [217]; Senegal (N) [212]; Sierra Leone (N) [212]; Somalia (N) [8, 2457]; South Africa (N) [8]; Sudan (N) [29]; Tanzania (N) [210]; Togo (N) [212]; Uganda (N) [210]; Zaire (N) [230]; Zambia (N) [28]; Zimbabwe (N) [8]

Fig. 6.1. Extract from ILDIS LegumeWeb, showing automated synonymic indexing, so that a request for *Abrus cyaneus,* has a page for *Abrus precatorius* returned.

at Missouri Botanical Garden. Some of these are data-dependent buttons: only if a species is recorded as present in Australia does a hyperlink button offer connection to the same species in APNI, the Australian Plant Name Index as in Fig. 6.2. The RAPHIS project (Shimura *et al.*, in press) provides links both ways between *Rhizobium* data in the Japanese world catalogue of *Rhizobium* cultures and the host plant species in LegumeWeb. To see this working, go to *Lathyrus pubescens* in LegumeWeb and click on the *Rhizobium* button at the foot of the page. Alternatively go to the *Rhizobium leguminosarum* collection No. IMYZA WDC31, in the *Rhizobium* database and click on the Host Plant – *Lathyrus pubescens*.

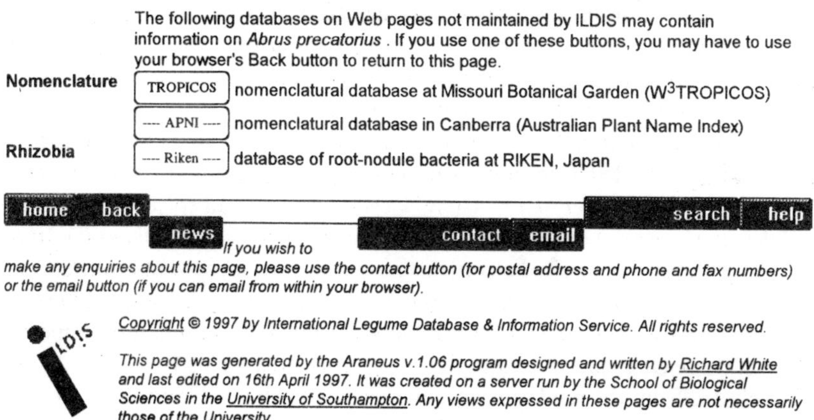

Fig. 6.2. Hyperlink buttons in ILDIS LegumeWeb, linking the page for this species with data on the same species in other databases – W3TROPICOS, APNI and Rhizobium Database at the Riken.

Keeping Track: Present Limitations

Despite the advances made in electronic indexing and the use of hyperlinks between data systems, there are still some real difficulties, beyond the obvious problem that not all resources are on the Internet:

1. The indexes do not cover all species or even all species available within online databases. This is because indexing engines such as ALTAVISTA search only on the fixed Web pages available, and not the full databases available from them. There is thus information available within the online databases on many more species and many more names. What is needed is indexed access to all of these.
2. Locating species by looking up names in indexes still does not work satisfactorily. There is no such thing as a correct taxonomy, and so some taxonomic indexing systems allow for the same species to be in different databases under synonyms. Without at least some synonymy quite common species may not be found, despite being present in a system. Try looking up *Lygos monosperma* (the name used in *Flora Europaea* (Tutin *et al.*, 1968) for a common Mediterranean plant usually known as *Retama monosperma*) or *Sarothamnus scoparius* (the name used in many editions of the *Flora of the British Isles* (Clapham *et al.*, 1962) for a plant also known as *Cytisus scoparius*) in systems such as the NCBI Taxonomy (1997). To be truly precise, systems ought to provide not only automated synonymic indexing, but also precise handling of the ambiguities caused by homonyms, misapplied names and pro-parte synonyms. To see how these function, try searching for *Lathyrus pubescens* or *Acacia senegal* on LegumeWeb.
3. There is the difficult problem of checking for variations in circumscription

between data sources. Linking database entries around the world by species names, as in the *Cytisus scoparius*/*Sarothamnus scoparius* or the *Abrus precatorius*/ *Abrus cyaneus* examples above, only works accurately if the species in question is of the same circumscription in all databases. Even when a species is known under the same name, there may be a major difference in which organisms are included or excluded. Common examples are polymorphic species with segregates that may be either included (e.g. *Vicia narbonensis* subsp. *johannis* included in *Vicia narbonensis*) or excluded (as a separate species *Vicia johannis*). In one database the two taxa can be contrasted as separate species, in another one is included in the other. Both databases have records for a species called *Vicia narbonensis*, but the range of plant populations included is very different. Important among these cases are the examples of domesticated forms variously included or excluded from the wild species (e.g. *Canis lupus* the wolf and *Canis familiaris* the dog). It would clearly be useful to test for such disparities so that users of hyperlinks between the records in different databases could be forewarned if there is a problem. I know of no system that currently performs these tests, but at Southampton a colleague and I are starting a project on rule-based taxonomic integrity checking.

Species 2000: Coordinated Catalogue of Life

'Species 2000: indexing the world's known species' is an ambitious programme to facilitate keeping track of all organisms on a global basis. Its principal objectives are: (i) to create a virtual index of all known plants, animals, fungi and micro-organisms; and (ii) to create a global framework by which species diversity databases in all parts of the world can be connected together (Bisby and Smith, 1996; Schalk *et al.*, in press). It will thus provide for any species both a resource in which its status can be verified and a gateway to further knowledge on that species from around the world.

The virtual index will use as its segments what are called global species databases (gsds) (Bisby, 1993a, 1994). Examples would be the Mammal Species of the World database and the ChronBact database of bacterial species. A gsd includes an indexed taxonomic checklist of all species in one taxon over the whole world. Clearly, such a system is limited to known species and can be enlarged as newly reported species are added. Important desiderata within the Species 2000 model are that a gsd should contain a coordinated taxonomic system which: (i) is responsibly created by taxonomic experts so that species from all parts of the world are classified uniformly in a single system; (ii) contains cross indexes to and from alternative current taxonomies; and (iii) has in place a mechanism for updating the taxonomy in response to new species as well as changing data and taxonomic opinion.

Meeting these criteria is quite demanding, so they are to be treated as long term goals rather than immediate requirements. To apply them too rigidly would exclude some of the best databases available! Complete coverage is already available for many small and medium sized groups, such as bacteria, birds, mammals, mosses, and the *Cactaceae*, *Magnoliaceae* and *Casuarinaceae* among flowering plants, although not all meet the full functionality needed. Two systems that go a long way to meeting the functional requirements are ILDIS LegumeWeb (Zarucchi *et al.*, 1993), with 15,500 of 18,000 species of *Leguminosae*/*Fabaceae* completed, and FishBase with 16,000 of 25,000 fish species available (Froese and Pauly, 1996). Three larger enterprises, each covering 120,000 to 280,000 species, are the IOPI Global Plant Checklist (1997), USDA World Database of Diptera (1997) and the UNESCO-IOC Register of Marine Organisms (1997). These are understandably still very far from completion, and there are overlaps with some smaller gsds.

Species 2000 (1997) is developing a central access system for the primary array of gsds, organized in such a way as to cover eventually all groups without overlaps. When completed with some 100–150 databases, this will provide the virtual index of all known organisms. A prototype, connected to just five databases, is available at the Species 2000 Web site (see Fig. 6.3). This design takes advantage of the very useful property of gsds: that they do not overlap or interact with each other. They can therefore be put together end-to-end without problem: for instance details of the taxonomy of cacti have no interaction with

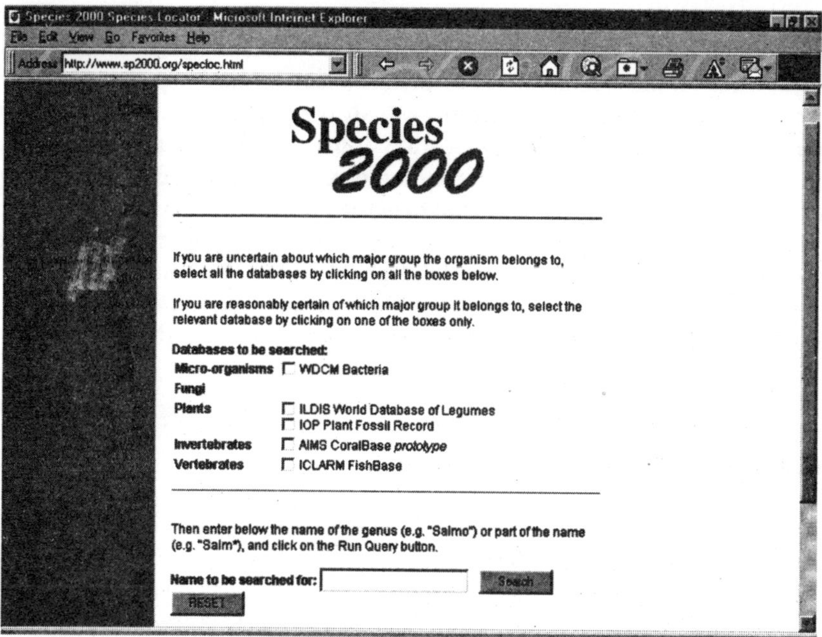

Fig. 6.3. The Species 2000 prototype Common Access System, presently linking five databases.

the taxonomy of *Leguminosae*. This is in stark contrast to the problems of trying to create a global list from regional flora and fauna databases, where the overlaps are multiple and the contents of all databases have to be reworked to bring them into agreement (Bisby, 1993b).

By itself, this initial concept of Species 2000 providing an index of all organisms could provide a major advance for verifying taxonomic details and keeping track of organisms. But, for many users of species diversity information systems, there is little direct interest in taxonomic details themselves. The more important objective is to use the taxonomic framework to connect with the world body of knowledge about these species. They would like to access descriptions, distributions, images, maps, uses, husbandry data, ecology, chemistry and so on, but all accurately accessed under the correct species. My experience has been that many users in applied biology would like the taxonomic indexing (the keeping track) to be completely invisible and yet perfect in its functioning, indeed as if performed by 'magic'! It is in this vein that we turn to the second objective of Species 2000, to provide the 'magic' that links users easily and automatically to rich sets of data about these species from many parts of the world.

In the medium to long term, what is needed is full interoperability among the many types of species diversity information systems. But, at present, a computer scientist would describe the databases available as dispersed, autonomous and heterogeneous, that is the most challenging sort! In the short term, Species 2000 proposes to interlink the databases using the species name hyperlinks of the sort initiated by ILDIS LegumeWeb and the *Rhizobium* database (Shimura *et al.*, in press). If the primary array of databases can be used to present a page of data for every species, then hyperlinks based on geography or topic area can be provided by the gsd operators to a wide variety of national and topic-based systems organized by species. In particular, if geographical distributions are available, these hyperlinks can lead to descriptive, specimen and germplasm databases in many countries. Imagine if, for pan-tropical species descriptions, images and germplasm sources might be available from Costa Rica, Mexico, Colombia, Thailand, Indonesia and the Philippines. For a Northern Hemisphere species, North American, European, Russian and Japanese sources might all be available.

Conclusions

1. Substantial progress has been made with creating species diversity knowledge systems for many groups of organisms, including those of interest in plant pathology and applied biology.
2. We now know a great deal about technical aspects of the taxonomic and indexing functions needed to keep track of organisms accurately and to access the world body of information about them.
3. The 'Species 2000; indexing the world's known species' programme

provides an ambitious framework within which many organizations are putting in place the segments of species diversity information services covering all organisms.

Acknowledgements

ILDIS is organized by James Zarucchi, Yuri Roskov, Brian Schrire, Richard White and myself, along with ten regional centres and a network of legume specialists. ILDIS has been supported by its participating botanical institutions, plus the former SERC, Chapman & Hall Publishers, the Leverhulme Trust, The Russian Science Federation, the Soros Foundation, the Commonwealth Science Council and UNESCO Botany 2000. Species 2000 is organized by Lois Blaine, Michael Dadd, Rainer Froese, David Gee, Peter Schalk, Junko Shimura, plus myself, and has been supported by IUBS, CODATA, the GEF, UNEP, *Zoological Record*, and ICLARM. I am grateful to the Royal Society and the Leverhulme Trust for the support of a Senior Research Fellowship in the current year. I thank Richard White for commenting on the text.

References

AltaVista (1997) Internet Search Engine. http://altavista.digital.com

APNI (1997) *Australian Plant Name Index.* gopher://www.anbg.gov.au:4320/7apnitaxon

Bisby, F.A. (1984a) Automated taxonomic information systems. In: Heywood, V.H. and Moore, D.M. (eds) *Current Concepts in Plant Taxonomy*. (Systematics Association Special Volume 25) Academic Press, London, pp. 301–322.

Bisby, F.A. (1984b) Information services in taxonomy. In: Allkin, R. and Bisby, F.A. (eds) *Databases in Systematics.* (Systematics Association Special Volume No. 26) Academic Press, London, pp. 17–23.

Bisby, F.A. (1993a) Progress with inventories. *Nature* 363, 11.

Bisby, F.A. (1993b) Botanical strategies for compiling a global plant checklist. In: Bisby, F.A., Russell, G.F. and Pankhurst, R.J. (eds) *Designs for a Global Plant Species Information System*. Oxford University Press, Oxford, pp. 145–157.

Bisby, F.A. (1994) Global master species databases and biodiversity. *Biology International* 29, 33–40.

Bisby, F.A. (ed) (1995) Characterisation of biodiversity. In: Heywood, V.H. (ed.) *Global Biodiversity Assessment*. Cambridge University Press for UNEP, Cambridge, pp. 21–106.

Bisby, F.A and Smith, P.M. (1996) *Species 2000: indexing the world's known species* (Project Plan Version 3). 44 p. UNEP, Nairobi (available from Species 2000 Secretariat, School of Plant Sciences, University of Reading, UK).

ChronBact (1997) *Chronicle of Bacterial Names*, Riken, Tokyo and DSMZ, Braunschweig (password protected). http://www.jcm.riken.go.jp/chronbact/

Clapham, A.R., Tutin, T.G. and Warburg, E.F. (1962) *Flora of the British Isles.* 2nd edn, Cambridge University Press, Cambridge.

IOPI Global Plant Checklist (1997) *Provisional Global Plant Checklist* of the Organization for Plant Information (IOPI). http://bgbm3.bgbm.fu-berlin.de/iopi/gpc/

ILDIS LegumeWeb (1997) Web access to the *World Database of Legumes* of the International Legume Database & Information Service. http://www.ildis.org

Mammal Species of the World Database (1997) The American Society of Mammalogists. http://www.nmnh.si.edu/msw/

NCBI Taxonomy (1997) (US) National Center for Biotechnology Information: *Taxonomy Homepage*. http://www.ncbi.nlm.nih.gov/taxonomy/taxonomyhome.html

Pankhurst, R.J. (1978) *Biological Identification: the Principles and Practice of Identification Methods in Biology*. Arnold, London, 104 pp.

Rhizobium Database (1997) *World Catalogue of Rhizobium Collections*, presently maintained at the Japan Culture Collection, Riken, Tokyo. http://fragrans.riken,go.jp:8000/cgi-bin/rhizobia.pl

Schalk, P.H., Bisby, F.A. and Smith, P.M. (in press) Species 2000: indexing the world's known species – towards a who is who directory of organisms. In: Mauro F. (ed.) *Mediterranean Biodiversity*. Proceedings of the International Symposium on 14–16 October 1996. ENEA, Rome, Italy.

Shimura, J., Ichyanagi, Y., Hollis, S., White, R.J. and Bisby, F.A. (in press) A Linked Rhizobium and Plant Host Information System (RAPHIS) on the World-Wide Web. *Microbiological Culture Collections*.

SINGER (1997) *System-wide Genetic Resources programme of the CGIAR*. http://noc1.cgiar.org/seartype.htm

Species 2000 (1997) *Species 2000: Indexing the World's Known Species*, a scientific programme of IUBS, CODATA and IUMS. http://www.sp2000.org.

Tutin, T.G., Heywood, V.H., Burges, N.A., Moore, D.M., Valentine, D.H., Walters, S.M. and Webb, D.A. (eds) (1968) *Flora Europaea* Vol. 2. Cambridge University Press, Cambridge.

UNESCO-IOC (1997) *Register of Marine Organisms*. UNESCO Intergovernmental Oceanographic Commission. http://145.18.162.199/database/urmo/default.html

USDA (1997) *World Database of Diptera*. Systematic Entomology Laboratory of the United States Department of Agriculture. http://www.nmnh.si.edu/entomology/diptera

W3TROPICOS (1997) Web access to the VAST (*VAScular plant Tropicos*) database at Missouri Botanical Garden. http://mobot.mobot.org/pick/search/pick.

Zarucchi, J.L., Winfield, P.J., Polhill, R.M., Hollis, S., Bisby, F.A. and Allkin, R. (1993) The ILDIS Project on the world's legume species diversity. In: Bisby, F.A., Russell, G.F. and Pankhurst, R.J. (eds) *Designs for a Global Plant Species Informations System*. Oxford University Press, Oxford, pp. 131–144.

7

Keeping Pathogens in their Place: International Plant Quarantine Databases

I.M. Smith

European and Mediterranean Plant Protection Organization,
1 rue Le Nôtre, 75016 Paris, France
Fax: +33 142 2489 43/E-mail: hq@eppo.fr

Introduction

Keeping pathogens (or plant pests in general) in their place is the objective of plant quarantine, which aims to prevent, by regulatory measures, the introduction of pests into new areas. Most of these measures concern consignments of plants or plant products moving in international trade, and the measures are therefore barriers, or at the very least a hindrance, to trade. The International Plant Protection Convention (IPPC), in force since 1952, provides for cooperation between governments in preventing introduction of plant pests and also limits the measures taken to those that are 'made necessary by phytosanitary considerations'. The IPPC also introduces the concept of a 'quarantine pest', as the principal target of phytosanitary measures. These provisions have restricted to a substantial extent the unjustified use of phytosanitary measures as a concealed barrier to trade. However, the IPPC hardly touches on the fact that the risks from quarantine pests vary considerably, and lets governments take measures which could be considered excessive in proportion to the risks. Also, though the IPPC provides for open exchange of information between trading partners on their phytosanitary measures, this is not very well respected in practice. These are some of the reasons why certain member governments of the GATT, during the Uruguay round of trade negotiations, thought it necessary to develop much more precise requirements for the justification of phytosanitary measures. Incorporated in the 'Agreement on Sanitary and Phytosanitary Measures' (SPS Agreement), these new requirements are now in force for members of the World Trade Organization (WTO, 1994).

The SPS agreement introduces in particular the concept of 'pest risk analysis' (FAO, 1995a), and of the relation between intensity of measures and degree of risk. Pest risk analysis (PRA) requires good knowledge of the identity of quarantine pests, their biology, their host plants, and the commodities which may carry them in international trade. The SPS agreement also introduces the term 'pest-free area' (FAO, 1995b), allowing for different measures according to the presence or absence of particular pests in the areas concerned. Governments may be called upon to justify their measures much more precisely than in the past. Thus, it becomes particularly important that accurate information on quarantine pests and their geographical distribution should be available. The computer databases developed in recent years are an excellent and indispensable medium for holding such information. However, the ease with which the information can be distributed, consulted, copied, transformed or condensed by electronic means creates new problems, which are considered in this article.

Plant Quarantine Databases

Data useful for plant quarantine have been available for many years in various paper publications, at national and international level (EPPO data sheets, IBPGR germplasm guidelines, CABI pest and disease maps, USDA data sheets, Australian data sheets, various compilations of the Regional Plant Protection Organizations). These were mostly pest- or host-specific. In the 1980/90s, a family of computerized relational databases has evolved, compiling the data from many of these sources and making them available equally easily on a pest, host or country basis. The FAO database PQdBase and the EPPO database PQR (Smith and Smith, 1991) shared a common origin in the 1980s. Originally constructed in dBase II, the two have been much developed since, in Clipper, Foxpro and related systems. With the addition of much new data and new data retrieval options, their content and structure has been used in data bases of the Caribbean Plant Protection Organization, of the Interamerican Institute for Cooperation in Agriculture, and of CAB INTERNATIONAL (CABI's Crop Protection Compendium). The same applies to a greater or lesser degree to national databases in France, Hungary, Russia, USA, and probably other countries. New data can rather easily be passed between these systems, and often is in practice. The practical aspects of the construction and maintenance of such databases will be considered with particular reference to the EPPO system PQR.

PQR

Figure 7.1 shows the relationship between the main tables of PQR. The main data tables hold information on pests, host plants, countries and their states (or other subnational units) and regional plant protection organizations (RPPOs).

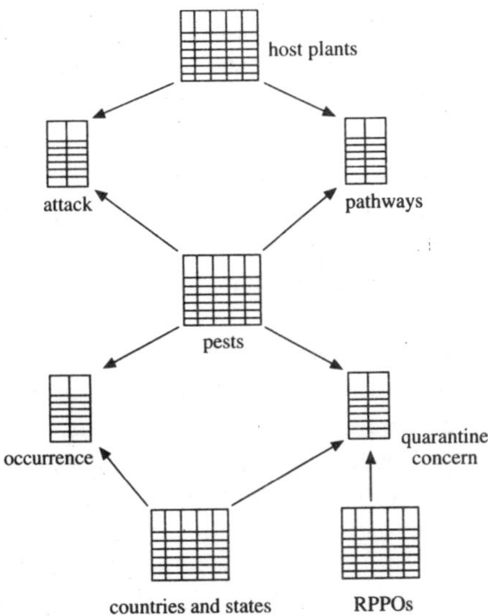

Fig. 7.1. Relationships between tables in the EPPO database PQR.

Each individual pest or host is uniquely represented by its 5- or 6-character Bayer code (see below), and each country by its 2-character ISO code. All lines containing information on a particular pest (e.g. synonyms, common names) contain the same pest code (and similarly for the host and country codes in the other tables). The main tables are related to each other by a second set of tables, giving appropriate relationships between pests and countries (presence or absence, quarantine concern), or pests and hosts (attack, pathways for introduction). The lines in these tables contain pairs of codes from the main files. For example, the occurrence of a pest in a country is represented by the combination of its Bayer code with the country's ISO code, together with a further set of fields providing detailed information about that pest record. An important feature of PQR is that the pest/country relationship takes two independent forms – presence or absence of the pest (in one table) and regulatory importance of the pest (in another). Also the pest/host relationship takes two forms – which hosts are attacked (with indices of the type of relationship – major/minor/incidental/artificial) and which host commodities can act as a pathway (with indication of the commodity concerned – plants/cut flowers/fruits/seeds). These distinct relationships are frequently not separated in other databases, and this can be a source of considerable confusion.

The pest and host coding system used in PQR is of more general interest. It is a subset, on licence, from the Bayer code system, developed by Bayer AG in Germany for use in agrochemical databases on pests, crops and plant protection products. Bayer codes are used by many databases not directly concerned with plant quarantine. The full set of Bayer codes provides a very convenient and comprehensive coverage of the organisms which are important for plant protection (extending also to storage pests, biological control agents, etc.). Each code is unique and non-modifiable (whatever changes in nomenclature may be made). So the set of codes in effect provides an international, language-independent system for harmonized plant protection databases, facilitating data exchange. Since the full database also contains a very full set of synonyms and common names, and of taxonomic relationships, it also constitutes an extensive international pest and host thesaurus (Table 7.1). In future, the Bayer codes will be managed by EPPO.

PQR holds information on quarantine pests worldwide. Its coverage is essentially that of the quarantine lists of the RPPOs, together with those of certain countries (Canada, Russia, Ukraine, USA). A basic minimum of information is held on every pest (over 1800 in all), but full data are mainly available for the quarantine pests of EPPO and of the European Union (about 330). This derives originally from the EPPO data sheets, and has recently been updated, completed and validated with the support of two EU-funded contracts carried our jointly with CABI – one for the production of new data sheets (EPPO/CABI, 1992, 1996) and one for the production of geographical distribution maps.

Geographical Distribution Data

For the purposes of accurate PRA and of justifying phytosanitary measures under the SPS agreement, there is no doubt that the geographical distribution of quarantine pests is a key point. Geographical distribution records should, ideally:

- concern pests of clear known identity.
- be as precise as possible.
- be backed up by adequate detail.
- come from a reliable source.
- be up to date.
- have official status.

Table 7.1. The Bayer code database as a pest and host thesaurus.

	Preferred names	Synonyms	Common names	Total
Animal pests	18565	8973	25446	52984
Microbial pests*	3263	3260	9568	16091
Plants, including weeds	27098	11364	54636	93098

*Many viruses are now being added.

Much of the data available in the past have been deficient in relation to one or other of these criteria, and it remains difficult to overcome these deficiencies even now. Old records, of phytopathogenic bacteria and viruses in particular, often refer only to symptoms, so that the true identity (in modern terms) of the pathogens concerned is not certain and mostly cannot be checked. There might even be something to be said for declaring all such records invalid before a certain date (say 1980), so that they have to be validated by more recent information. More generally, many records are of simple presence in a country, which may correspond to a wide variety of different situations. Such a record may correspond to anything from a pest which is ubiquitous to one which has only been found once. Single findings, similarly, may correspond to an interception (i.e. detection on imported material), to genuine rarity, to a very local presence, or even to a wide distribution if there are no experts looking. Because much pest recording is done from a biogeographical standpoint, many publications are only interested in first records, and collect no information on what happens afterwards. Yet pests disappear, or are eradicated, or the record is shown to have been erroneous in the first place. For plant protection purposes, it is the situation today which is important.

For a database such as PQR, the consequence of this is that the status of a pest in a country should preferably be supported by several items of data. These items may come from different sources, and there is no essential reason why different sources should agree. Indeed, our experience is that different sources quite often conflict, at least in detail. In PQR, geographical distribution status in a country has been codified into four categories of presence and five categories of absence (Table 7.2). These allow for different degrees of precision between sources, and in particular for changes over time. Thus, an old record of local presence may stand next to a recent record of absence (as a result of eradication). It may be mentioned that there is no category of 'Absence' as such. If there is no information, then there is no record in the database. Only if the status of absence can be officially backed (i.e. the country can be considered as a pest-free area in the terms of the SPS agreement) can specific information about it be put into the database. In addition to these coded entries, PQR provides for free text

Table 7.2. Categories for the geographical distribution of pests used in the EPPO database PQR.

Present	Absent
A widespread	D found in the past but not established
B local	E found in the past but eradicated
C few reports	F intercepted only
X no details	G mistaken, unconfirmed or doubtful
	H officially declared absent

describing the status of the pest in the country in as much detail as necessary, with bibliographic references or other indications of source.

The Problem of Sources

Reliability of the sources of geographical distribution information (Table 7.3) is obviously critical. Yet sources are often not very reliable. In the plant quarantine context, it is important to have information from official sources, or to have official confirmation of status. However, officials of one country are frequently sceptical about information from officials in other countries, and scientists are often sceptical of all of them. From a scientific standpoint, the traditional good source is a published refereed article. The reality is that few articles deal critically with the question of distribution. The facts provided are often incidental to the main subject. There is little reason to suppose that referees take much notice of the veracity of geographical distribution information in the articles they review.

Database compilers are mostly searching the literature with the help of computer systems, and are actually finding abstracts, not articles. Modern technology has not yet solved the problem of how to obtain full texts of articles other than in the good old-fashioned way. So, records go into databases without critical examination of the full source. For example, an abstract carried the title 'Six nematodes dangerous for Italy' and went on to outline the damage caused. In fact, none of the six nematodes were present in the country, and the purpose of the article was to highlight the risk of introduction. Abstracting systems succeeded in giving exactly the opposite information (that the nematodes are present in Italy) and will go on, for ever, telling searchers that these very dangerous nematodes are associated with Italy.

Another important point is that computer searching systems allow access to previously unavailable 'remote' material, published in distant countries, in other languages. The reliability of the sources involved is often unknown to the database compiler. The same applies to privately transmitted information

Table 7.3. Primary sources of reports of pests in countries.

Official declarations
Local publications
Scientific papers
Their abstracts
Herbarium specimens
Consultants
Word of mouth

(consultants, word of mouth) and even to herbarium specimens (if the circumstances of collection are not known). All these types of unreliability have always existed, of course, but they were less important in the past. Nowadays, the collection of information from multiple sources into computer databases, which can easily print out pest lists for countries, or country lists for pests, or on-screen maps, acquires a spurious authority because of the medium itself. The user is impressed by the output, and fails to notice that the original sources are no longer available to him. It is thus an important responsibility of the database manager to be sure of his sources, to validate information as far as possible, and to 'proof-read' for accidental errors.

History of a Record

Another specific problem of geographical distribution records is that they are frequently copied from one database to another. This has of course happened from the very beginning of pest recording, but has become much easier in computerized systems, and in particular can now happen uncritically. Figure 7.2 shows a case history which can teach some lessons. It starts from a leaflet published in Morocco in 1938, describing a formalin treatment for pea seeds to protect them from *Pseudomonas pisi*. This was abstracted in *Review of Applied Mycology*, and was interpreted in the 1950s as a record of the pest in Morocco. This record then passed through a series of publications and databases, by the pathways shown in Fig. 7.2. The status of the record changed over the years, mostly for no obvious reason. Morocco was consulted twice. On the later occasion, the Plant Protection Service declared that it had no direct knowledge of any occurrence of *P. pisi*, but could only be categorical about the relatively recent past (20 years). The earlier statement (held in the EPPO archives) suggested that the pest was previously present (but probably only because of that CMI record which could not be ignored). At the end of the operation, it appears superficially that the presence of *P. syringae* pv. *pisi* (as it has now become) in Morocco is well documented by reliable sources. However, our conclusion is that it almost certainly never occurred in Morocco at any time.

Conclusions

A few examples can be used to show that such cases are not only of academic interest. Import of apples to Japan from other countries (especially USA) has been prohibited because of the risk of introduction of *Erwinia amylovora* (the cause of fireblight of pome fruit). For many years, a shoot blight of pear has been known in Hokkaido (the northern island of Japan). It was naturally supposed that this might be due to *E. amylovora* but local publications attributed it to another bacterium. Under the SPS agreement, such situations have to be

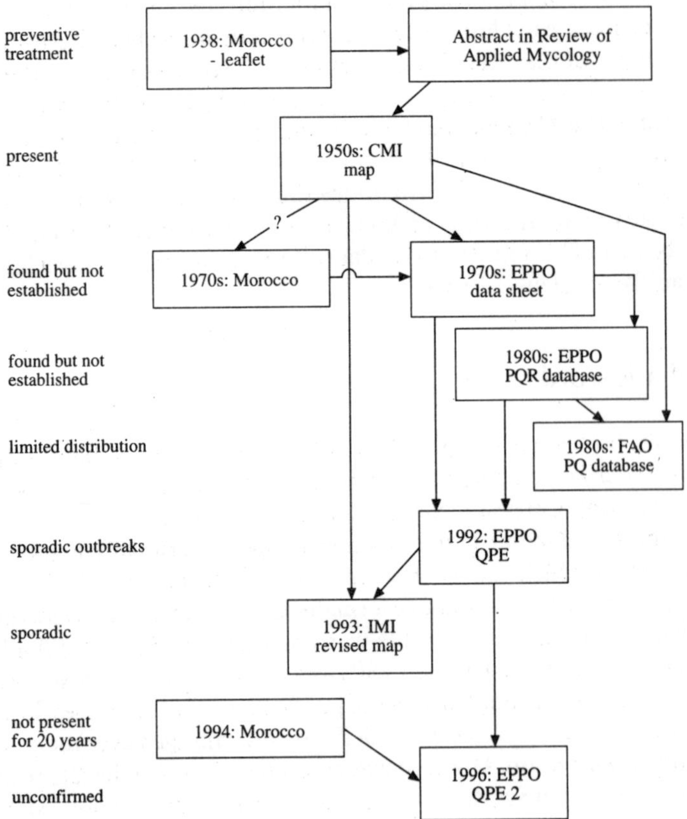

Fig. 7.2. History of a record: *Pseudomonas syringae* pv. *pisi* (bacterial blight of pea) in Morocco.

resolved. Isolates of the bacterium have been identified as *E. amylovora* by North American scientists, and there is now ongoing dispute between Japan and USA on access of American apples to the Japanese market. Within Europe, *Clavibacter michiganensis* subsp. *sepedonicus* (the cause of ring rot of potato) is the subject of severe requirements affecting trade in potatoes. Algeria used to export early ware potatoes to European Union countries and was one of a relatively short list of countries allowed to do so. The publication of an article indicating the presence of the bacterium in one area in Algeria was sufficient reason for the removal of its name from that list in EU Directive 77/93. Considerable work was needed to establish that the bacterium is no longer present, or in other words re-establish pest-free area status.

So, records of pests in countries can have great consequences on trade. Similarly, records of pests attacking hosts can lead to severe restrictions of trade in commodities of those plants. Plant health authorities look to international databases as their immediate source of accurate information. If the information they find is inaccurate or simply insufficient, there can be considerable consequences in terms of unjustified quarantine measures. So the managers of international plant quarantine databases have a considerable responsibility to provide well documented, complete and accurate information.

References

EPPO/CABI (1992) *Quarantine Pests for Europe.* CAB International, Wallingford.

EPPO/CABI (1996) *Quarantine Pests for Europe.* 2nd edn. CAB International, Wallingford.

FAO (1995a) *Guidelines for Pest Risk Analysis.* International Standards for Phytosanitary Measures No. 2. FAO, Rome.

FAO (1995b) *Requirements for the Establishment of Pest-free Areas.* International Standards for Phytosanitary Measures No. 4. FAO, Rome.

Smith, I.M. and Smith, B.C. (1991) PQ – the EPPO data base on plant quarantine pests. *Bulletin OEPP/EPPO Bulletin* 21, 211–218.

WTO (1994) *Agreement on the Application of Sanitary and Phytosanitary Measures.* World Trade Organization, Geneva.

Handling Facts to Produce Information – Emerging Trends in Biological Databases

8

S.B. Jones

CAB INTERNATIONAL, Wallingford, Oxon OX10 8DE UK
Fax: +44 (0)1491 833508/E-mail: s.jones@cabi.org

Introduction

Biology has often in the past been seen as one of the least computerized of the sciences. However, in recent years the development of large databases has brought biological disciplines to the forefront of some of the development of modern computer technology. Examples are the databases underlying projects to characterize the human genome, large databases of taxonomic systems, and the computerization of large collections. This is in addition to the growing trend towards electronic publishing, in which the full-text of journal articles and books is available electronically as well as the more traditional literature databases holding bibliographic data, abstracts and indexing. In biology, 'full-text' often crucially includes graphical information, especially photographic material, which has major consequences for storage and retrieval.

This paper looks at the computer science implications of these quantities of data and looks forward to the emerging technologies which will help to meet the challenge of managing the databases in the future. In each case, some early experience is examined and pointers drawn which may help those considering the technology.

Data Warehousing

Data Warehousing is one of the current 'hot topics' in computing, generally applied to management information systems in commercial enterprises (Gill and Rao, 1996). The idea is to create a very large data repository encompassing all the information available to a company, and then to explore this for trends which can be used to improve the business. The analysis techniques are often those of data mining, discussed later, but in this section the mechanisms for storing the data are discussed. An important feature of a data warehouse is that it is separated from the associated data entry systems and is designed for efficient information retrieval rather than for on-line transaction processing.

To give a non-biological example, consider a supermarket. The Electronic Point of Sale system creates and stores raw data on all the purchases made by every customer. Once transferred to a data warehouse system, this can be compared with stock levels to enable just-in-time ordering. If there is a 'loyalty card', the purchasing pattern of each customer can be tracked and appropriate incentives given. The performance of each store in a chain can be analysed. In addition, external data such as the weather and local demographics can be included to enhance forecasting of purchases. The data for these analyses is likely to be in several separate databases, some very large.

The simplest retrieval systems from data warehouses are closely related to those of 'ordinary' relational databases, although they must in general be capable of simultaneously handling data stored under different management systems. The term 'On-line Analytical Processing' (OLAP) is often used for analysis on a data warehouse, generally expressed as a contrast to 'On-line Transaction Processing' (OLTP) which is the name conventionally used for modern data entry systems.

The data in a data warehouse is often stored in large relational databases, although flat files are sometimes used. An important question is whether the data has to be physically co-located, or can it be stored in several different places, a so-called 'federated' model? In the latter case, network links will be required to combine the data, and performance of the system can crucially depend on the quality of the network.

Data warehouses have started to be used in the biological sciences. For example, in the field of biotechnology there are several large data repositories which may be of use to researchers: factual data such as genetic and protein sequences; bibliographic summaries of the research literature; the full text of literature; relevant patents and regulatory data from several countries; data on the availability of resources such as cell lines.

Large sequence databanks, such as the one held at the European Bioinformatics Institute, generally have an internal production version in a relational database management system, and an 'issue' version which is a flat file representation of data ready to be released for wider use. There will be data in the production system which is not released, for example sequences which are deposited with a requirement that they be kept confidential until a related publication is completed.

The need to develop a coherent infrastructure for biotechnology in Europe has been recognized and the ADLIB project (Advanced Database Linkages in Biotechnology) will be a significant step towards proving the worth of an infrastructure of this nature for biotechnology researchers (Jones, 1996). The aim of the ADLIB project is to demonstrate an environment in which three major commercial scientific literature databases can be linked to a variety of factual and full-text databases in biotechnology. Although ADLIB is not explicitly intended as a data warehouse project, it shares many characteristics with this technology.

ADLIB will serve both academic and industrial users and will involve academic and commercial database hosts. The environment will be used to include small and specialized databases. User groups covering a very wide range of specialities and site locations will be offered access to the linked databases under controlled conditions. Commercial, copyright and legal issues related to the potential future operation of the linked service will be examined.

The ADLIB team initially examined the possibility of pooling all the data into one database on a single server (Fig. 8.1). This was discounted because the field structures and other aspects of the data vary greatly between the different databases, and it was felt that valuable data could be lost by making a permanent pool. Instead the team is building a 'tank' of databases (Fig. 8.2) in which the structure of each database is retained but powerful parsing software (Etzold and Argos, 1993) is used to link the data 'on the fly' as and when required. ADLIB is also looking at the feasibility of holding some of the data in a 'virtual tank' whereby databases are linked from remote sites over networked connections. The experience to date is that this can present significant performance problems, especially when network connections across national boundaries are concerned.

Other key issues being examined by the ADLIB team centre on the commercial realities of exposing such a mixed group of databases as a product. The means of recovering revenue in an equitable way for database producers and for

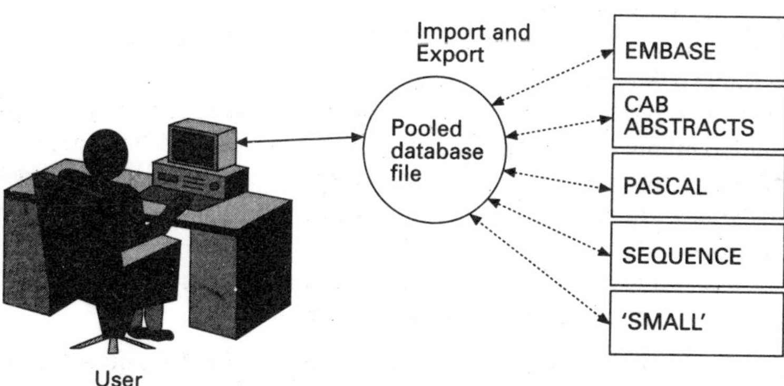

Fig. 8.1. The pooled database concept in ADLIB.

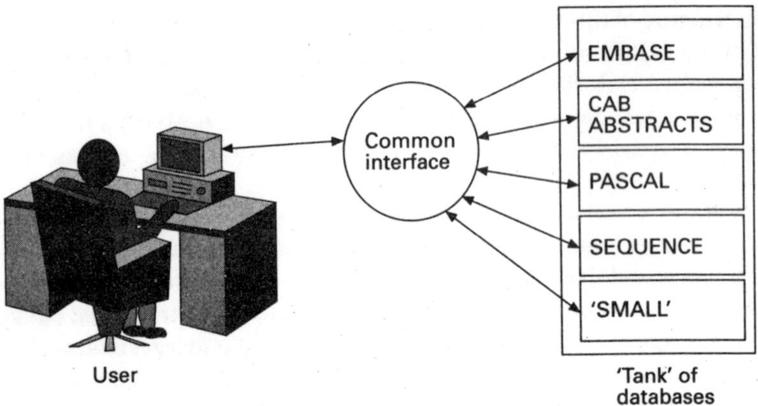

Fig. 8.2. The database tank concept in ADLIB.

customers needs to be resolved. In addition, there would be complex copyright and other legal issues attached to such a product.

Data Mining

Data mining is a concept closely tied to data warehousing. It concerns the processing of repositories of information to provide unsuspected relationships. In general, data mining is a process a step beyond basic OLAP, in that it does not presuppose models for the data and is often based on non-parametric statistics or on technology associated with artificial intelligence.

Typical techniques used in data mining are:

- *Clustering algorithms*, based on statistics and geometry to identify like and unlike members of a set.
- *Association rules*, or decision trees which are used to break down the data into ever more detailed and related subsets.
- *Neural networks*, in which a very general model is built using a system which is said to mimic a simplified model of the human brain. The system can be 'trained' on known data to carry out pattern recognition and other tasks.
- *Genetic algorithms*, which are used to identify groupings with specific characteristics.
- *Data visualization*, in which computer graphics techniques are used to view the data and facilitate manual searching for patterns.

One problem with the use of these techniques is that the sheer quantity of data may overwhelm the technology: it can take a very long time to train a neural network on a very large dataset.

The key problem with data mining, though, is that the models used are essentially 'black boxes' and the reasoning behind a decision may not be clear. Also, in common with any non-deterministic mathematical or statistical system, any conclusions are only valid within the calibration range of the system. This is a severe disadvantage for a system which is being used to predict new patterns. The conclusion has to be that these techniques are quite acceptable when used in some business applications, such as finding patterns in supermarket purchase records, but they are not safe to be used on their own in scientific research. If new scientific 'discoveries' are made by these means then they must always be fully checked by conventional methods.

There appears to be a trend towards the use of the term 'data mining' in a much looser way, meaning any techniques used to extract new information from existing large datasets. While this is essentially an attractive colloquialism, it may be dangerous and counter-productive if the strict use of the term becomes associated with unreliable conclusions in other contexts.

Object-oriented Databases

Today's databases are generally being built using the relational database model, often in a client server mode. This means that the data is expected to conform to a predetermined data structure which has often been established as the end result of lengthy analysis and optimization. The function of a server is generally to conduct searches and to pass the results back to the client in a defined way. Subsequent processing will be done by the client. It is possible to have many servers and many different databases in this model, but the structure of each database must be known in advance to the client so that it can use the data.

In scientific research, and in other fields, this can be a very restrictive model in practice. It is often very difficult to prescribe a data structure for experimental data in advance, because a line of research may show that additional data items should be collected. This tends to mean that data structures need to change and therefore designs become inefficient and there is a major management problem for widely distributed systems.

These issues are resolved with the emerging technology of object-oriented databases, particularly in distributed systems. In computing terms, objects are self-contained data and program structures which provide services to larger programs. They enable systems to be built in a highly modular fashion, because the internal workings of an object do not need to be known to the outer program. The attraction of applying this model to data storage can be seen by looking at an example. In a conventional model, a database storing a genetic sequence would have a defined structure, including for example information about the person and institution depositing the sequence, the date of deposition, information on relevant literature, the host species, the experimental conditions and the sequence itself. This structure will generally need to be known by any system retrieving the data for onward processing. In an object-oriented system, the structure would be hidden and flexible, and the data could be returned in the form requested by the client.

Use of distributed object-oriented systems requires a standard for the way in which clients interface with objects. This will remove the need for a client to 'know' about the individual characteristics of a particular object. A conforming object can then supply the data to the client in the requested form. If the structure of the data in the object changes, the object-side interface may need to be re-written, but the client need not know this and should be able to remain unchanged.

The emerging standard for this architecture is the Common Object Request Broker Architecture (CORBA) (Orfali *et al.*, 1996) developed by the Object Management Group. The core of the standard is the Object Request Broker (ORB) which includes the communication infrastructure necessary to identify and locate objects, handle connection management and deliver data. There are competing standards for ORBs, notably the Microsoft OLE standards, but there are signs that CORBA will be fairly widely adopted for large data management systems. However, the track record for the establishment of open standards against proprietary opposition is not good. There are already operational CORBA-based systems in molecular biology (Cameron, 1995).

Conclusions

It is very difficult to predict even the medium term future in computing. However, trends in software architectures are often slower than those in hardware, and the three key technologies discussed in this paper appear to represent the next generation of data management technology. In some ways they are specialist: they concentrate on the hard problems of managing large data sets, of knowledge discovery from data, and of managing dispersed and structurally unstable data. It is unlikely that they will replace entirely the current database management systems based on relational and even flat-file technology. These older software technologies will continue to benefit from increased computational power and falling costs of data storage.

References

Cameron, G. (ed.) (1995) *Linking Biological Databases, Report of a Meeting Supported by the Commission of the European Union*, European Bioinformatics Institute, Hinxton, Cambridge, 17 pp.

Etzold, T. and Argos, P. (1993) SRS – an indexing and retrieval tool for flat file data libraries. *Computer Applications in Bioscience* 9, 49–57.

Gill, H.S. and Rao, P.C. (1996) *The Official Guide to Data Warehousing*. 1st edn. Que, Indianapolis, 382 pp.

Jones, S.B. (1996) Linking databases in biotechnology. *Neuroimage* 4, S59–S60.
Orfali, R., Harkey, D. and Edwards, J. (1996) *The Essential Distributed Objects Survival Guide*. 1st edn. John Wiley & Sons, New York, 604 pp.

9

Effective Management and Delivery of Biodiversity Information

R. Allkin

'Plantas do Nordeste', Centre for Economic Botany, Royal Botanic Gardens, Kew, Richmond, Surrey, TW9 3AE
Fax: +44 (0)181 332 574/E-mail: r.allkin@rbgkew.org.uk

Introduction

Accurate identification and a reliable, coherent, stable and predictive classification are fundamentally important to the effective use and conservation of natural resources. While this is obvious to systematists, it is not as apparent to the general public, to other scientists or even to biologists in other disciplines. This is one factor limiting new funding for systematics despite the current vogue for 'biodiversity'. Chemists working with medicinal plants, for example, know that they must identify accurately the plants they work with but are less aware of their dependence on the nomenclatural reliability of their own and others' records when searching the literature or exchanging data with others.

Given current levels of funding, systematists may quite possibly be publishing as much research, of as high a quality as possible. Questions remain, nevertheless, about how effectively systematists serve society. Through what mechanisms do we listen to, and seek to satisfy, the demands of society? What is done to explain the fundamental importance of systematic information to others? Does society have access to coordinated systematic information? Should it not do so? Through what routes are systematic data transformed and delivered to non- technical audiences? To what extent should systematists be responsible for delivering biodiversity information at all? Surprisingly little attention appears to have been given to such issues by the systematic community (Bisby, 1988).

This chapter does not attempt to provide answers to these generic questions but explores, in the context of a multi-disciplinary research programme, some of the issues surrounding the delivery of biodiversity information for use by society.

'Plantas do Nordeste: Local Plants for Local People'

Background

Plantas do Nordeste (PNE) is an Anglo-Brazilian collaborative programme in the semi-arid northeastern region of Brazil. PNE was established in 1992 and involves the Royal Botanic Gardens, Kew in the UK and more than 25 institutions in Brazil. It is an interdisciplinary programme generating, through pure and applied research, new information about the native plant resources of the region and their use. In 1994 the PNE Association was established in Recife as a Brazilian non-profit organization to coordinate the programme.

Northeast Brazil covers 1.5 million square km (*c.* the combined size of France, Germany, Spain and Italy) and is home to 30% of the Brazilian population (*c.* 44 million people). The region is characterized by social and environmental problems resulting partly from decades of unsustainable exploitation of natural resources and inappropriate land use. About 22 million of these people live in poverty, slightly more than half of whom still live in rural areas. Rural poverty in the Northeast has led to large-scale migrations to major cities (particularly in the south of Brazil) and to the Amazon exacerbating social and environmental problems there.

PNE is targeting the native vegetation and communities of the 'drought polygon' (an area demarcated by the Brazilian Government as that in which the rural economy is most affected by periodic droughts) and aims to bring socio-economic benefits to these communities, whilst conserving biodiversity, through the promotion of appropriate and sustainable use of the region's native plant resources. Eleven projects have been initiated to date in two research programmmes: 'Biodiversity' and 'Economic Botany'. A third programme 'Information, Dissemination and Training' supports, and builds upon the results of, the other two.

Biodiversity

The Northeast has the greatest variety of ecosystems of any region in Brazil and is one of the world's major centres of biodiversity, with an estimated 20,000 plant species. Much has still to be learnt about the flora and new species are still commonly found. PNE's biodiversity subprogramme currently includes projects researching and documenting the native vegetation of three biomes: 'Caatinga' (dry thorn scrub forests) in the states of Ceará, Rio Grande do Norte, Piauí and Pernambuco; 'Upland Moist Forests' (or 'Brejos') in Pernambuco and Paraíba; and the 'Chapada Diamantina' table mountains in southern Bahia.

Economic botany

The economic botany programme includes more applied projects. These are dependent upon accurate systematic information from the biodiversity programme and elsewhere and themselves generate new knowledge about the use and management of the region's resources. Some projects carry this research through to practical conclusions with direct socio-economic impacts in those communities involved. Principal areas of study have been forage plants, including systems of agro- and silvi-pastoral production (four projects) and medicinal plants, including capturing local knowledge, chemical and pharmacological evaluation and medicinal plant gardens that serve hospices for malnourished mothers and children and urban slums (four projects). Projects approved or under consideration for future funding, include wood utilization, bee plants, native fruits, dried flowers and root crops.

Information, dissemination and training

The third PNE subprogramme, 'Information, Dissemination and Training' (IDTS), is that through which PNE achieves its wider goals and is of primary interest in this chapter. This subprogramme:

- facilitates the collation, coordination and exchange of information about the region's plant resources.
- promotes standardization and quality control of information gathered.
- facilitates the flow of information within PNE.
- ensures greater access to reliable information through extensive information dissemination services.
- multiplies benefits arising from individual PNE projects throughout the region.
- ensures that these information services are demand-led and respond to user needs.
- strengthens technical and managerial capabilities.

Funding restrictions have particularly limited IDTS's dissemination activities, but funding is sought from the UK's Overseas Development Administration for an ambitious five year programme which it is anticipated will start in 1998.

Information Management

The need for an effective information system

The vegetation of northeast Brazil is relatively poorly known, but information does exist as undocumented knowledge, informal records (card reference files, personal databases or internal reports) or published information. This information remains, nevertheless, mostly inaccessible to those that could use it. Even published data are scattered amongst journals of limited availability and, as a consequence, foresters and agronomists can often not identify local plants nor list, even using common names, the distribution, habitat preferences and potential uses of relatively well known trees. This is illustrated by a recent forestry project, funded by the World Bank, which was based upon the use of a local tree, known throughout the region by the same common name ('Joazeiro') and widely recognized as having economic potential. The project was flawed since the foresters were unaware that systematists in a local university had published studies showing that this 'tree' was in fact three related species with different properties. The economic impact of the project was clearly compromised for want of access to accurate information.

Poor communications in the past and the distances involved have restricted information exchange within the region with consequent duplication of effort. Scores of small organizations offer medicinal remedies to those in lower income groups, that use plants for example. Many of these initiatives are empirical in approach and do not benefit from extensive chemical and pharmacological studies carried out in neighbouring states indicating which plants bring real benefits and can be used safely.

To support and coordinate the collation, management, exchange and dissemination of available knowledge about the plants of the region, PNE is creating a Plant Information Centre (CNIP) in the Federal University of Pernambuco in Recife.

Support for those generating information

Managing data about biological organisms is not as straightforward as it might first appear (Allkin, 1988; Allkin *et al.*, 1992). Sadly, more botanical database projects have failed than have succeeded; most commonly through poor planning rather than technological limitations. CNIP offers technical support and an information management consultancy. The service is free of charge to PNE participants and aims to enable all stakeholders (including those receiving copies of PNE's data sets and other agencies managing information about the region's natural resources) to manage their own information more effectively and to better achieve their goals (Allkin and Maldonado, 1991). This service will include training courses and workshops and regular site visits, to monitor progress and offer advice on:

- assessing existing information assets.
- establishing attainable data management goals.
- database design to meet the project's own research goals and, where appropriate, to be compatible with PNE's broader objectives.

CNIP will:

- include a reference library to improve access to international literature.
- select and install software for use throughout the network and train participants in its use.

- provide the infrastructure, skills and tools to permit a regular exchange of information between participants.
- assist projects to disseminate their own research results.

Software used within PNE were chosen (Allkin, 1997) using criteria such as the quality and security of data and the ability to exchange data with other software (Allkin and Winfield, 1993a, b). Systems in use include ALICE, DELTA and BRAHMS. ALICE is used for all databases containing information about species: their use, morphology, ecology and distribution (Alice Software, 1996). ALICE was also selected for its ability to provide adequate nomenclatural indexing (by handling the complexities of synonymy, homonymy and misapplication of names), its data independence (permitting each PNE project to define its own database differently and add any number of new data classes), and because it includes programs designed for database users as well as for those building databases. The DELTA system (Dallwitz, 1993) is used to create identification manuals and on-line identification using data exported from ALICE. BRAHMS (Filer, 1996) is to be used by systematists for specimen collection records which ultimately will be summarized as taxon records and passed to Alice databases.

Regional coordination of information

An important task is to review information about the region's plants which is already available and to select data sets for inclusion in a central information pool. Data from the literature or from databases available elsewhere are being brought together in a core biodiversity database and a number of linked thematic databases (e.g. 'Medicinal plants', 'Forage plants') all built using the ALICE system. CNIP is building and maintaining these databases to provide the basis of PNE's information service. A UK support unit is to repatriate collections, information and images about plants of Northeast Brazil from Kew Gardens and other European institutes. One important source of information, for example, is the SEPASAL database (Davis *et al.*, 1995) which contains information about useful plants from semi-arid regions of the world.

The PNE projects themselves are another source of information. Each project designs database(s) for its own purposes which, consequently, differ from those built by other projects. All project databases, however, share some common fields and data are entered into these fields using agreed standards (Allkin, 1997). Once project results have been published, and are already in the public domain, they will contribute to the central information pool. Otherwise, projects can contribute as much or as little of their data as they choose. The ALICE system allows them to select any subset of the taxa and/or any subset of the data classes to be exported in a choice of formats. In return for contributing data each project receives copies of PNE's central databases. Collaborative management of biodiversity information through a central exchange (or 'information hub') is not new. The International Legume Database and Information Service

(ILDIS, Zarucchi *et al.*, 1993) was possibly the first example. The PNE model differs in that data ownership is explicitly distributed amongst contributors and that the satellite databases are planned from the outset to be different from one another (Allkin and Winfield, 1993a; Allkin, 1997). More recent initiatives also taking a more federated approach include 'IOPI' and 'Species 2000' (Bisby, Chapter 6, this volume).

Distributed data ownership and the benefits of pooling data

Delivery of accurate up-to-date information requires the cooperation of the scientists who generate new information. They wish to have access to published data and that of their colleagues, to control their own data and to manage it effectively. Each PNE project, therefore:

- has ownership and control over any data collated locally.
- decides which of those data are to be shared with others.
- receives assistance in managing, publishing and disseminating their research.
- adopts PNE data standards where appropriate and exchanges data through agreed exchange formats.
- has free use of the library and information service and receives copies of central datasets.

Intellectual property rights and data access are complex issues about which debate is often emotional and ill-informed: to the cost of those most disadvantaged. Multinational pharmaceutical companies can access information published by Brazilian scientists in ways which are not possible for rural Brazilian NGOs or even those scientists themselves. IDTS seeks to promote more informed debate within the region.

While access to information encourages scientists to participate in PNE, another incentive is the opportunity to place their own data within a common framework with data from other sources and disciplines. This allows the scientists to explore their own data in a broader context asking questions, generating reports and undertaking analyses which would be impossible in their own database, or even were they to have copies of all of this information but in a number of separate databases. As PNE's information resources grow, this will become increasingly attractive.

An important, though often neglected, advantage of coordinating data centrally is the improvement that this can bring to the quality of our knowledge base. Intercalating data from diverse sources enables inconsistencies between different authorities to be detected (generating useful debate) and enables gaps in our knowledge to be identified more readily.

The challenges of data exchange

While it is technically simple to pass data from one computer or computer program to another, it can be very difficult to achieve a meaningful exchange of information between complex databases, even when these use the same software. Such exchange requires careful planning, coordinated effort and sophisticated data exchange tools (White and Allkin, 1992; Zarucchi *et al.*, 1993). It cannot be stressed too strongly that the causes of incompatibility between biological data sets are the responsibility of biologists rather than computer programmers (Allkin, 1997).

PNE has already invested substantially in facilitating information exchange between projects, disciplines and institutions. Technical working groups, consisting of PNE participants and invited specialists have created data standards for priority data classes. Much of the data collated within current PNE projects follow these data standards which have been developed for disciplines of current interest: medicinal and forage plant use and specimen collection. Use of these standards point to changes desirable in subsequent versions. Existing data standards (e.g. Bisby, 1994; Brummit and Hollis, 1992; Cook, 1995) are used and adapted, where necessary, to PNE's needs.

The software used to merge data sets is being developed further to make management of data exchange with PNE more flexible. As data from diverse sources are merged, inconsistencies or differences of opinion must be detected and resolved (Zarucchi *et al.*, 1993; Allkin, 1997). If this is not done, the resulting data set will behave unpredictably (lose data integrity) and be much less useful. Many inconsistencies, and some of the more complex to resolve, are associated with scientific nomenclature. They derive from many causes ranging from simply using different abbreviations for authorities to genuine differences of opinion about taxonomic relationships. When merging two databases, the software detects all classes of logical incongruency between them and, in creating the merged database, follows a set of pre-defined behavioural rules according to the type of conflict found. Inconsistencies are corrected automatically where possible. Any remaining conflicting records are filtered out and reports generated to help scientists resolve them. One of the original data sets can be given priority so that when certain types of conflict are found, the program will accept the opinion of the priority set. Future versions of the software will allow priority to be assigned separately for each data class. Thus nomenclatural records from author A could be given priority over those of author B while B's chemical records took priority over those of A, for example.

Information Flow within PNE

Information generated by systematists (from within PNE or elsewhere) is critical to PNE's applied research and to the reliability of our information services.

Two examples illustrate the importance of the flow of this information within PNE.

Medicinal plants

Professor Matos, from Fortaleza, Ceará leads one of PNE's four medicinal plant projects (Plantas do Nordeste, 1996). His team have built databases that contain information about *c.* 600 plant species recorded, within Northeast Brazil, as having medicinal value. Many of these species are grown in the University's botanical garden and have been chemically and pharmacologically analysed. To date, more than 50 species have been shown to have genuine medicinal benefits and to cause no deleterious side effects. Many plants in popular use have no measurable effect or can even be hazardous. Plants with proven efficacy are grown in 'living pharmacies' (medicinal plant allotments) within local communities. Hospices for malnourished infants and community groups within large urban slums are amongst those that already benefit from 'living pharmacies' which provide medicines which could otherwise not be afforded and a modest income as plant extracts (syrups, teas, etc.) or living plants are sold in wealthier neighbourhoods. To replicate such experiences elsewhere requires that the knowledge about which plants to use becomes more widely available.

Professor Matos takes great pains to identify accurately the plants with which he works since there are many examples of medicinal plants being misused through misidentification or misplaced faith in common names which vary between states. When using information from the literature or other laboratories, Matos also depends on the nomenclatural accuracy of his own and others' records and on effective nomenclatural cross indexing. One of the principle attractions of participation in PNE for Matos was gaining access to reliable systematic data. Figure 9.1 illustrates how systematic information, from a variety of sources, is critical to the extension component of Matos' work.

Forage plants

Figure 9.2 illustrates the benefits of information exchange between two PNE projects. The left hand panel describes a project in Feira de Santana, Bahia led by Professor Paganucci. Specialist monographic work, field observation and herbarium studies of legumes from Caatinga vegetation in Bahia (*c.* 250 species) (A) underlie creation of a database containing morphological, geographical, ecological and phenological data with behavioural traits relevant to the use of the plants for forage. The database benefits from data passed from the ILDIS database (Zarucchi *et al.*, 1993). An illustrated field manual, generated from the database, contains details of *c.* 30 species judged, through consultation of the database, to have greatest potential as forage crops (B).

Professor Filho leads a forestry management project (right hand panel), which has significantly increased yields of meat and milk from goat herds

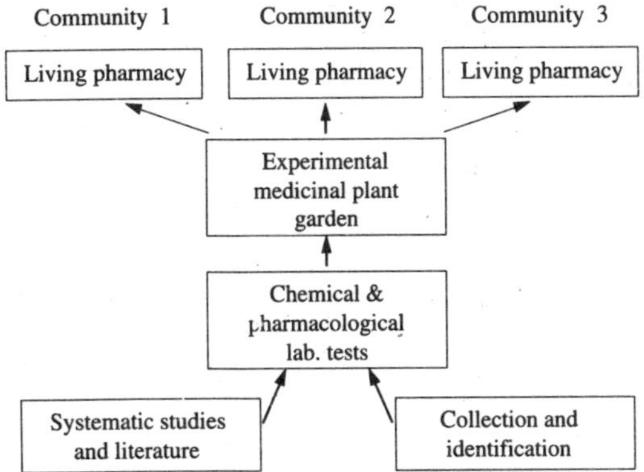

Fig. 9.1. Living pharmacies: the importance of systematic knowledge to health care.

belonging to communities close to his research institute (EMBRAPA) in Sobral, Ceará (C). Eighteen tree species were selected as being major components of the native vegetation. At regular intervals during three years, phenological observations were made and detailed laboratory tests of biomass production, palatability and nutritional value of various plant organs undertaken for individuals of these 18 species at a number of sites (A). The increased yields derive from improvements in both the quality and seasonal availability of forage through management of the native Caatinga forest. Techniques employed include selective thinning (according to palatability, nutritional value, etc.), partial coppicing (to provide new growth, leaf litter and shade) and management of the herb layer (B). Species with relatively low nutritional values can still play an important part in the management cycle if they provide forage during the dry season, offer more than one source of forage (e.g. bark, leaves and fruit) or are suitable for coppicing.

The Bahian project permits accurate identification of Caatinga legumes and suggests new species for the Ceará project to exploit. The database in Bahia and field manual, in turn, can benefit from including palatability and nutritional data from laboratory analyses undertaken in Ceará.

Information Delivery

A broad regional information service

IDTS aims to disseminate PNE's accumulated information though a range of information services and to provide mechanisms by which users can feed their views back into PNE as a whole (Fig. 9.3). Each PNE project already disseminates its own research results but since IDTS is coordinating information from

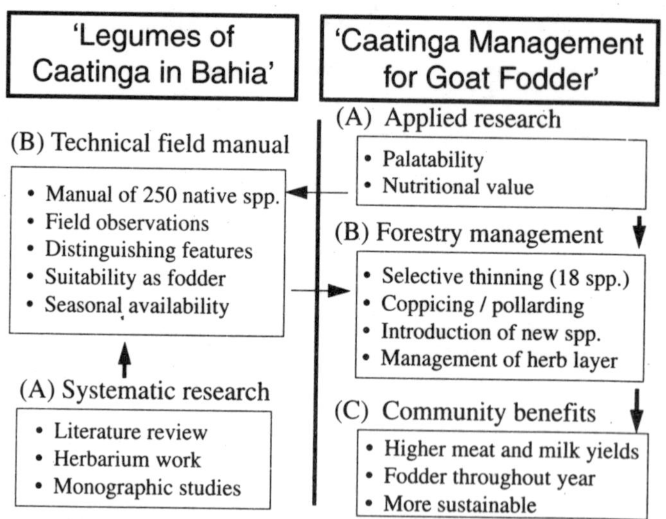

Fig. 9.2. Exchange of systematic knowledge and its relevance to goat husbandry.

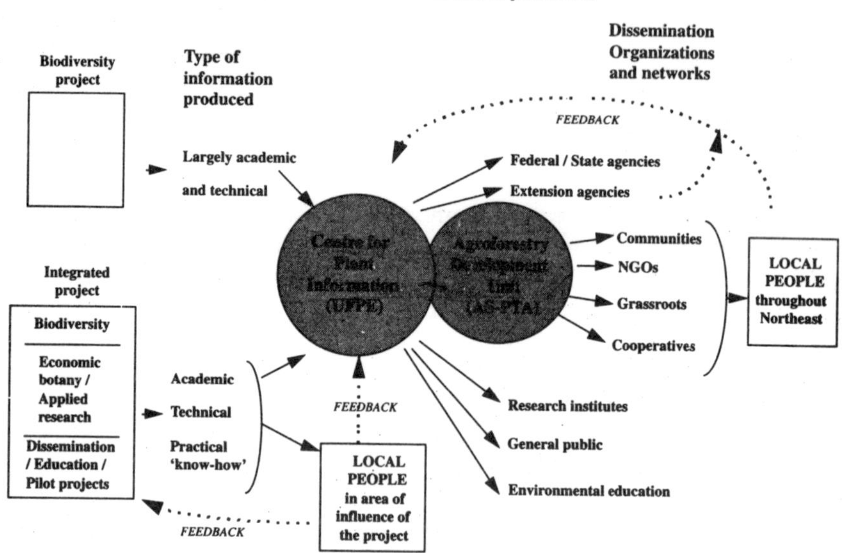

Fig. 9.3. Information flow within and beyond Plantas do Nordeste.

many sources, it is able to deliver information products of a broader scope than can any one project alone. It is neither practicable nor desirable to include all of the data from specialist project databases within central data sets, but the latter do contain pointers to the sources of further detail. Finding the right balance between depth of information and universality is one of the challenges that we are addressing.

Target audiences

The potential beneficiaries of our information service are numerous and include rural communities, poor urban communities (improved health and nutrition), children (through environmental education) and the scientific and technical community. Not all beneficiaries can be given equal priority, however, and target audiences were selected using:

- PNE's stated objectives.
- the reliance of potential beneficiaries upon the sorts of information that PNE is generating.
- the capabilities and contacts of those participating in IDTS.

Small-holder families in the 'drought polygon' were identified as the audience of highest priority. Another priority are the scientists and technicians working within those agencies responsible for the conservation, development and rural extension of the region, who, in the longer term, can bring far reaching benefits. These two audiences have different information needs and our dissemination programme reflects this. Activities are assigned to two separate units: the CNIP (described above) and an Agroforestry Development Unit (ADU) run by an NGO (see Fig. 9.3). CNIP will target more technical audiences within federal, state or non-government agencies while ADU will coordinate dissemination aimed at rural communities. The activities of the two units are coordinated by a joint committee and, while each deals primarily with certain audiences, some information services will be produced jointly.

Intermediaries and a two-way flow of information

Small-holder families in Northeast Brazil include some 3 million people and IDTS can only have direct contact with a small percentage of them. The scientific and technical community, though smaller in number, is widely distributed geographically and dispersed amongst many types of organization. To reach its audiences and have a lasting effect on the region's development, IDTS must therefore operate, at least partly, through intermediate organizations. The classes of intermediary through whom IDTS might operate include:

- non-governmental organizations (NGOs).
- governmental environmental and planning agencies.

- extension agencies.
- universities and schools.

Formal user needs analyses are planned during the next two years to help define target audiences and intermediaries more precisely. We expect this to be an ongoing debate but initially, intermediaries were selected using the following criteria:

- their ability to address the needs of priority target audiences.
- their track-record.
- their ability to communicate with IDTS members.

The relationships between information producer, intermediary and end-user are rarely simple and for many information products, several intermediaries will be involved sequentially, as raw data are brought together, analysed and presented (Fig. 9.4). In Northeast Brazil, IDTS might collaborate with any number from hundreds of organizations who interact in complex ways. Extension agencies in neighbouring states, for example, follow different procedures from one another and have quite different success rates. Links between Government and Non-Government sectors are generally weak, although there are important exceptions. Hundreds of small NGOs exist (with a staff from three to a dozen people) amongst whom a few have had notable success in rural development. Establishing a dissemination strategy requires that these many agencies be classified in some way or other. Conservation and environmental education organizations (relatively few in number) can be separated from those with a primarily developmental focus for example. Recognizing sub-classes amongst

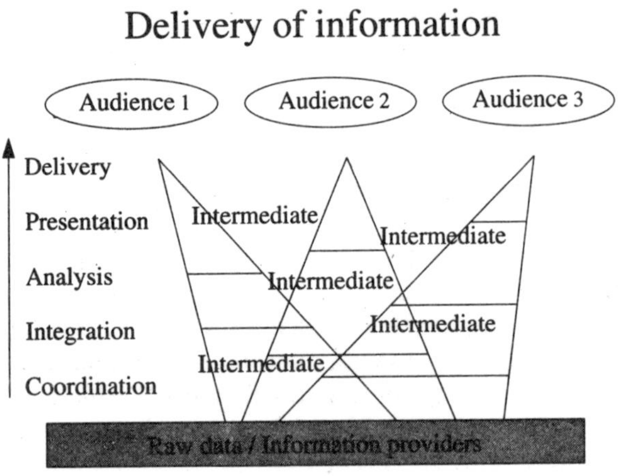

Fig. 9.4. Delivering information to diverse target audiences. (This figure is adapted from a diagram in World Conservation Monitoring Centre, 1996.)

NGOs in the latter group, however, is less straightforward since linkages between them tend to be less rigid. Informal groupings exist, however, which generally reflect similar historical, ideological or cultural backgrounds.

Particular PNE projects, such as those of Professors Matos and Filho, have been particularly successful at putting their ideas into practice to the benefit of communities with whom they work. IDTS wishes to adapt these experiences in pilot field projects in other communities and demonstrate those which are successful to people from all parts of the region. Technical information, even that with obvious economic potential such as 'high nutritional value' or 'drought hardiness', is seldom directly useful to a farmer. Such information often requires considerable elaboration before it can be used in developmental projects when questions arise such as: How might this species be most easily propagated? How can sufficiently large volumes of seed be made available cheaply? and How might its use be combined with existing land-use practices? Social, cultural and economic factors become important and the role of the scientist diminishes. Implementation of field pilot projects with local farmers requires skills and experience beyond that normally found within the research community. It also requires familiarity with, and the confidence of, those communities in which the projects are to be carried out.

IDTS includes an Agroforestry Development Unit (Fig. 9.3) run by AS-PTA, a Brazilian NGO with alternative agroforestry experience whose regional headquarters is in Recife (Guijt and Sidersky, 1996). AS-PTA will implement IDTS's pilot projects in communities where a dialogue already exists through AS-PTA's own projects or those of other NGOs that collaborate with them. AS-PTA's presence, as an integral part of IDTS, is particularly valuable because it shares technical advice and experiences with a network of similar developmental NGOs: the 'PTA' network. The network contains 27 organizations nationally of whom nine are in the Northeast. Each of these nine organizations itself forms part of a secondary network of smaller organizations within their own State. Through PTA, therefore, IDTS has access to approximately 40 other NGOs and grassroots organizations (e.g. Cooperatives, Rural Unions) and the communities with whom these work. The organizations involved with IDTS will evolve as the information available grows and as our experience and dissemination policies mature.

AS-PTA's experience and contacts are important in ensuring communication between, on one hand, representatives of our primary target audience and the intermediaries that serve them and PNE (producing information), and IDTS (managing information) on the other. CNIP will play a similar role for more academic and technical audiences. All stakeholders will participate in the design of future dissemination projects and in monitoring our services.

Dissemination mechanisms

Individual PNE projects disseminate their work through academic and technical publications (e.g. illustrated field guides), popular leaflets, videos and more practical implementation within communities. Innovative approaches such as street theatres involving street children have helped popularize the use of medicinal plants for example.

For the broader IDTS information services, traditional media, including popular pamphlets, other grey literature and videos will remain important media for both CNIP and ADU. TV and local radio which is particularly effective for rural families are also possible. CNIP is to run an information desk to respond to queries and will distribute information subsets to technical audiences and to intermediaries for further applications. Data will be published electronically in several formats: as 'read-only' databases with search software in Portuguese, as CDs and on the World Wide Web. The Internet is an increasingly important vehicle as the Brazilian Government extend the geographical reach of the 'National Research Network' and broadens its use beyond the research community to include non-government and commercial sectors. Exciting possibilities exist for presenting and monitoring information services to technical audiences within and outside of Northeast Brazil since recent additions to the ALICE system mean that information subsets from Alice databases, including illustrations, can be automatically made available on internet or intranet sites (Dalcin, 1997; White, 1997).

Conclusions

The requirements of different audiences for information about biodiversity are very different and information products and services need to be designed explicitly to target particular audiences. To be sustainable, information systems must balance the needs of all stakeholders including the researchers who generate information and the intermediaries who manage and interpret it as well as the end users. Successful information systems involve two-way communication between information providers and their users so that the users' opinions can be fed back into the system. Such communication can be facilitated through intermediaries such as organizations or individuals with: (i) skills that are complementary to those of the information providers; (ii) a familiarity with the information needs of particular target audiences; and (iii) the ability to talk with both parties.

PNE offers benefits to participating scientists and intermediaries including access to information and the means to exchange data with other institutes and disciplines. Intellectual property rights and local interests are protected through distributed data ownership, a broad constituency and transparency.

Acknowledgements

The PNE data management model derives from previous collaboration and discussions with Peter Winfield, Richard White and Frank Bisby. Implementation of IDTS in Brazil and details of the information service owe much to the energies and ideas of many colleagues from Brazil and Kew, of whom I am particularly indebted to Everardo Sampaio and Pablo Sidersky.

The PNE/IDTS programme acknowledges funding from the Reuters Foundation for developing our information exchange capability, from the Brazilian National Research Council (CNPq) for training grants and from the Weston Foundation and the Royal Botanic Gardens Kew.

References

ALICE Software (1996) ALICE Software Homepage: http://dspace.dial.pipex.com/alice.

Allkin, R. (1988) Taxonomically intelligent database programs, In: Hawksworth, D.L. (ed.) *Prospects in Systematics*. Systematics Association Special Volume No. 36. Clarendon Press, Oxford, pp. 315–331.

Allkin, R. (1997) Data management within collaborative projects. In: Dransfield, J., Coode, M.J.E. and Simpson, D.A. (eds) *Plant Diversity in Malesia III*. Royal Botanic Gardens, Kew, London (in press), pp. 55–74.

Allkin, R. and Maldonado, S. (1991) Recommendations for the application of botanical databases within Latin America. *Taxon* 40, 527–529.

Allkin, R., White, R.J. and Winfield, P.J. (1992) Handling the structure of biological data. *Mathematical and Computer Modelling* 16(6), 1–9.

Allkin, R. and Winfield, P.J. (1993a) Cataloguing biodiversity: new approaches to old problems. *Biologist* 40(4), 179–183.

Allkin, R. and Winfield, P.J. (1993b) Software development strategies for Global Plant Information Systems. In: Bisby, F.A., Pankhurst, R.J. and Russell, G.R. (eds) *Designs for a Global Plant Species Information System*. Clarendon Press, Oxford, pp. 304–318.

Bisby, F.A. (1988) Communications in taxonomy. In: Hawksworth, D.L. (ed.) *Prospects in Systematics*. Systematics Association Special Vol. 36. Clarendon Press, Oxford, pp. 277–291.

Bisby, F.A. (1994) *Plant Names in Botanical Databases*. Plant Taxonomic Database Standards No. 3 Version 1.00. Hunt Institute for Botanical Documentation, Carnegie Mellon University, Pittsburgh for the International Working Group on Taxonomic Databases for Plant Sciences.

Brummit, R.K. and Hollis, S. (1992) *World Geographic Scheme for Recording Plant Distributions*. Plant Taxonomic Database Standards No. 2 Version 1.00. Hunt Institute for Botanical Documentation, Carnegie Mellon University, Pittsburgh for the International Working Group on Taxonomic Databases for Plant Sciences.

Cook, F.E. (1995) *Economic Botany Data Collection Standard*. Royal Botanic Gardens Kew. 146 pp.

Dalcin, E. (1997) 'Surfing' your data with AliceWeb: a new program for Alice users. ALICE *Software News* 5, 8–9.

Dallwitz, M.J. (1993) DELTA and INTKEY. In: Fortuner, R. (ed.) *Advances in Computer Methods for Systematic Biology: Artificial Intelligence, Databases and Computer Vision*. Johns Hopkins University Press, Baltimore and London. pp. 287–296.

Davis, S.D., Sinclair, N.J. and Cook, F.E.M. (1995) The work of Kew's Centre for Economic Botany and the Survey of Economic Plants for Arid and Semi-Arid Lands (SEPASAL). In: West, N.E. (ed.) *Rangelands in a Sustainable Biosphere: Proceedings of Fifth International Range Congress*. Society for Range Management, Colorado, pp. 111–112.

Filer, D. (1996) *BRAHMS – Botanical Research and Herbarium Management System* Version 4.0, Department of Plant Sciences, Oxford.

Guijt, I. and Sidersky, P. (1996) Agreeing on indicators. *ILEIA Newsletter* 12(3), 9–11.

Plantas do Nordeste (1996) *Three Year Annual Report*. Royal Botanic Gardens, Kew, London for the Plantas do Nordeste Association.

White, R.J. (1997) The ILDIS 'LegumeWeb' service. ALICE *Software News* 5, 6–7.

White, R.J. and Allkin, R. (1992) A language for the definition and exchange of biological data sets. *Mathematical and Computer Modelling* 16(6), 199–224.

World Conservation Monitoring Centre (1996) *Guide to Information Management in the Context of the Convention on Biological Diversity*. United Nations Environment Programme, Nairobi, Kenya. 69 pp.

Zarucchi, J.L., Winfield, P.J., Polhill, R.M., Hollis, S., Bisby, F.A. and Allkin, R. (1993) The ILDIS project on the world's legume species diversity. In: Bisby, F.A., Pankhurst, R.J. and Russell, G.R. (eds) *Designs for a Global Plant Species Information System*. Oxford University Press, Oxford. pp. 131–144.

10

Keeping Track of Where Pathogens Are: Geographic Information Systems

P. Blaise

Institute of Plant Sciences, Section Phytomedicine/Pathology, Swiss Federal Institute of Technology, Universitätstr. 2, CH 8092 Zürich, Switzerland
Fax: +411 632 1108/E-mail: Blaise@ipw.agrl.ethz.ch

Introduction

An important part of a scientist's work is to think of ways to represent information so that underlying processes can be better understood. This is particularly important in the analysis of spatial data which epidemiologists are faced with as soon as they leave the two dimensions of time and severity and go on to consider dispersal and spread of pathogens. Unfortunately, the complexity and difficulty of handling multi-dimensional data has impeded research on spatial processes in phytopathology. The range of such research extends from small scale studies such as the analysis of patterns of spread of soil-borne pathogens, to large scale studies dealing with the spread of airborne pathogens that can travel over hundreds of kilometres. Although we are familiar with most representation methods for standard data in phytopathology, problems arise when dealing with the representation of spatial data. Unless we sacrifice a part of the information, e.g. to reduce an area to a single location, special tools are required. Geographers, as the main scientific community concerned with spatial data, had already developed special tools in the 1960s called geographic information systems (GIS). These are distinguished from other data management systems by their inherent ability to maintain the spatial relationships of variables as well as their attributes. However, GIS have been traditionally regarded as very high-tech and very expensive, requiring vast hardware, software and data resources in order to be useful, so that until recently, their use was restricted to very specialized applications. Such systems still exist but fortunately, thanks to the increasing power of microcomputers, GIS are becoming a common tool on desktop computers, and so are available to plant pathologists.

Geographic Information Systems

Although GIS seem to mean different things to different people, the most accepted definition is a set of software tools that is used to input, store, manipulate, analyse and display geographic information. Although most GIS deal with geographic data, not all are restricted to this, so that 'spatial information systems' would be a more appropriate term. Spatial data is information that is linked to a specific location in space, e.g. the area planted with potatoes in a county, the slope of a field or the diseases present on a wheat plant.

Data models

Reality is much too complex to be represented on a computer and so has to be simplified. Geographers call this simplification a data model and represent the real world by four spatial entities: the point, the line, the area and the surface. In agricultural sciences a point may represent a single tree; a line might represent a road between two fields, each being represented by an area; while a surface may represent the soil surface of the field where the plants will be sown. Each of these spatial entities has attributes which are attached to it, e.g. the species and the age of the tree, the previous culture on the field or the type of road.

The storage and representation of the spatial entities is generally done according to one of two main spatial data models: the raster and the vector models (Peuquet, 1984).

The raster model divides the reality into a grid of cells of equal size (Fig. 10.1 right), to each of which is associated an attribute value, e.g. a value saying: this is a wheat field. The number of cells necessary to represent a given area depends on the resolution of the grid.

In the vector model, objects are represented by dots (vertices) joined by straight lines (segments) (Fig. 10.1 left). In addition to the attributes associated with the objects, the GIS usually also stores the relationship between the single elements (topology).

Data manipulation and analysis

In addition to its function as an information, visualisation and representation tool, the utility of a GIS comes from the ability to manipulate the geo-referenced data and generate new data from analyses. The most important manipulation is probably the overlay process, in which two or more map layers are either superimposed or combined to produce a new map (Fig. 10.2). Functions used during the combination of several layers can be calculations on specific attributes over several maps, geometrical operations like intersecting polygons, or clustering based on distances or on attributes of objects. Depending on the package used, the amount and sophistication of functions available to perform data

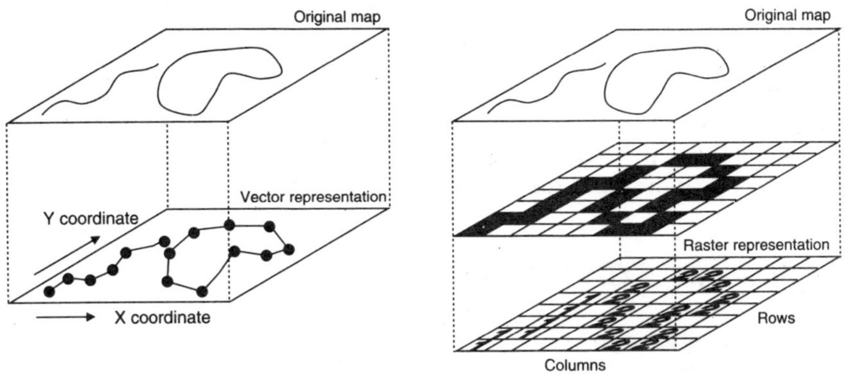

Fig. 10.1. The two main spatial data models of GIS. Left: the vector model, and right: the raster model.

Fig. 10.2. The concept of layers allows the overlaying and combination of several maps to generate new maps. In this case, the top layer map results from the superimposition of the other three layers.

analysis may vary, but most packages allow general statistics, classification and interpolation. The most sophisticated packages also feature modelling capabilities allowing not only simple data analysis, but also the generation of complex simulation models.

More information on GIS can be found in specialized literature, or on the World Wide Web (WWW). A good starting point is the 'GIS WWW Resource List', an index of WWW servers which are related to GIS (http://www.geo.ed.ac.uk/home/giswww.html). It is maintained by B.M. Gittings at the Department of Geography in the University of Edinburgh, in collaboration with the Association for Geographic Information.

Applications of GIS in Pest Management

GIS software is most correctly described as 'enabling technology' (Liebhold *et al.*, 1993), in that it provides the researcher with the ability to store, retrieve, process and display their spatially referenced data. Depending on the data available and the goal sought, the GIS can be used in two ways: (i) the handling and visualization of spatial data; and/or (ii) in analysis and modelling. Although the use of GIS is well established in fields such as land-use, natural resources and environment management, landscape ecology and remote sensing, biological sciences have only discovered these tools recently and applications in pest management are sparse. As in modelling and population dynamics, entomologists applied this new technology to their data earlier than phytopathologists. A major use of GIS technology by entomologists has been to relate insect outbreaks to biological and physiographic features of the landscape (Van Sickle, 1989), and other applications have included the compilation and analysis of insect-census data (e.g. Schaub *et al.*, 1995), forecasting pest outbreaks (e.g. Johnson, 1993) and pesticide resistance (Coop *et al.*, 1994).

Although there have been some spatial studies in plant pathology (e.g. Jeger *et al.*, 1987; Madden *et al.*, 1987; Nelson, 1996) none of them appear to have used a GIS. When plant pathologists have used GIS, it has been mainly for their power as visualization tools. Nelson *et al.* (1994) used the ARC/INFO system (ESRI, Redlands, California) to display the results of geostatistical analyses of a multi-virus, multivector, disease complex on tomatoes in Mexico and to design a plant virus management programme. The ability to visualize the regional situation and the association of high risk factors with high disease incidence resulted in modification of the management approach and the spatial/temporal arrangement of plantings. While Gessler and Blaise (Blaise and Gessler, 1988; Gessler and Blaise, 1990, 1992) used a GIS to visualize epidemiological data (see below), Seem *et al.* (1989) and Royer *et al.* (1989) tried to combine mesoscale weather forecasts with digital elevation models and disease models to generate localized plant disease predictions for apple scab and potato late blight.

Three Examples

In 1988, confronted with the need to handle a large amount of spatial data and unable to find an adequate tool on personal computers, we initiated the development of a simple GIS suited to our needs (Blaise and Gessler, 1991). In the following years, we have used this tool for several studies at our Institute, three of which are presented here as illustrations of possible applications in plant pathology.

Epidemic of canker stain of plane trees in southern Switzerland

Canker stain of plane trees (*Ceratocystis fimbriata* f.sp. *platani*) was first diagnosed in 1984 in southern Switzerland (Gessler and Mauri, 1987) and caused significant damage in eliminating the urban trees representative of this region. Due to the importance of the plane trees in this part of Switzerland, there was a strong need for frequent information about the progression of the disease. A survey of the disease distribution was made in 1988 in the whole canton, and the coordinates and health status of over 8000 plane trees were recorded and entered in a GIS (Blaise and Gessler, 1988). In the following years, the database was updated regularly with information on new cases of the disease or the removal of dead trees. Although the main application was the management of the census-data and the generation of maps showing the state of the epidemic (Fig. 10.3 left), some analyses revealed interesting aspects. Clustering confirmed what had been suspected regarding the dispersal of the disease, in that the main disease vector was man. This can be seen by clustering trees which are less than 500 m apart from each other (Fig. 10.3 right). The fact that numerous clusters contained diseased trees excluded the sole transmission by roots or by insects. A very nice demonstration of the power of visualization.is the case of Ceresora (Fig. 10.4), where the data showed that within a small creek, which was only flooded from time to time, all diseased trees were infected by spores from a single tree situated upstream, which had subsequently died and been removed.

Survey of plant protection measures taken in apple orchards

In an advisory service, client data are, by their nature, spatially oriented. Daily, weekly or monthly data elaboration is required, e.g. for the release of regional forecasts or for the planning of efficient itineraries for client visits. Evaluation of data is best made in relation to the geographical situation of the client, especially in Switzerland where climatic conditions can change radically within a few kilometres. The problem of pin-board maps with different colours representing different information, apart from the amount of time necessary to set them up, is that they are limited to one criterion at a time. In a case study, we used a GIS to manage data from a survey made by an advisory service on the plant protection measures taken in apple orchards in Zürich canton. One hundred and sixteen farmers responded to the survey and their answers were registered together with their geographical location. The information collected included guidelines followed, pesticide applications, equipment, date of first observation, severity and incidence of the diseases present. The GIS allowed us to visualize the data

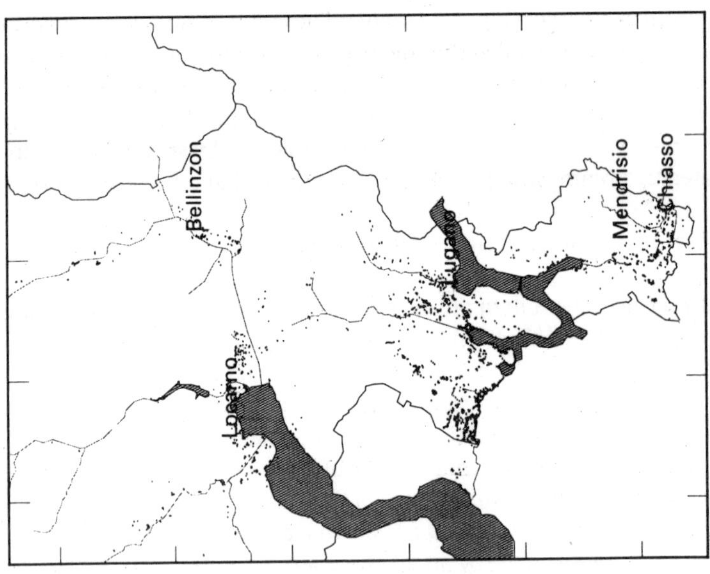

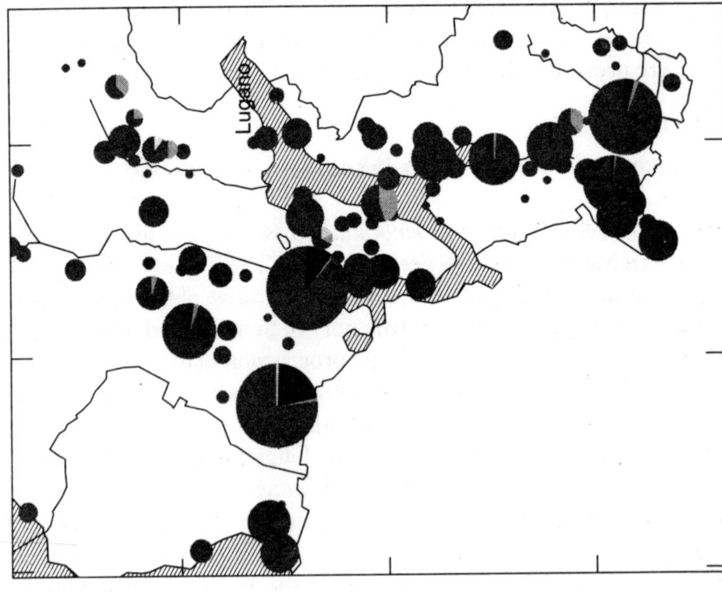

Fig. 10.3. Data on the epidemic of the canker stain of plane tree in southern Switzerland. Left: overview, each dot represents a tree; right: after clustering all trees less than 500 m apart. Size of the pies is proportional to the logarithm of the number of trees in the cluster. Grey scales correspond to the health status of the trees: Healthy ■; Suspect ■; Diseased ■; Removed ■.

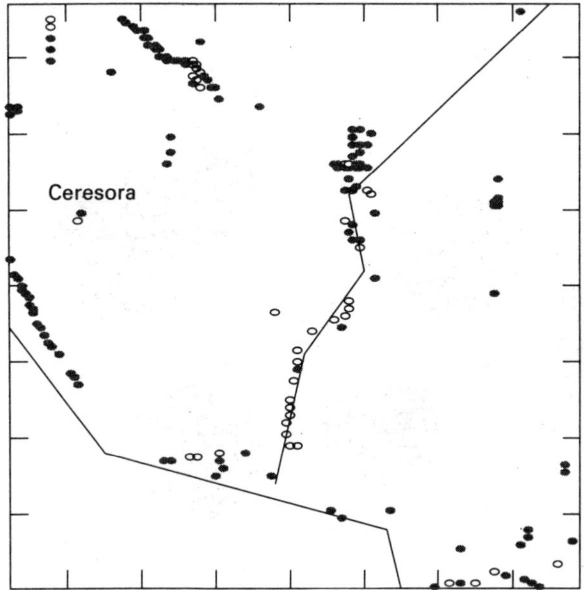

Fig. 10.4. Detailed analysis of the Ceresora region, in southern Switzerland. Each ellipse represents a plane tree, the lower line is a small river, the upper line a creek flooded from time to time. Healthy trees are stippled and diseased trees open ellipses. The black ellipse marks the tree that had died and been removed.

collected by generating thematic maps, where colours were assigned to the orchards according to selected criteria (Fig.10.5 left). After clustering the orchards in six relatively homogeneous geographical zones, one could compare the plant protection data between those regions (Fig. 10.5 right) and observe that, for example, although some regions were more exposed to favourable conditions for apple scab than others, no significant differences were observed in either the spraying intensity or in the scab incidence (Fig. 10.6).

Phytophthora *blight epidemic in the Rhein valley*

A late blight epidemic on potatoes caused by *Phytophthora infestans* was tracked in time and space (Rossi, 1990) in a geographically isolated agricultural zone (the Rhein valley leading into Lake Constance, limited on both sides by mountains, Fig. 10.7). The following information was registered for 145 commercial potato fields: area, cultivars (up to six different cultivars per site), estimated infection date, type of infection as primary or secondary (a field infection was classified as primary when it appeared to arise from a single plant), cultivar infected and growth state of potato at time of infection. Each field was identified

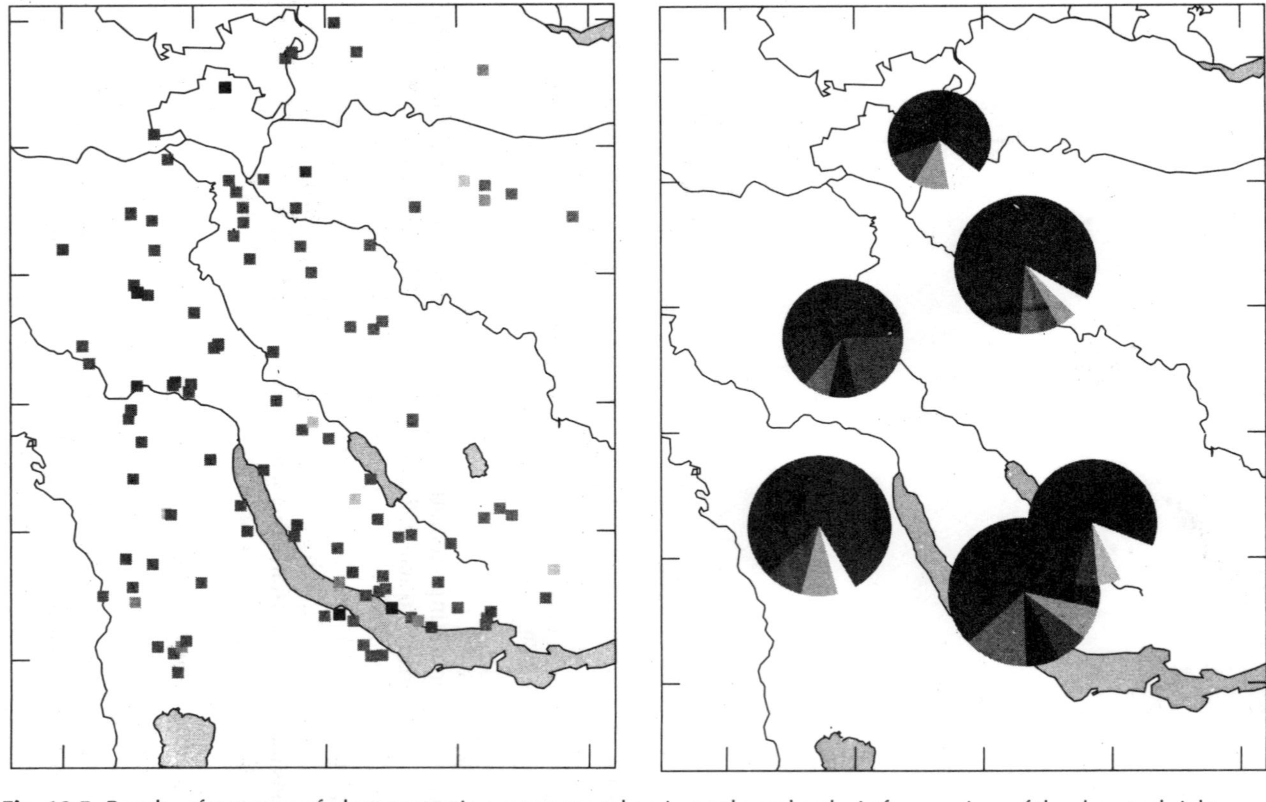

Fig. 10.5. Results of a survey of plant protection measures taken in apple orchards. Left: overview of the data and right: clusters of orchards into six geographical zones. The grey scales correspond to the plant protection guidelines followed: integrated production ■; Migros ■; Organic ■; Own ■ ; Other ■.

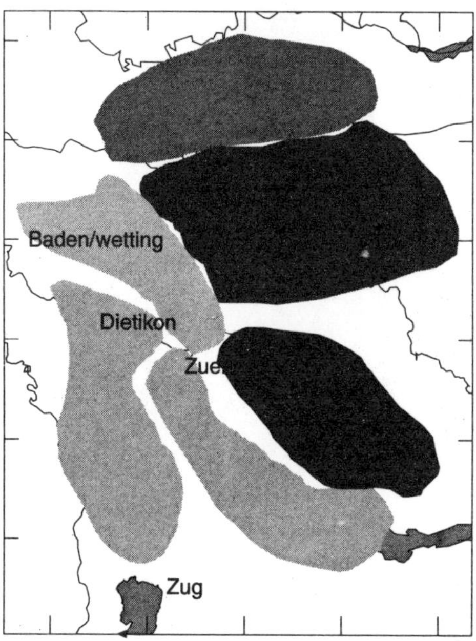

Fig. 10.6. Results of a survey of plant protection measures taken in apple orchards in the region of Lake Zurich. The six zones correspond to six geographically distinct regions with homogeneous conditions for apple scab. Average number of treatments against apple scab: 9–10 ■; 10–11 ■. Average percentage of fruit scab: 5–6 ■; 6–8 ■; 8–10 ■.

by the farmer's name and at least four sets of coordinates from official maps. Due to the difficulty in visualizing detail from 116 fields spread over 40 km on a sheet of paper (Fig. 10.7), only a selected example of the evaluation is presented here as an illustration of the analyses possible: a group of 22 fields was selected in the northern part of the valley (Fig. 10.8). By combining the information on the infection type (fill pattern) and on the date at which the first infection was observed (colour), it can be seen that the criteria used to differentiate primary from secondary infection is probably inadequate in this case for the following reason: when considering the two fields where infection appeared in the week from 8/6 to 14/6 (black colour), one can exclude the possibility that the primary infection from the eastern field was responsible for the secondary infection in the western one. Considering the subsequent development of the epidemic and the fact that no other primary infection occurred in the western group, one can conclude that: (i) the secondary infection from the week 8/6–14/6 was a primary infection; or (ii) both infections in that week were secondary infections, and the source was not detected.

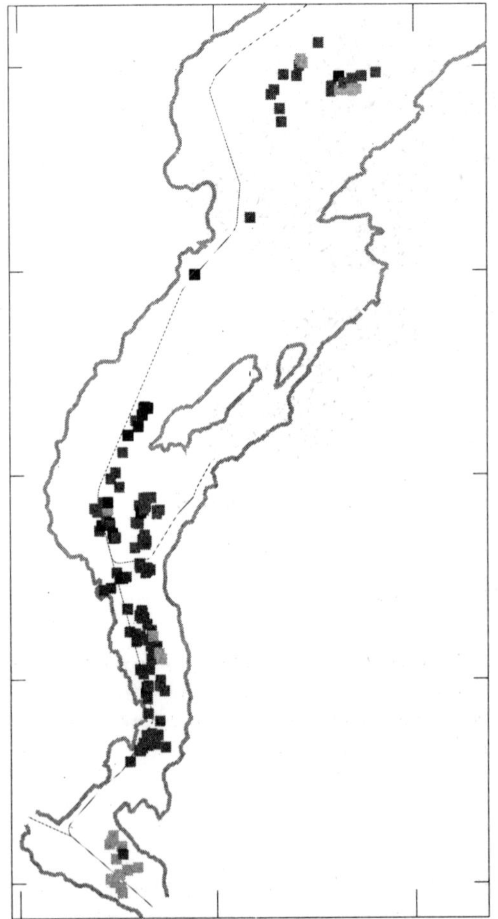

Fig. 10.7. The Rhein Valley with the 145 commercial potato fields where the epidemic of *Phytophthora infestans* was followed.

Conclusions and outlook

Plant pathologists often have to deal with geo-referenced data. It is therefore surprising that GIS are not widely used tools in this discipline, and this is hopefully only due to the fact that until recently GIS were considered very expensive, required powerful computers and often had an unfriendly user interface. Although the few existing applications of GIS in plant pathology have principally used their visualization features, examples from applied insect ecology show that a much wider range of applications could be used in plant pathology.

The combination of GIS with other technologies such as remote sensing, global positioning systems (GPS), and geostatistics is likely to bring about important changes in disease forecasting and pesticide application recommendations.

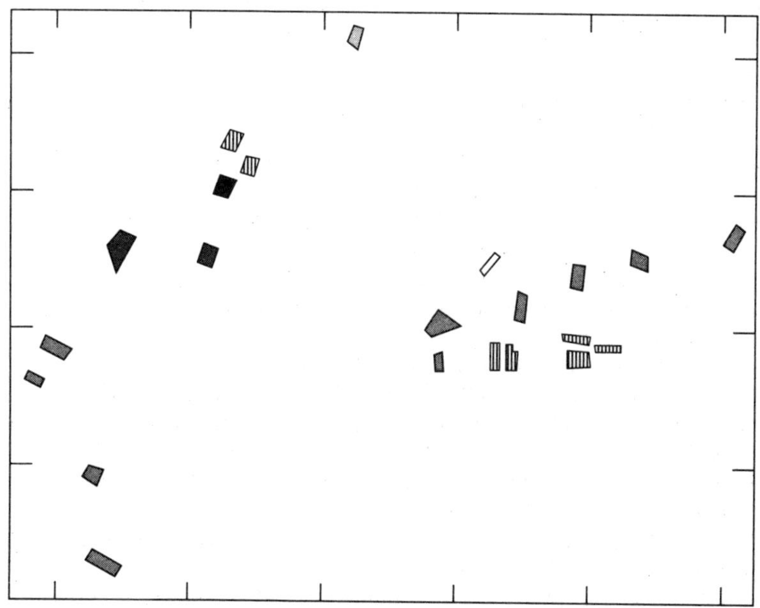

Fig. 10.8. Detailed study of potato fields infected by *Phytophthora infestans* in the Rhein Valley. The displayed area corresponds to the upper part of Fig. 10.7. The infection type in a field is represented by its fill pattern: primary □; secondary ■; and no infection ▥; The grey level corresponds to the week where the first infection occurred: 8.6–14.6 ■; 22.6–28.6 ■; 29.6–5.7 ■; 6.7–12.7 ■; later than 20.7 □.

Brown and Steckler (1995) illustrated a possible direction for the future by preparing weed maps for fields of no-till corn from digitized aerial photographs, importing them into a GIS and combining them with decision models. They suggested that the resulting prescription maps could be used to control a field sprayer so as to apply the appropriate herbicide combination to each weedy area.

Acknowledgements

I wish to thank B. Koller for the invaluable help in generating the drawings, U. Merz and C. Gessler for critical reading as well as Paul Bridge for the careful editing of the manuscript.

References

Blaise, P. and Gessler, C. (1988) Disease mapping on PC: Application to the canker stain of plane trees in southern Switzerland. In: *Abstracts 5th International Congress of Plant Pathology*. Kyoto, Japan, p. 288.

Blaise, P. and Gessler, C. (1991) Geobase: a simple geographical information system on a personal computer. *CABIOS* 7, 155–160.

Brown, R.B. and Steckler, J.P.G.A. (1995) Prescription maps for spatially variable herbicide application in no-till corn. *Transactions of the ASAE* 38, 1659–1666.

Coop, L.B., Croft, B.A., Riedl, H. and Burts, E.C. (1994) Geographic information systems and simulation in regional orchard management: pesticide resistance in Hood River Valley, Oregon. *Acta Horticulturae* 367, 383–390.

Gessler, C. and Blaise, P. (1990) Application of a PC-based GIS in plant pathology. In: Di Cola, G. and Gilioli, G. (eds) *Proceedings of the International Workshop on Computer Science and Mathematical Methods in Plant Protection*. Quaderni del Dipartimento di Matematica n. 15, Università di Parma, Parma, Italy, pp. 42–50.

Gessler, C. and Blaise, P. (1992) Using a PC based GIS for advisory services information management. *Acta Horticulturae* 313, 263–271.

Gessler, C. and Mauri, G. (1987) Le Malattie e parassiti del platano. Situatione nel Ticino. *Botanica Helvetica* 97, 349–356.

Jeger, M.J., Kenerley, C.M., Gerik, T.J. and Koch, D.O. (1987) Spatial dynamics of *Phymatotrichum* root rot in row crops in the blackland region of north central Texas. *Phytopathology* 77, 1647–1656.

Johnson, D.L. (1993) Geographic information systems. In: Norton, G.A. and Mumford, J.D. (eds) *Decision tools for Pest Management*. CAB International, Wallingford, pp. 209–218.

Liebhold, A.M., Rossi, R.E. and Kemp, W.P. (1993) Geostatistics and geographic information systems in applied insect ecology. *Annual Review of Entomology* 38, 303–327.

Madden, L.V., Pirone, T.P. and Raccah, B. (1987) Analysis of spatial patterns of virus-diseased tobacco plants. *Phytopathology* 77, 1409–1417.

Nelson, M.R., Felix-Gastelum, R., Orum, T.V., Stowel, L.J. and Myers, D.E. (1994) Geographic information systems and geostatistics in the design and validation of regional plant virus management programs. *Phytopathology* 84, 898–905.

Nelson, S.C. (1996) A simple analysis of disease foci. *Phytopathology* 86, 332–339.

Peuquet, D.J. (1984) A conceptual framework and comparison of spatial data models. *Cartographica* 21, 66–113.

Rossi, A. (1990) Epidemiologie von *Phytophthora infestans* auf Kartoffeln im St. Galler Rheintal. Diploma work, Institute of Plant Sciences, Section Phytomedicine/Pathology, Swiss Federal Institute of Technology Zürich. Unpublished.

Royer, M.H., Russo, J.M. and Kelley, J.G.W. (1989) Plant disease prediction using a mesoscale weather forecasting technique. *Plant Disease* 73, 618–624.

Schaub, L.P., Mani, E., Bloesch, B. and Schwaller, F. (1995) Distribution of *Quadraspidiotus perniciosus* (San José scale) in Switzerland based on interpolated pheromone trap data. *Bulletin OEPP* 25, 631–636.

Seem, R.C., Russo, J.M. and Kelley, J.G.W. (1989) High resolution weather forecasts to predict local apple scab infection. In: Gessler, C., Butt, D. and Koller, B. (eds) *Second IOBC Workshop on Integrated Control of Pome Fruit Diseases. Brissago, Switzerland.* vol. II, pp. 66–87.

Van Sickle, G.A. (1989) GIS – a tool in forest pest management. In: *A Wider Perspective, Proceedings of GIS – 89*, Forestry Canada. Victoria, BC, pp. 213–219.

Integrated Information Management: a Multimedia System for Crop Protection

A. Sweetmore*, C.Y.L. Schotman, B.C. Zhang*, S.A. Rudgard and P.R. Scott

CAB INTERNATIONAL, *Wallingford, Oxon OX10 8ED, UK*
Fax: +44 (0)1491 833508/E-mail: cabi@cabi.org

Introduction

Crop protection research has traditionally been discipline-based. This begins with the structure imposed by the academic institutions in which researchers are trained: the component disciplines of crop protection fall within the schools of botany or zoology, which are often administratively separate and may also be in different locations. By its very nature, research work tends to involve a narrowing of focus: on a particular order, family or genus, on a particular aspect of a species' ecology or physiology, or on detailed diagnostic techniques which demand intimate knowledge of very small variations. This detailed and intricate work is all vital to the enhancement of knowledge. However, there has been a tendency for the structure and hierarchies imposed by the academic knowledge base to be carried through to practical crop protection research organizations, with a consequent loss of cross-fertilization and a tendency to compartmentalization.

In particular, the move towards integrated pest management (IPM), which takes a more holistic approach to the complex business of crop protection, has demonstrated the limitations of a discipline-based approach. This problem was identified by an IPM Working Group providing advice to the CGIAR:

> The holistic approach to IPM has created difficulties at the technical level. The bulk of IPM researchers are the product of a narrow background. They identify themselves as an entomologist or pathologist; perhaps a virologist, weed scientist or nematologist; even an ornithologist or rodent expert. It follows that each will

* See Contributors List for present address.

 Information Technology, Plant Pathology and Biodiversity (eds P. Bridge, P. Jeffries, D.R. Morse and P.R. Scott)

> provide a different perspective on the word 'pest'. Most will be confident that they are able to take a view on how a variety of management techniques can be integrated to provide an effective solution in the context of their specialism; few would claim to have the skills to anticipate, or even detect, negative impacts on other components of the pest complex. There are, for example, precious few rice pest experts, though there are many rice entomologists and rice pathologists. It is generally accepted that, ideally, IPM seeks an approach that takes account of interactions between all biological agents that cause loss, as well as an appropriate mix of control technologies. Only rarely will one pest so dominate a situation that IPM is an achievable target for workers from a single crop protection discipline.
>
> (IPM Working Group, 1992)

Recognizing and responding to this shift in perspective, a number of international and national institutions have revised their internal structures. To take just two examples: the International Rice Research Institute (IRRI), in their report *Towards 2000 and Beyond*, recognized that 'IRRI's present organisational structure makes it difficult to establish problem-focused interdepartmental programmes ... wide dispersal of resources over a large number of departments and support units constrains formation of the critical mass of scientific knowledge needed for breakthroughs', and has since attempted to overcome compartmentalization through teams of IRRI scientists from various research departments focusing on nine interdisciplinary, mission-oriented research programme areas (IRRI, 1989). The UK Overseas Development Administration has also reviewed its approach: in the early 1990s, in the crop protection area, priorities were identified for mycology and bacteriology, virology, nematology, entomology and weed science against a crop and commodity background. However, by 1994, this approach was replaced by a crop protection strategy area, which adopts a production systems approach and takes a more holistic view of crop protection, taking full account of the agronomy of cropping and intercropping systems (Smith, 1994).

The end-users of the products of research, particularly farmers and those advising them in small-scale, intercropped systems in developing countries, have long been taking a more holistic approach. This difference in outlook has led in many cases to a gulf between the activities of the research community and the needs of farmers and their advisers.

The electronic Crop Protection Compendium, a multi-media relational database on CD-ROM, goes some way towards enabling scientists to take a more multi-disciplinary approach, by providing interlinked information on pest, natural enemy and crop species covering a broad range of disciplines. It is not claimed that the Compendium is likely (at least in the near future) to provide in-field assistance for subsistence-level farmers. But its interdisciplinary approach, along with the infinite variety of links within the system, will assist researchers to take a broader, more holistic view and thus to provide more useful results which will, in the longer term, be of assistance to those working in the field.

The Compendium Concept

The concept of a relational database for crop protection, which became the Crop Protection Compendium, was formulated at an international workshop on information needs for crop protection, held in 1989 by CABI in partnership with the Food and Agriculture Organization of the United Nations (FAO) and the Technical Centre for Agricultural and Rural Cooperation (CTA) (Harris and Scott, 1989). There was a strong sense of expectancy at this workshop that information technology, correctly harnessed, could be on the point of delivering crop protection information far more efficiently, and that the benefits would be felt particularly in developing countries where the dearth of information was greatest. For demonstration at the 1989 workshop, CABI developed a prototype information system consisting of a database of names of pests linked to a series of independent files. These files contained different types of information about the pests – text descriptions, abstracts of literature, illustrations, and distribution maps. On selecting a pest by name, access was provided to the appropriate files and these various resources could be examined.

CABI discussed the concept with many interested people during the period 1989–1991 and, with the support of the Australian Centre for International Agricultural Research (ACIAR) carried out a feasibility study and held a small international workshop to expose the concept to thorough discussion. The Workshop was attended by 14 participants from ten countries, representing CABI, the Caribbean Agricultural Research and Development Institute (CARDI), the Cooperative Research Centre for Tropical Pest Management (CRCTPM), the European and Mediterranean Plant Protection Organization (EPPO), the Food and Agriculture Organization of the United Nations (FAO), the South Pacific Commission (SPC), the United States Department of Agriculture (USDA), and the University of the West Indies (UWI).

Much discussion centred on the balance to be struck between the factual core of the Compendium on the one hand, and the utilities to interpret data, to make predictions and to aid decision-making on the other. It was concluded that priority should be attached to construction of the core, which was a major task and which had a high value in its own right as a generic information resource. The core would be regarded as a platform on which utilities for specific applications could be built.

The Crop Protection Compendium

Module 1 – South-east Asia and Pacific

The 1992 Workshop recommended that work should proceed stepwise towards global implementation. With support from ACIAR, CABI carried out a feasibility study to determine the scope and requirements of a working Compendium for South-east Asia, southern China and the Pacific region.

Potential users of this regional Crop Protection Compendium were visited in 1994. Contact was made with more than 40 institutions in Malaysia, Indonesia, the Philippines and Thailand. A clear need for the proposed development was expressed by researchers, lecturers, plant health officials, quarantine officers, extension specialists, policy makers and the agrochemical industry. In support of the visits, a questionnaire was designed to identify potential users, determine their needs, and gather opinions on desirable features and content, and sent to approximately 300 recipients in the region. Responses were received from more than 40% of those mailed, with positive responses from 92% of these (Scott and Schotman, 1994).

Module 1 of the Crop Protection Compendium was released in 1997. Information contained in the Compendium is of two types: (i) detailed data sheets on *c.* 1000 pests, beneficials, crops and countries/regions; and (ii) basic data comprising *c.* 20,000 species names, and links with hosts and geographic data. A detailed description of the content and operation of the Compendium is provided below.

CABI's intention, which has been explained to members of the Consortium, is to develop a global Crop Protection Compendium by early 1999, subject to adequate resourcing. Meanwhile, Module 1 of the Compendium will be updated annually, maintenance costs being offset by revenue earned from sales in the developed world.

Relational database system

The core component of the Crop Protection Compendium is a relational database system containing names of pest organisms, beneficial organisms, crop plants and countries, all represented by codes. Files holding relational links between organisms and crops and between organisms and countries allow the derivation of host and geographic ranges of any organism. This simple system of links between pests, hosts and countries held in a relational database system was the basis for the PQ plant quarantine database developed by the European and Mediterranean Plant Protection Organization (EPPO) (Smith and Smith, 1991). It has provided the pattern for several other crop protection databases, such as the plant protection information system of the Caribbean Plant Protection Commission (Schotman, 1990) and the FAO Global Plant Quarantine Information System.

In the database of names and codes, organisms are included not only under their preferred scientific names, but also under synonyms and common names in English and other languages. There are thus several ways to retrieve information about a given organism. In addition to species, names of higher taxonomic groups are included in this database to allow retrieval of information at any appropriate level.

Data sheets

After selection of an organism from the name database, information about it is presented on the screen in the form of a standard data sheet. First, a cover page is presented, giving brief text, a quarter-screen illustration, and a small distribution map. A menu then gives access successively to: details of its names, taxonomic position, host range, and natural enemies; a detailed description; control methods; a user-notepad entry; additional illustrations; and a world distribution map, with options to zoom-in to chosen regions and to view the literature citations on which each occurrence is based. There is also an option to examine abstracts of the relevant literature. An illustration is included in this volume by Scott (Chapter 1).

The description is a key part of this information. It adopts the format of the FAO model data sheet with certain modifications, including the following main headings: Identity, Hosts, Geographical Distribution, Biology and Ecology, Natural Enemies, Economic Significance, Identification, Control, Bibliography. Words within the text may be used to access other data sheets or glossary entries, as explained under 'Soft links' below.

Data sheets for host crops and for countries have comparable formats appropriate to their application, consisting in each case of text, and a small illustration. In addition, both crop and country data sheets allow access to FAO data from the FAOSTAT database on production of major crops and agricultural land use. Software has been developed to present FAOSTAT data in a user-friendly graphical format. For Module 1, pesticide usage data from the Landell Mills database are included for Malaysia, Thailand, the Philippines, Indonesia, Vietnam and China, classified by crop and by pest group.

Name codes and taxonomic hierarchies

Relational links between items in the databases are made through codes. For countries and other geographic regions, a system based on the standard two-character isocodes is used. For organisms, the system used is a development of that proposed by the Bayer company (Faust, 1979), in which species of animals and plants are allocated a six- or five-character code.

Relational files are used to record the upward and downward relationships of every taxon, so that the Compendium contains an inherent taxonomic hierarchy. It is possible to display an organism's taxonomic position by listing the levels above it, and to expand the contents of a given taxon as a hierarchical tree.

Retrieval

There are numerous routes by which the data may be accessed, and the Compendium is designed to provide a user-friendly basis for browsing as well as for performing specific searches. Because a relational database system is used, the initial results of a search can be refined on the basis of other criteria.

The main access points are through pests, crops, countries and abstracts. A common type of retrieval would be through the name of a country (or other geographic region). The data sheet for a given country presents brief text, a map, and a small global location-map. A menu then offers a list of pests recorded for that country. This may be long, so options are then offered to narrow down the list. The criteria offered include pest group (e.g. insects, fungi, bacteria, viruses, weeds), host crop (through access to a listing of all crops), plant part attacked (e.g. leaves), stage affected (e.g. seed) and symptoms. This is likely to lead to a short list of pests, whose individual data sheets can then be examined.

Narrowing down from a long list of candidate pests according to criteria such as these provides a simple but effective way of reaching at least an approximate identification of an unknown pest. It requires no specialist knowledge or diagnostic keys.

The data can also be accessed through crops, narrowing down as appropriate according to country, pest type etc. Alternatively, a taxonomic group can be selected as the starting point. Abstracts can also be searched, and the names of organisms or other items selected from the text can be used to call data sheets.

Illustrations

Several types of illustrations have been incorporated into data sheets, including pictures with 8 bits per pixel (256 colours), 4 bits per pixel (16 shade grey-scale images) and 2 bits per pixel (monochrome line drawings). If no illustration is available for a particular species or other taxonomic group, the system loads where possible an illustration of the closest available group to which it belongs (genus, family or order). Additional pictures are available for a large number of data sheets.

Light Lib Images from DFL Software, Inc. is used as an add-in software component to display the pictures. This component provides a means of displaying a wide range of image formats, such as PCX, GIF and JPG, thus allowing the pictures to be stored in a compressed format.

Distribution maps

A simple, purpose-built Geographic Information System (GIS) has been developed to plot the geographical distribution of an organism on an outline map of the world. The plotting symbol is varied according to the frequency of the pest in the country concerned. The viewpoint can be changed and it is possible to zoom-in to a chosen region. At this larger scale, more detail can be shown, including the distribution in subnational divisions for large countries (such as Australia, Brazil, Canada, China, India, Russia, USA) and for those with detached parts (such as France, Indonesia, Italy, Japan, Malaysia, Portugal, Spain, UK).

The plotted distribution of one organism can be overlaid with that of another or compared side by side; for example a pest distribution can be overlaid with that of its host or its natural enemy. A distribution can also be overlaid with a climatic map.

Literature abstracts

References cited in data sheets which have been abstracted by CABI are available within the Compendium, and can be searched by keyword. Words within a record can be used for further searching or to access data sheets or glossary entries.

Glossary

A glossary of technical terms used in crop protection is available within the Compendium. The entire list of terms can be browsed and definitions displayed as required. Alternatively, any term within a section of text (which may be a data sheet description, an abstract, another glossary entry, or even a recently-entered user note) may be marked and used to call a glossary definition. Illustrations may be displayed next to the glossary text. Within this list of technical terms are data obtained from the *Pesticide Manual* of British Crop Protection Enterprises on active ingredients of pesticides, including names, products, applications and environmental impact.

Soft links

The user can highlight any word or group of words within a section of text, in order to obtain additional information. The highlighted text is looked up in the database of names (scientific names, synonyms and common names of pests, beneficial organisms, crops and countries) and in the glossary. If it is found, the appropriate data sheet or glossary entry is displayed. This provides a powerful free-text-linking facility without the need to set up hypertext links in advance.

User notepads

The user may enter personal notes related to any pest, beneficial organism, crop or country. This may be done directly from the keyboard or from an external text file. The result is stored separately from the Compendium data, in a location of the user's choice; if required, users can opt for both a local, personalized notepad and an institution-wide, networked notepad. This utility will enable networked users to share their own information in context, and could also play an important role in keeping the Compendium updated with information from users.

Diagnostic utilities

There are several possible ways to determine the identity of an unknown pest. The most easily accessible to a non-specialist is to use the relational structure of the Compendium to narrow down from many possibilities to few, based on knowledge of one or more of the following: pest group, location, host crop, plant part attacked, symptom type, and commodity (see 'Retrieval' above).

The first Module of the Compendium includes several different types of diagnostic tools, including two CABIKEY applications (White and Scott, 1993): to the Asian fruitflies, and to the important beetle families and subfamilies. In addition, simpler dichotomous TAXAKEY applications are provided, to the arthropod orders, families of Heteroptera, Homoptera, and adult and larval Lepidoptera. Two further applications, aids to identifying the weed and nematode species included in this Module of the Compendium, are also available. In all cases, the result is returned to the Compendium system and the appropriate data sheet is displayed on request.

Open architecture

The system is designed with open architecture, allowing interfaces to be made with add-on or external systems such as the diagnostic tools described above, bibliographic databases, expert systems and the Internet. An interface with the World Wide Web is being developed, with links to a selection of useful pest management Web sites around the world and the option to add user-defined links. External software and databases can in principle be called from anywhere in the system, and the result can be returned to the Compendium for further processing.

In addition, the CAB-IPM (SE Asia) database will be an integral part of Module 1. This is a collection of abstracts of publications, both conventional and non-conventional ('grey literature'), relating to pest management in South-east Asia. It focuses particularly on IPM, and is being developed by CABI under a separately funded project. Material is derived not only from CAB ABSTRACTS, but also from the AGRICOLA database compiled by the USDA National Agricultural Library, and from regional sources in South-east Asia including a number of national institutions that have agreed to act as National Focal Points for the CAB-IPM (SE Asia) database.

Data types and sources

The data in the Compendium fall into two distinct sets: (i) manually compiled detailed data sheets; and (ii) semi-automated compiled 'basic data', consisting of preferred names, synonyms, taxonomic position, host range, natural enemies, geographic distribution and citations.

Compilation of data for the Crop Protection Compendium has been a major task, even for the first module. The most significant component of the work is the preparation of detailed data sheets, which must be done by specialists with up-to-date knowledge of the organisms and their economic significance. For Module 1 of the Compendium, pest and natural enemy data sheets have been provided by 322 specialist authors in 44 countries. The project has maintained close links with institutions in the South-east Asia region, via a Regional Co-ordinator based at CABI's Asian Regional Office in Kuala Lumpur. Twenty-five per cent of the pest data sheets were generated in Asia. Data sheets received are then validated by experts in each discipline.

The basic data has been extracted mainly from existing databases through software routines. Many relationships between pests and their natural enemies have been obtained, from sources such as the Natural Enemy Databank (NED) of CABI's International Institute of Biological Control (IIBC) (Fry, 1989), and by electronic scanning of descriptors in the field of CAB ABSTRACTS used to generate printed indexes in abstract journals over the last 20 years. Frequency counts of references to these relationships will be used to eliminate infrequent items that require validation. Similar routines can be developed to extract host relationships and geographical occurrences from databases such as CAB ABSTRACTS. While the results may be incomplete, they represent an invaluable starting point for validation and extension from other sources.

In addition, information from Bayer, the VIDE (Virus Identification Exchange) data bank (Boswell *et al.*, 1986) and the EPPO PQ database (Smith and Smith, 1991) have been used to enhance the Compendium data set.

Technical Requirements

The minimum hardware requirement will be an IBM PC or compatible, with 486DX2 or Pentium processor (minimum 486DX), Super VGA monitor, running Microsoft Windows 3.1x or Windows 95, preferably with 16 Mb RAM (a minimum of 8 Mb) and 20 Mb hard disk space, and a 2x CD-ROM drive. While such facilities are in only limited use at present, CABI will do all it can to support the wider availability of this valuable technology, which is already spreading at a remarkable rate as prices drop and awareness extends. The Compendium has been developed for Windows 3.1, and will also run through later versions including Windows 95.

Future Developments

Information technology and modes of communication are developing so rapidly, especially through the personal computer, that technological factors are not expected to limit the useful application of the Compendium. At present,

CD-ROM is expected to be the practical distribution medium for the Compendium, whose content will be 400–500 megabytes in extent. CD-ROM technology is already spreading fast in developing countries, in the wake of the essentially similar CD technology used for music and other applications. For the receipt of feedback from the user community, and for the dissemination of updates by CABI, other media may be appropriate, including for example diskette and telecommunications. Developments in global networking, especially through the Internet and the World Wide Web graphical interface, are proceeding rapidly: it appears certain that their penetration will extend quickly, and will spread increasingly to developing countries. CABI will pay close attention to the changing technical scene, and will ensure that useful developments are harnessed for the Compendium project. CABI's first interest will be the useful application of the results of the project, especially in developing countries; the use of widely accessible technology will therefore remain a high priority.

Conclusions

The interdisciplinary nature of the Compendium will aid users to gain a greater understanding of the complex interrelations within agronomic systems, and to take a more integrated approach to crop management and crop protection. It will thus provide strong underpinning for pest management programmes, both as an information resource and as a training aid. It will be helpful to plant protection practitioners, advisers, researchers, students, quarantine officers, national and regional plant protection organizations. It will represent a major step forward in the application of modern information technology to pest management. Furthermore, the concept lends itself to many other applications in agriculture, forestry, health, and natural resource management, and has much potential for further elaboration and development.

The Compendium's successful deployment will depend on effective international cooperation and the two-way flow of information. Thus, the Compendium will be a resource developed through international partnership to the mutual benefit of those involved. Its maintenance will depend in turn on the flow of information from the community of users, to keep it up to date and relevant and to ensure its development in line with trends in crop protection.

Acknowledgements

We acknowledge the support of the Australian Centre for International Agricultural Research (ACIAR) for the original feasibility study for the Crop Protection Compendium.

We acknowledge the financial support of the following, who are members of the Crop Protection Compendium Development Consortium, or who are assisting its development in other ways: Asian Development Bank (ADB); Australian Centre for International Agricultural Research (ACIAR); Canadian International Development Agency (CIDA); Cyanamid, USA; Danish Government Institute of Seed Pathology/Danish International Development Agency (DGISP/DANIDA); DowElanco, USA; DuPont, USA; Gesellschaft für Technische Zusammenarbeit (GTZ), Germany; Hoechst Schering AgrEvo Gmbh, Germany; International Development Research Centre (IDRC), Canada; International Rice Research Institute, the Philippines; Monsanto, USA; Novartis, Switzerland; Overseas Development Administration (ODA), UK; Pioneer Hi-Bred, USA; Rohm and Haas, USA; Sumitomo Chemical Company Limited, Japan; Swiss Development Cooperation (SDC); United Nations Development Program (UNDP); United States Department of Agriculture: Animal and Plant Health Inspection Service (USDA-APHIS); Zeneca Agrochemicals, UK.

We also acknowledge with gratitude support from many institutional and individual experts, too numerous to list, who are contributing data sheets, images and other data to the project. All are acknowledged by name within the electronic Compendium.

The support of many CABI staff has been invaluable, and the advice of Dr Clive James is especially appreciated.

References

Boswell, K.F., Dallwitz, M.J., Gibbs, A.J. and Watson, L. (1986) The VIDE (Virus Identification Data Exchange) Project: a data bank for plant viruses. *Review of Plant Pathology* 65, 221–231.

Faust, W. (1979) Ein universelles System zur Codierung von Tier-, Pflanzen- und Pilznamen für die elektronische Datenverarbeitung im Pflanzenschutz. *Zeitschrift für Pflanzenkrankheiten und Pflanzenschutz* 86, 396–399.

Fry, J.M. (1989) *Natural Enemy Databank* (1989). CAB International, Wallingford, UK.

Harris, K.M. and Scott, P.R. (1989) *Crop Protection Information: an International Perspective.* CAB International, Wallingford, UK, pp. 237–291.

IPM Working Group (1992) *A Synopsis of Integrated Pest Management in Developing Countries in the Tropics.* Synthesis Report commissioned by the Integrated Pest Management Working Group. Natural Resources Institute, Chatham, UK.

IRRI (1989) *IRRI Toward 2000 and Beyond.* International Rice Research Institute, Manila, Philippines.

Schotman, C.Y.L. (1990) Plant protection information system of the Caribbean Plant Protection Commission (CPPC). In: *Final Report, Workshop on Plant Health Information and Monitoring in Latin America and the Caribbean, San José, Costa Rica, 24–26 April 1990.* Instituto Interamericano do Cooperación para la Agricultura, Coronado, Costa Rica. pp. 165–174.

Scott, P.R. and Schotman, C.Y.L. (1994) The concept of an Electronic Compendium and its application to crop protection. *Quarterly Bulletin of the International Association of Agricultural Information Specialists* 39, 166–173.

Smith, I.M. and Smith, B.C. (1991) PQ – the EPPO data base on plant quarantine pests. *EPPO Bulletin* 21, 211–218.

Smith, R.W. (1994) Changes in crop protection strategies for less-developed countries: an ODA perspective. In: Black, R. and Sweetmore, A. (eds) *Crop Protection in the Developing World*. BCPC Monograph No. 61. British Crop Protection Council, Farnham.

White, I.M. and Scott, P.R. (1993) Computerized information resources for pest identification: a review. In: Hawksworth, D.H.L. (ed.) *Identification and Characterization of Pest Organisms*. CAB International, Wallingford, UK.

12

Interpreting Information to Produce Knowledge: the Role of a Professional Society

A.C. Newton

Scottish Crop Research Institute, Invergowrie, Dundee DD2 5DA, UK
Fax: +44 (0)1382 562426/E-mail: a.newton@scri.sari.ac.uk

Introduction

This chapter addresses some questions about the role of information technology in plant pathology, and in particular the role of a professional organization such as the British Society for Plant Pathology (BSPP) in serving and anticipating the needs of its members in this area.

BSPP's Information Dissemination Role

In the past, BSPP has been instrumental in disseminating information through publishing books (largely conference proceedings), publishing its peer-reviewed journal *Plant Pathology* and its newsletter *BSPP News*, holding conferences, and representing plant pathologists, for example by membership of national and international committees. Now the Society has established, through the Internet, a presence on the World Wide Web known as BSPPWeb (http://www.bspp.org.uk). The contents of *Plant Pathology* are also available online (Blackwell Science, 1997). BSPPWeb contains general information about BSPP and its activities, the contents of *BSPP News*, and a new peer-reviewed electronic journal *Molecular Plant Pathology On-Line*. BSPPWeb also contains conference information including pre-conference material, abstracts and other notes, and facilities for on-line booking. There is also a searchable database of members' interests which they can keep up to date interactively. A link is maintained with a comprehensive list of pathology-related Internet resources (Kraska, 1997), and various other information resources are included such as

 Information Technology, Plant Pathology and Biodiversity (eds P. Bridge, P. Jeffries, D.R. Morse and P.R. Scott)

summaries of the annual reports of the UK Cereal Pathogen Virulence Survey. Together these represent a valuable information resource collected under one umbrella. However, whilst this may be a collection of information, it could not be described as an information system.

Current Developments

Every user comes into a network when trying to access information. The objective of an information system is to provide comprehensive but easy-to-follow routes through the network to all points the user needs to access. The objective is not to impress the user with the size of the network and their location in it, as this will either confuse, bewilder or daunt. A practical example of this is being implemented in some on-line journals where papers are published in a network of information readily available on-line. This includes not only database-type information of direct relevance to the paper's contents, but also backwards and forwards links to other citations. References can be hypertext-linked from text to reference list to on-line bibliographic database, where the abstract or even the full paper may be accessed. The paper is placed in a dynamic information network through forward citation indexing, by regular updating of links to all publications that cite the paper.

The Challenge

As with all new developments, we must continually assess whether they meet their objectives. For example: is this development of information technology for its own sake, because it is new and gives us new toys? Does it put up more barriers to people outside our field, or does it make information more accessible? What about education of the next generation of plant pathologists? Will the text book be obsolete soon? Should there be a new type of 'electronic text book'? If, for example, someone required information on potato late blight (*Phytophthora infestans*), potential sources of information and resources of interest may be: books, scientific papers, bulletins, crop intelligence reports, conferences, published statistics and surveys, information about control methods, practical and anecdotal information from agricultural advisors, agricultural policy documents, lists of current research projects, lists of researchers and advisors, databases of germplasm collections of potato and its wild relatives, and of *P. infestans* and its relatives, genetic maps and databases of host and pathogen, research employment opportunities, and many other types of information. There is scope for a structured information system in which such sources are linked together in a logical and helpful format, and such that providers of information make it available in suitable formats.

An Information System Specification

A key quality of such information systems is their user-friendliness. The list of requirements could be:

- a Windows-based package.
- intuitive presentation, as few bother to read manuals.
- rapidly set to the user's level of competence.
- rapid access to up-to-date information.
- self-training and automatic prompting.
- natural language input, where practical.
- able to handle uncertainties in queries.
- sustainability, through an interested group such as a learned society.

The Internet effectively mimics distributed computing on a local area network (LAN), where information and programs from multiple sources can be accessed from anywhere on the network. In each case a degree of management is required, or at least an index to the material that is available. A good information system should be balanced, giving due prominence to significant content and avoiding technology-driven developments if they are not appropriate. This is where the input of a professional society can be valuable.

One of the biggest problems in developing information networks is the compatibility and accessibility of different information resources. The World Wide Web interface allows hitherto incompatible formats to be made compatible without having to access the software directly. The user need not know where the information is stored, what electronic format is used, or what software is being run to access it. For example, the user may see buttons or sliders to operate, lists to select from, or boxes to make entries in, all of which are familiar to PC or Mac users. Options, preferences and permissions can be tailored to users' requirements, allowing different levels of access for different needs.

Existing Systems

Systems that go some way in this direction within a specific application include the US National Library of Medicine's Unified Medical Language System (NLM UMLS) (NLM, 1997), whose goal is to make it easy for users to link information from patient record systems, bibliographic databases, factual databases, expert systems, and other resources, without the need to negotiate diverse telecommunications or information retrieval procedures. In the field of crop protection the CABI *Crop Protection Compendium* (Sweetmore *et al.*, Chapter 11, this volume) aims to present the user with diverse types of information within a single system. Software packages such as *Reference Manager* can integrate bibliographic information from a wide range of sources including locally constructed databases, network-mounted CD-ROMs such as CAB ABSTRACTS and *PlantPathCD* (CABI, 1997), and on-line databases such as BIDS (Bath Information and Data Services) (BIDS, 1997), and they can be tailored to a single area of interest.

An additional dimension that can be integrated where professional societies such as BSPP are involved is the provision of access to active researchers in a field. No information system can replace experience, and there is no satisfactory way of accessing anecdotal information or making the mental connections that an active researcher makes spontaneously. It is essential to involve people in the continuous development of information systems to maintain and develop the usefulness of the systems. Furthermore, once a coherent structure is in place, many other suppliers of information will be encouraged to make their resources available in the most useful format.

Features of an Information System

Information exchange will be encouraged by a framework which is not only accessible and easy to use, but which also accepts and integrates information rapidly and in a user-friendly manner, without the need to go through intermediaries. If the authority of the information is clearly identifiable, and credit is attributable, then this improves the prospects of acceptance and use. Authority status, for example, ranging from personal opinion through preliminary research report or database entry to peer-reviewed publication, must be transparent.

Most researchers regularly access very few journals – just their favourites. Even then they look at very few papers, scan the content of even fewer, and rarely read any papers from beginning to end. Just as we often read the back page of the newspaper first, so we often only scan authors, title and abstract in a research paper, then maybe figures and tables, and the rest only if we are really interested. An electronic information system should be designed in the same way. Hypertext should enable this to be accomplished relatively easily. On the World Wide Web, 'cookies' can be used to record a user's previous selections and preferences, making it possible to adapt presentation to needs without the user even realizing. However, this should not be to the exclusion of intelligent prompting, which should enable irrelevant material to be easily rejected without having to navigate through layers of information. It should also enable special priority to be given where appropriate to information sources normally given minimal status.

Despite their memory-hungry nature, graphics are one of the great attributes of electronic information systems, where file size and not cost is their major penalty. Images can allow multiple levels of technical description to be bypassed, providing access by the non-expert into specialist fields. This is especially valuable for pest and pathogen identification, as discussed in many chapters of this volume. If images are linked with verification data such as geographic location, host, etc., many of the pitfalls of such 'picture book' pathology can be avoided. Whilst communications restrictions (bandwidth especially) currently limit the usefulness of graphics on the World Wide Web, considerable progress in access to multimedia can be expected through the intelligent use of a combination of locally mounted information on CD-ROM, improved data reading devices, reduced CD-ROM production costs, downloading and updating during quiet times, use of automatically updated mirror sites, file compression technology, and restraint in the use of graphics for purely presentational purposes.

Conclusions

Information technology needs investment, management, publicity and training. Probably all areas of science will benefit greatly from its development over the next few years. In plant pathology, epidemiology will benefit especially, owing to the need to handle either computationally demanding models and analyses, or to capture, process, store and retrieve the sheer volume of spatial and temporal data.

Danger lies in the ease with which information can be stored, without necessarily being read, analysed and interpreted before being put to further use. A good information system provides only enough data to enable initial sorting into categories such as 'discard', 'store for later appraisal', or 'make full appraisal'. Information in the last category simply requires more to be revealed for further assessment. However, information classified as 'store for later appraisal' needs to be carefully labelled, sorted and stored. It is better to discard than to store inappropriately.

There is a clear role for a professional society such as BSPP in initiating and coordinating the development of information systems to serve the interests of members, and in actively involving them in the process of linking and associating information from different sources. The organization should be providing rapid and mutually beneficial channels of communication, discussion, interpretation and information dissemination for members, at their required level of technology access or expertise. The tools for presenting information are clearly being developed. They now need to be applied and evaluated in a continuous development plan.

References

BIDS (1997) *BIDS on the Web.* University of Bath, Bath Information and Data Services, Bath, UK. World Wide Web page at http://www.bids.ac.uk/

Blackwell Science (1997) *Blackwell Science.* Blackwell Science, Oxford, UK. World Wide Web page at http://www.blackwell-science.com/

CABI (1997) *PlantPathCD 1973–1996*. Bibliographic database on CD-ROM. CAB International, Wallingford, UK.

Kraska, T. (1997) *Plant Pathology Internet Guide Book*. University of Hannover, Hannover, Germany. World Wide Web page at http://www.ifgb.uni-hannover.de

NLM (1997) *UMLS Knowledge Sources*. US National Library of Medicine, Bethesda, USA. World Wide Page at http://www.nlm.nih.gov/research/umls/umlsdoc.html

Building Models of Epidemics to Help Take Decisions

13

M.J. Jeger

Department of Phytopathology, Wageningen Agricultural University, POB 8025, 6700EE Wageningen, The Netherlands
Fax: +31 317 483412/E-mail: mike.jeger@medew.fyto.wau.nl

Introduction

It was a great pleasure and honour for me to give the Garrett Memorial Lecture, on which this chapter is based. I first encountered the writings of S.D. Garrett a few months into my PhD research when I was wrestling with the problems of infection–inoculum relationships and the question of independent action of spores. I was much struck by the way in which Garrett reviewed the mechanisms of infective dose for animal pathogens. In an apparently effortless verdict he wrote '... an outstanding contribution towards the solution of this problem has been made ...'; and then reviewed the evidence for plant pathogens (Garrett, 1970, pp. 13–19). At this time, in 1977, epidemiology as a quantitative discipline was rapidly developing and I remember being much less impressed by his concepts of inoculum potential which appeared to me to be essentially unmeasurable and thus inoperable as definitions. Considerable controversy was generated by these concepts with a series of criticisms (e.g. Vanderplank, 1975, pp. 84–87) and ripostes (e.g. Baker and Drury, 1981) which more than anything attest to the acidity and (sometimes) wit possible within the English language. Nonetheless concepts of inoculum potential continue to preoccupy epidemiologists concerned with soil-borne pathogens and I will return to them at the end of this chapter.

Epidemiology is the essence of my lecture. In a recent Presidential Address to the British Society for Plant Pathology, Royle (1994) presented the dual aspects of plant disease epidemiology that make it the 'cornerstone of good disease control' – understanding and prediction – and argued that a balance must be struck between these two aspects. I certainly echo his views that fundamental

 Information Technology, Plant Pathology and Biodiversity (eds P. Bridge, P. Jeffries, D.R. Morse and P.R. Scott)

understanding of epidemiological processes within an appropriate framework is necessary but not sufficient for improved disease prediction and control. Further, it is my view that understanding can contribute at different levels relating to the tactics, strategies, and even policies, of disease control. It is my purpose in this chapter to illustrate how epidemiological modelling can contribute to establishing this appropriate balance.

It is appropriate to do this in a book on the application of information technology to plant pathology. While the main advances that I shall describe are conceptual rather than computational, epidemiological modelling is one of the classic fields in which information technology has supported plant pathological thinking. Furthermore, I shall give examples where advances in the approach to modelling have depended on enhanced computational power.

Models in Plant Disease Epidemiology

Much has been written on the typology of models used in plant disease epidemiology that I will not attempt to review here. In Table 13.1, I have tried to show how different modelling techniques have been developed for different purposes and how these in some cases have been related to practical aspects of disease management. I do not intend to overemphasize the horizontal alignments in this Table but it seems to me that these serve to illustrate key features of the contribution of modelling to disease management.

Exploratory models

Descriptive statistics have long been used to explore often comprehensive data sets with a view to establishing relationships between weather and disease variables that may be used for forecasting future disease development. Based on these relationships and provided other criteria are satisfied (Jeger, 1987), decisions on the necessity for spray treatments can be made. Because the disease cycle can be broken down into successive components of infection, colonization, sporulation and dispersal and because each of these components is strongly influenced by weather variables (Schroedter, 1983; Jeger, 1987), there is clearly scope for exploring the idiosyncrasies of disease progress curves (Royle, 1994) by modelling these individual components using techniques such as multiple regression analysis (Butt and Royle, 1990). Again, there is much literature on this topic – my only comment is that the practical use of forecasting is often limited more by other constraints (perception of risk, logistical problems, and economics) rather than epidemiological insight.

Theoretical models

Theoretical models form part of an enormous literature from population ecology, combined with and enriched by more recent modelling in human/animal

Table 13.1. Types of models developed in epidemiology and examples of applications in disease management. Horizontal alignments indicate a personal view of the most productive relationships at the 'type' level not for the specific examples given.

Epidemic modelling	Disease management
Exploratory analysis	Tactics
• statistics	• forecasting
• multivariate techniques	• infection periods
• time series/spatial pattern	• spray schedules
Theoretical models	Strategy
• differential/difference eqns	• cultivar deployment
• population genetics/dynamics	• fungicide resistance
• physical theory	• control efficacy
Spanning hierarchies	Policy
• computer simulation	• risk assessment
• neural networks	• quarantine
• crop physiology/loss	• funding priorities

epidemiology. In plant disease epidemiology, the contributions of Vanderplank (1963, 1975) fall very much into the category of theoretical models, although his approach was sufficiently individualistic to effectively create barriers between plant disease epidemiology and the broader mainstream of population ecology, as will be discussed in later sections of this chapter. It is only recently that these barriers are breaking down. Because of their nature, theoretical models can seldom be used in support of tactical disease control decisions. Their contribution has been mostly in support of experimental epidemiological research, and to address strategic questions of control.

Hierarchical models

Disease simulators

Models that span biological hierarchies (cell, organ, individual, population, etc.) have had a somewhat chequered history. In 1977, with the increased power and accessibility of computers, computer simulation was seen as the great leap forward in plant disease epidemiology arising from the development of simulators of specific plant diseases (Waggoner, 1978, 1990), based on the underlying biological processes and mechanisms. A large number of such simulation models had been developed by the mid 1980s (Teng, 1985). Unfortunately, to my mind, insufficient attention was given to exactly to what uses simulation models could or should be put (Jeger, 1986b). Shrum (1978) predicted that, in the long run, simulation or systems analytic models would predominate in the practical forecasting of plant disease. Whether it is still premature to comment on this prediction I'm not sure but there is little evidence to support it at this point in time.

*Apple scab (*Venturia inaequalis*)*

To take one disease as an example, apple scab (caused by *Venturia inaequalis*) is a classical example of a plant disease where the relationships between weather variables and infection seem obvious; the infection periods defined by Mills (Mills and La Plante, 1951) seem ideally suited for forecasting, as do aspects of the overwintering of the fungus (Jeger and Butt, 1983). A theoretical approach was taken to model the risk of primary infection by ascospores in spring (MacHardy and Jeger, 1983) and this has led to the development of practical calculation procedures as described fully in MacHardy (1996). Several mechanistic simulators of apple scab have been developed starting with EPIVEN (Kranz *et al.*, 1973: Kranz, 1979). Since then a number of simulators have been developed and compared (Seem, 1986). The most comprehensive simulator developed by Seem *et al.* (1989) includes management decisions and economic optimization as well as the underlying biological processes and relationships (see MacHardy,1996, pp. 504–509 for a thorough discussion of the uses of this complex model).

An intermediate approach has been taken in the UK where a PC-operated decision support system, based in part on simulation of mechanistic processes, has been developed (Xu and Butt, 1993; Xu *et al.*,1995). The model proved successful in detecting critical infection periods in early spring and subsequent disease development which were not predicted using the simple Mills criteria. However, one criticism of the original Mills criteria, at least in the UK, was that too many infection periods were identified. The ability of any model to screen out these false positives would be a key test of its performance. Further, it can be argued that any decision support system must integrate with the whole range of disease and pest problems in an orchard for practical use by farmers. For example Van den Ende *et al.* (1996) list some 19 arthropod pests of apple included in GABY, a computer-based decision support system in use in the Netherlands.

Uses of hierarchical models

My own view on these sometimes quite complex hierarchy-spanning models is that they are unlikely to be used directly for determining the tactics or strategy of disease management, but in the future are more likely to be used more in policy decisions such as risk assessment (e.g. introduction of genetically modified microorganisms or disease resistant plants), issues relating to the movement of important quarantine pathogens, or in combination with crop physiological or even land-use models to assist in decisions on funding priorities.

Theoretical Epidemiological Models and Strategic Decisions

Relationship with a research programme

It is perhaps restrictive to view the use of models in decision making as geared only to practical applications in decision-making. Sometimes models can prove an invaluable tool in clarifying the issues involved in a research programme and assisting decisions made by researchers. This use has much in common with the view of Levin (1981) who envisaged the relationship between theoretical ecology and real-world ecology in terms of an analogy with a DNA molecule. Seen similarly, modelling and experimentation are two mutually reinforcing strands in plant disease epidemiology. To illustrate this I will use some collaborative work done on nematode overwintering and epidemiology. This started with a conceptual view of how *Meloidogyne* spp. eggs and juvenile stages might overwinter (Fig. 13.1) (Starr and Jeger, 1985) on different host crops in Texas. A model was proposed which was certainly able to predict the qualitative form of the conceptual relationship (Jeger and Starr, 1985). More than that, the model also predicted other shapes depending on the relative values of parameters describing rates of egg and juvenile mortality and hatching. Subsequently, the model was fitted to data from different locations, crops and years and, despite the inherent variability in such data, was successful in describing the major trends and differences observed (Jeger *et al.*, 1993).

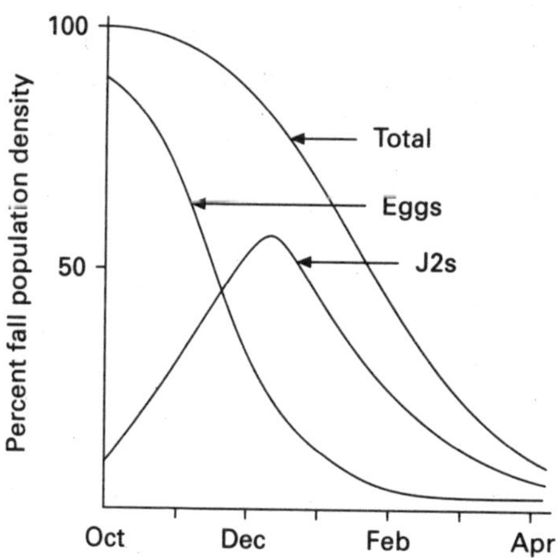

Fig. 13.1. Schematic representation of winter survival dynamics of root knot nematode eggs and J2 juveniles in Texas, after Starr and Jeger (1985).

Vanderplank's differential-delay equation

The model describing the nematode overwintering was derived in terms of two (or in some cases more) differential equations describing the rates of change in the nematode egg and juvenile populations. This type of model is quite standard in population ecology but was slow to be adopted in plant disease epidemiology largely because of the pervasive influence of Vanderplank's seminal book on plant disease epidemics (Vanderplank, 1963). Starting with what was in effect the well known logistic equation, Vanderplank derived an equation giving the increase of disease as a function of infectious tissue (e.g. sporulating lesions) and the amount of healthy tissue. This equation can be written as:

$$dy(t)/dt = R[y(t-p) - y(t-i-p)][1-y(t)] \tag{13.1}$$

where: $y(t)$ is disease at time t measured on a proportion scale; p is the length of the latent period (time between infection and infectiousness); i is the length of the infectious period (time a diseased unit remains infectious); and R is a rate parameter. The first term in square brackets on the right hand side then gives the amount of disease that is infectious; the second term gives the proportion healthy. The two most important aspects of this equation are that: (i) disease will eventually level off to approach an asymptote strictly less than 1; and (ii) for an epidemic to occur the product iR (what Vanderplank called his progeny–parent ratio) must be greater than 1.

Further analyses of this equation have been made by Jeger (1986a), May (1990), Jeger and van den Bosch (1994a,b), and Swinton and Anderson (1995). Some of the earlier ambiguities concerning the equation have been clarified but it is clear that there are more flexible ways of formulating epidemic models using linked differential equations as proposed first for plant disease epidemics by Jeger (1982).

Epidemic models based on linked equations

The basic idea in these types of model is to recognize that the host population can be broken down into categories of:

- healthy units – not yet infected.
- latently infected units – infected but not yet infectious.
- infectious units – infected and infectious.
- post-infectious – infected but no longer infectious.

Differential equations are then written down for each of these categories, with parameters describing the infection rate (sometimes termed contact or transmission rate) and the rates of progression from the latent to the infectious and from the infectious to the post-infectious categories. In human/animal epidemiology these categories are often termed Susceptible, Exposed, Infectious, and Removed, giving rise to the so-called SEIR epidemic. The major advantage of this modelling approach is that it is much easier to analyse effects of host and pathogen attributes and the effect of control practices in such a formulation. It is usually possible to derive an expression for the asymptotic (or final) size of the epidemic and also for the epidemic threshold. In particular, a key composite quantity, the basic reproductive number, can be derived in terms of the model parameters. In the simplest case, this is conceptually the same as Vanderplank's progeny–parent ratio.

The basic reproductive number (often denoted by R_0) is of fundamental importance in determining qualitative aspects of an epidemic behaviour. Its derivation allows the analysis of strategic control decisions such as sanitation (destruction of diseased plants), deployment of resistant cultivars, the effect of mixing fungicides with different epidemiological modes of action, and the requirements for a successful biocontrol agent. The basic reproductive number is defined as: the number of new infected units that arise following the introduction of one infected unit into an otherwise healthy population. If this quantity is greater than 1 then an epidemic will occur (which may still be at a relatively low final intensity); if less than 1 then there will be no epidemic. The difference between these two outcomes is shown schematically in Fig. 13.2. During the early stages of an epidemic (i.e. $R_0 > 1$) disease increase appears exponential, although it will eventually level off as healthy host tissue becomes limiting. However, if $R_0 < 1$ then although there is a small increase in disease it will level off even with an abundance of healthy host tissue present. The point

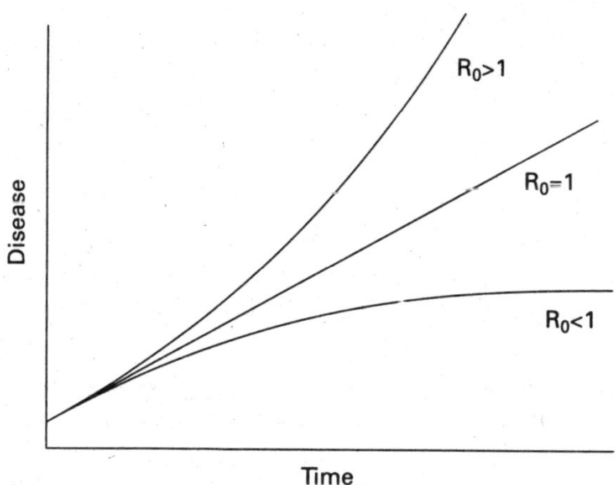

Fig. 13.2. Schematic representation of early disease development dependent upon the value of the basic reproductive number (R_0): if $R_0 > 1$, there is unbounded, apparently exponential, increase in disease; if $R_0 < 1$, there is a bounded increase in disease only, even when healthy host tissue is not limiting.

of any control option is of course ideally to reduce the basic reproductive number to less than 1.

Control of virus diseases by roguing

To illustrate the use of the basic reproductive number, I will refer to work on plant virus diseases of perennial crops, including cocoa swollen shoot, citrus tristeza, banana bunchy top and plum pox (Chan and Jeger, 1994). An SEIR type model was formulated in terms of the basic epidemiological processes, but also included natural host mortality, disease-induced mortality, and roguing of diseased plants at varying efficiencies. The basic equations and definitions of parameters are given in Table 13.2A, with the full details in Chan and Jeger (1994). Analysis of these equations yielded expressions for the final size of the four host categories (Table 13.2B), the basic reproductive number (C), and the critical roguing rate (D) necessary to prevent an epidemic. From these analyses and from parameter values estimated from the literature, it was concluded that roguing could be effective in controlling the diseases. Moreover it was possible to evaluate trade-offs (constructed in terms of effort, but could equally be done in terms of economics) between roguing and replanting to obtain optimal solutions for the size of the healthy plant population (Jeger and Chan, 1995).

Of course, plant virus epidemics involve the interaction of host, virus and vector, with a wide variation in transmission characteristics possible, especially between non-persistent and persistent transmission. We have recently developed a model that combines the epidemiological approach of Chan and Jeger (1994) with transmission characteristics of the vector and also birth, death and migration parameters. The resulting model (Jeger *et al.*, 1997) is of course more complex (shown schematically in Fig. 13.3) than the case without explicit consideration of vectors, but does nevertheless allow the same types of qualitative analysis. In particular, we were able to derive a basic reproductive number including the vector transmission and population parameters, and show clearly the difference in epidemic patterns between non-persistently transmitted viruses and propagative viruses. In the latter case, much higher vector populations (or vector activity) are required to give the same basic reproductive number observed for an epidemic. Consequently, reduction in vector population densities would be the key control option for propagative viruses. For non-persistent transmission, roguing could be effective but only at very low population densities. Holt *et al.* (1997) have recently linked a disease and vector model for the specific case of African cassava mosaic virus to look at roguing/replanting strategies in relation to the deployment of resistant cultivars in Uganda and possible virulence shifts in the virus and/or vector.

Long term dynamics

Our ability to examine the consequences of the long-term dynamics of non-linear systems such as the epidemiological models above has been much

Table 13.2. Analytical model of plant virus disease dynamics (Chan and Jeger, 1994; Jeger and Chan, 1995) based on linked differential equations for each plant category. A. Basic model and parameter definitions; B. Asymptotic or final sizes of the host populations; C. Determination of the basic reproductive number; D. Critical roguing rate for prevention of an epidemic.

A.

$$\frac{dH}{dt} = r(K - P) - \mu H - k_1 H \frac{S}{K}$$

$$\frac{dL}{dt} = k_1 H \frac{S}{K} - \mu L - k_2 L$$

$$\frac{dS}{dt} = k_2 L - \mu S - k_3 S$$

$$\frac{dR}{dt} = k_3 S - (\mu + \alpha) R$$

where *H*, *L*, *S* and *R* are the categories of healthy, latently infected, infectious and post-infectious plants, respectively, k_1 is the contact rate, and k_2 and k_3 are rates of disease progression from latent to infectious and from infectious to post-infectious, respectively. The rates k_2 and k_3 are effectively the inverses of the mean length of latent and infectious periods, assumed to be exponentially distributed. Plants are replanted at a fraction *r* of the difference between the number of plants present ($H + L + S + R$) and *K*. Plants die at a rate μ; with disease causing an extra mortality rate, α, in the post-infectious category.

B.

In the presence of disease

$$H^* = K(\mu + k_2)(\mu + k_3) / k_1 k_2$$

$$S^* = \frac{K\left[r\,k_1 k_2 - (\mu + r)(\mu + k_2)(\mu + k_3)\right]}{k_1\left[r(\mu + k_2 + k_3) + (\mu + k_2)(\mu + k_3) + r k_2 k_3 / (\mu + \alpha)\right]}$$

$$L^* = \left[(\mu + k_3)/k_2\right] S^*$$

and

$$R^* = \left[k_3 / (\mu + \alpha)\right] S^*$$

C.

For disease to persist

$$r\,k_1 k_2 / (\mu + r)(\mu + k_2)(\mu + k_3) > 1$$

Rearranging

$$H^*(k_1 / K)\left[1/(\mu + k_3)\right]\left[k_2/(\mu + k_2)\right] > 1$$

(1) (2) (3) (4)

where (1) is the number of trees in the absence of disease, (2) is the number of new infections per unit time per tree, (3) is the approximate duration of the infectious stage and (4) is the probability a tree reaches the infectious stage.

D.

Critical roguing level

$$\eta_c = \left[r\,k_1 k_2 - (\mu + k_2)(\mu + k_3)(\mu + r)\right] / (\mu + k_2)(\mu + r)$$

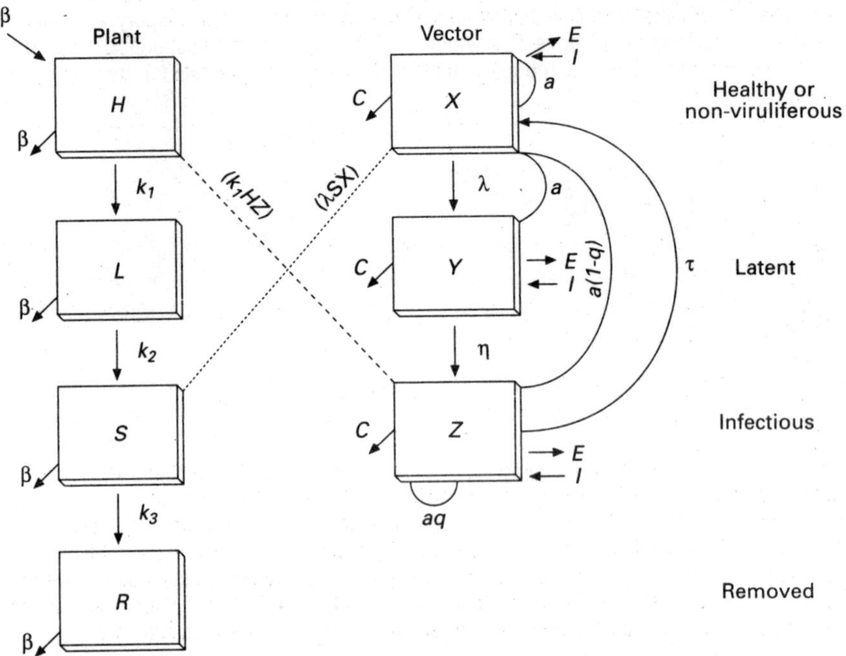

Fig. 13.3. A model of plant virus disease dynamics linked to a vector population. The host population is partitioned into healthy, latently infected, infectious and removed (postinfectious) categories; the vector population is partitioned into non-viruliferous, latent and infectious categories. Parameters are as follows: β, plant mortality (= birth) rate; c, vector mortality rate; a, vector birth rate; I and E, immigration and emigration rates; k_1, inoculation rate; λ, aquisition rate; k_2, host latent period; k_3, host infectious period; η, vector latent period, τ, vector infectious period; q, proportion of offspring viruliferous. Graphical representation is by Madden (unpublished).

improved by the availability of powerful computing facilities. Analysis of the long-term outcome of host and pathogen dynamics following a gene-for-gene interaction in a natural system can be examined under a range of assumptions and allowing for population dynamic as well as genetic considerations to be included in the underlying model. For example Jeger (1997) proposed an epidemiological model based on gene-for-gene interactions and derived conditions for persistence of the resistant host and avirulent pathogen alleles in natural populations. In some cases, this took the form of single point equilibria but stable limit cycles were also possible. In general, inspection of the short term dynamics (within year) gave no indication of the long-term (20-year) behaviour. Proprietary software is now commonly available for dynamical analysis and for generating graphical output that reveals much of the complexity of these systems. In some cases these nonlinear deterministic systems give rise to chaotic dynamics in which the long term behaviour is ultimately unpredictable. Shaw (1994) was the first to consider the possibilities of such behaviour in plant disease epidemiology. He considered two models in which fungal pathogens were infected by: (i) a mycovirus; or (ii) a hypoparasite, and with seasonal harvesting in each case. Chaotic dynamics were readily induced in these models but there were clear patterns in the regions of the phase plots occupied by the dynamics.

Soil-borne plant pathogens

The modelling of epidemics caused by soil-borne plant pathogens has followed rather different lines from that for foliar pathogens. One possible reason is the sheer technical problems of relating the above-ground symptoms of infected plants with the dynamics of infection and pathogen activity below ground. Another may be the misconception that because of their situation, pathogen propagules simply lie in wait for any passing root that comes along. Yet as discussed by Wallace (1978) dispersal in soil can be important for pathogens including bacteria, nematodes, fungal spores and mycelium (Table 13.3). Gilligan in a series of contributions has shown conceptually and using theoretical and simulation approaches that differences are perhaps more of scale and fine detail than of a fundamental nature (Gilligan, 1987, 1990, 1995; Gilligan *et al.*, 1994). For some soil fungi such as *Phymatotrichum omnivorum*, growth of hyphal strands through soil appears to be a major component of disease development especially along the rows of crops such as cotton (Kenerley and Jeger, 1992). This results in characteristic patterns of disease development as shown in Fig. 13.4. At this point in time there are no theoretical models that have been developed to deal with mechanisms such as fungal growth in soil that are then spatially expressed in the data presented by Jeger *et al.* (1987). Computer simulation on spatially explicit grids is possible but highly demanding in terms of computation. The alternative is to resort to space–time statistics for analysis of

Table 13.3. Rate of spread of soil-inhabiting pathogens* (after Wallace 1978).

	Artificial conditions	Natural conditions
Bacteria	0.5–2.5 cm*day^{-1}	0.25 cm day^{-1}
Fungi – zoospores	14 cm h^{-1}	0.10 cm day^{-1}
Fungi – mycelia	20–25 cm month^{-1}	1–3 m year^{-1}
Nematodes	20 cm h^{-1}	0.1–1.0 cm day^{-1}

*Values are not standardized to equivalent time units as estimates were made over different time periods and conditions.

Fig. 13.4. Early disease development of *Phymatotrichum* root rot in cotton. Characteristically, runs (or sequences) of diseased plants spread along the rows, probably associated with growth of fungal strands from plant-to-plant.

such data (Stein *et al.*, 1997); again modern software has revolutionized our ability to undertake such analyses. It is possible that models of fungal growth in soil could assist in understanding and describing epidemics caused by *Armillaria* spp. forming rhizomorph systems.

Biological control and inoculum potential

As stated at the beginning of this chapter Garrett did more than anyone to attempt a definition of inoculum potential, whether of infection – 'inoculum potential is defined as the energy of growth of a parasite available for infection of a host, at the surface of the host organ to be infected'; or of competitive saprophytic colonization – 'inoculum potential is defined as the energy of growth of a fungus available for colonization of a substrate at the surface of the substrate to be colonized' (Garrett, 1970). It still remains difficult to accept that these are operational definitions in any real sense of the word. And yet there is undoubtedly an underlying reality to the concept of inoculum potential, at least for soil fungi forming a mycelial system for making contact with their host (or substrate) and then infecting (colonizing) it. For some fungi, both saprophytic colonization and infection are important in the life history. We have recently developed a model (Stolk *et al.*, 1997) in which a mycoparasite plays a role in controlling the resting structures (e.g. sclerotia) of a soil fungus, which may itself be a plant pathogen. The model is formulated in terms of the population density of host resting structures, the energy content per propagule of the mycoparasite, and the volume of soil colonized by the mycoparasite. A system of linked differential equations results which can be subjected to both qualitative and quantitative analysis. An interesting feature of the qualitative analysis is that a criterion similar in nature to the basic reproductive number can be obtained. It follows from this that a mycoparasite is only able to establish itself in the soil if the initial density of host propagules (H_0) per unit volume soil exceeds the quantity:

$$H_0 > w/e_H b \tag{13.2}$$

where e_H is the energy reserve per host propagule (J), b is the probability that a host propagule is infected if it occurs in the volume of soil colonized by the mycoparasite, and w is the amount of energy required for one unit volume growth of the mycoparasite (J/m^3). The units of the parameter w are of course those of stress or potential indicating perhaps that the concept of inoculum potential can be put on a firm physical basis. The model has also been fitted to data on *Sporidesmium sclerotivorum*, a mycoparasite of various *Sclerotinia* species including *S. minor* (Adams *et al.*, 1984, 1985). Some difficulties were apparent in obtaining the set of optimal parameter values but with appropriate *a priori* constraints good correspondence to field data on the depletion of soil populations of *S. minor* by the mycoparasite with time was obtained. The estimated vales of the parameters used in the criterion above were: H_0, 8.21×10^{-3} mm^{-3}; w, 3.15×10^{-5} $J.mm^{-3}$; e_H, 0.299 J; and b, 0.644. It can readily be checked that the criterion for persistence was satisfied.

Conclusions

In this chapter I have tried to illustrate the different types and uses of epidemic models for decision-making. For decision support systems in general, it is not so much the lack of basic understanding but rather the lack of an appropriate conceptual and logistical context that is limiting adoption for tactical decision making. This applies mostly to the exploratory type of model rather than complex simulation models where again similar constraints to implementation apply. Perhaps the most productive linkage of models is that between theoretical models and the types of decisions that arise in evaluating strategic questions of disease control. The basic reproductive number and its derivation is a major methodological tool for this purpose. The ready availability of software packages for qualitative and quantitative analysis of dynamical systems has improved our ability to explore patterns of epidemic development. Finally the modelling of epidemics caused by soil-borne pathogens continues to present a major challenge and it is very timely to re-investigate the validity of concepts such as inoculum potential in this setting.

References

Adams, P.B., Marois, J.J. and Ayers, W.A. (1984) Population dynamics of the mycoparasite *Sporidesmium sclerotivorum* and its host *Sclerotinia minor*, in soil. *Soil Biology and Biochemistry* 16, 627–633.

Adams, P.B., Ayers, W.A. and Marois, J.J. (1985) Energy efficiency of the mycoparasite *Sporidesmium sclerotivorum in vitro* and in soil. *Soil Biology and Biochemistry* 17, 155–158.

Baker, R. and Drury, R. (1981) Inoculum potential and soilborne pathogens: the essence of every model is within the frame. *Phytopathology* 71, 363–372.

Butt, D.J. and Royle, D.J. (1990) Multiple regression analysis in the epidemiology of plant diseases. In: Kranz, J. (ed.) *Epidemics of Plant Diseases: Mathematical Analysis and Modeling*. 2nd edn. Springer-Verlag, Berlin, pp. 143–180.

Chan, M.S. and Jeger, M.J. (1994) An analytical model of plant virus disease dynamics with roguing and replanting. *Journal of Applied Ecology* 31, 413–427.

Garrett, S.D. (1970) *Pathogenic Root-infecting Fungi*. Cambridge University Press, Cambridge, 294 pp.

Gilligan, C.A. (1987) Epidemiology of soil-borne plant pathogens. In: Wolfe, M.S. and Caten, C.E. (eds) *Populations of Plant Pathogens: Their Dynamics and Genetics*. Blackwell Scientific Publications, Oxford, pp. 119–133.

Gilligan, C.A. (1990) Mathematical modeling and analysis of soilborne pathogens. In: Kranz, J. (ed.) *Epidemics of Plant Diseases: Mathematical Analysis and Modeling*. 2nd edn. Springer-Verlag, Berlin, pp. 96–142.

Gilligan, C.A. (1995) Modelling soil-borne plant pathogens with special emphasis on spatial aspects of disease: reaction-diffusion models. *Canadian Journal of Plant Pathology* 17, 96–108.

Gilligan, C.A., Brassett, P.R. and Campbell, A. (1994) Computer simulation of early infection of cereal roots by the take all-fungus: a detailed stochastic, mechanistic simulator. *New Phytologist* 128, 515–527.

Holt, J, Jeger, M.J., Thresh, J.M. and Otim-Nape, G.W. (1997) An epidemiological model incorporating vector population dynamics applied to African cassava mosaic disease. *Journal of Applied Ecology*, 34, 793–806.

Jeger, M.J. (1982) The relation between total, infectious and postinfectious diseased plant tissue. *Phytopathology* 72, 1185–1189.

Jeger, M.J. (1986a) Asymptotic behaviour and threshold criteria for model plant disease epidemics. *Plant Pathology* 35, 355–361.

Jeger, M.J. (1986b) The potential of analytic compared with simulation approaches in plant disease epidemiology. In: *Plant Disease Epidemiology: Volume 1, Population Dynamics and Management*. Macmillan Publishing Company, New York, pp. 255–281.

Jeger, M.J. (1987) Meteorology and plant disease. In: Prodi, F., Rossi, F. and Cristoferi, G. (eds) *Agrometeorology*. Fondazione Cesena Agricultura Publ., Cesena, pp. 255–276.

Jeger, M.J. (1997) An epidemiological approach to modelling the dynamics of gene-for-gene interactions. In: Crute, I.R. and Holub, E.B. (eds) *The Gene-for-Gene Relationship in Plant-Parasite Interactions*. CAB International, Wallingford, pp. 191–209.

Jeger, M.J. and Butt, D.J. (1983) The effects of weather during perennation on epidemics of apple mildew and scab. *EPPO Bulletin* 13, 79–85.

Jeger, M.J. and Chan, M.S. (1995) Theoretical aspects of epidemics: uses of analytical models to make strategic management decisions. *Canadian Journals of Plant Pathology* 17, 109–114.

Jeger, M.J. and Starr, J.L. (1985) A theoretical model of the winter survival dynamics of *Meloidogyne* eggs and juveniles. *Journal of Nematology* 17, 257–260.

Jeger, M.J. and van den Bosch, F. (1994a) Threshold criteria for model plant disease epidemics. I. Asymptotic results. *Phytopathology* 84, 24–27.

Jeger, M.J. and van den Bosch, F. (1994b) Threshold criteria for model plant disease epidemics. II. Persistence and endemicity. *Phytopathology* 84, 28–30.

Jeger, M.J., Kenerley, C.M., Gerik, T.J. and Koch, D.O. (1987) Spatial dynamics of *Phymatotrichum* root rot in row crops in the Blacklands region of North Central Texas. *Phytopathology* 77, 1647–1656.

Jeger, M.J., Starr, J.L. and Wilson, K. (1993) Modelling winter survival dynamics of *Meloidogyne* spp. (Nematoda) eggs and juveniles with egg viability and population losses. *Journal of Applied Ecology* 30, 496–503.

Jeger, M.J., van den Bosch, F., Madden, L.V. and Holt, J. (1997) A model for analysing plant virus transmission characteristics and epidemic development. *IMA Journal of Mathematics Applied in Biology and Medicine* 14, 1–18.

Kenerley, C.M. and Jeger, M.J. (1992) Fungal diseases of the root and stem. In: Hillocks, R.J. (ed.) *Cotton Diseases*. CAB International, Wallingford, pp. 161–190.

Kranz, J. (1979) Simulation of epidemics caused by *Venturia inaequalis* (Cooke) Aderh. *EPPO Bulletin* 9, 235–242.

Kranz, J., Mogk, M. and Stumpf, A. (1973) EPIVEN – ein Simulator für Apfelschorf. *Zeitschrift für Pflanzenkrankheiten und Pflanzenschutz* 80, 181–187.

Levin, S.A. (1981) The role of theoretical ecology in the description and understanding of populations in heterogeneous environments. *American Zoologist* 21, 865–875.

MacHardy, W.E. (1996) *Apple Scab: Biology, Epidemiology and Management*. APS Press, The American Phytopathological Society, St Paul, Minnesota, 545 pp.

MacHardy, W.E. and Jeger, M.J. (1983) Integrating control measures for the management of primary apple scab, *Venturia inaequalis* (Cke.) Wint. *Protection Ecology* 5, 103–125.

May, R.M. (1990) Population biology and population genetics of plant-pathogen associations. In: Burdon, J.J. and Leather, S.R. (eds) *Pests, Pathogens and Plant Communities*. Blackwell Scientific Publications, Oxford, pp. 309–325.

Mills, W.D. and LaPlante, A.A. (1951) Diseases and insects in the orchard. *Cornell Extension Bulletin* 711, 100 pp.

Royle, D.J. (1994) Understanding and predicting epidemics: a commentary based on selected pathosystems. *Plant Pathology* 43, 777–789.

Schroedter, H. (1983) Meteorological problems in the practical use of disease-forecasting models. *EPPO Bulletin* 13, 307–310.

Seem, R.C. (1986) Methodological comparison of three apple scab simulators. *Acta Horticulturae* 184, 33–40.

Seem, R.C., Shoemaker, C.A., Reynolds, K.L. and Eschenbach, E.A. (1989) Simulation and optimization of apple scab management. In: Gessler, C., Butt, D.J. and Koller, B. (eds) *Integrated Control of Pome Fruit Diseases, Vol. II*. IOBC Bulletin, pp. 66–87.

Shaw, M.W. (1994) Seasonally induced chaotic dynamics and their implications in models of plant disease. *Plant Pathology* 43, 790–801.

Shrum, R.D. (1978) Forecasting of epidemics. In: Horsfall, J.G. and Cowling, E.B. (eds)

Plant Disease, An Advanced Treatise: Volume II, How Disease Develops in Populations. Academic Press, Inc., New York, pp. 223–238.

Starr, J.L. and Jeger, M.J. (1985) Dynamics of overwintering of eggs and juveniles of *Meloidogyne incognita* and *M. arenaria*. *Journal of Nematology* 17, 252–256.

Stein, A., van Groenigen, J.W., Jeger, M.J. and Hoosbeek, M.R. (1997) Space-time statistics for environmental and agricultural related phenomena. *Environmental and Ecological Statistics* (in press).

Stolk, C., Van den Bosch, F., Termorshuizen, A.J. and Jeger, M.J. (1997) Modelling the dynamics of a mycoparasite and its host: an energy-based approach. *Phytopathology*, (submitted).

Swinton, J. and Anderson, J.M. (1995) Model frameworks for plant-pathogen interactions. In: Grenfall, B.T. and Dobson, A.P. (eds) *Ecology of Infectious Disease in Natural Populations.* Cambridge University Press, Cambridge, pp. 280–294.

Teng, P.S. (1985) A comparison of simulation approaches to epidemic modeling. *Annual Review of Phytopathology* 23, 351–379.

Van den Ende, E., Blommers, L. and Trapman, M. (1996) Gaby: a computer-based decision support system for integrated pest management in Dutch apple orchards. *Integrated Pest Management Reviews* 1, 147–162.

Vanderplank, J.E. (1963) *Plant Diseases: Epidemics and Control.* Academic Press, New York, 349 pp.

Vanderplank, J.E. (1975) *Principles of Plant Infection.* Academic Press, New York, 216 pp.

Waggoner, P.E. (1978) Computer simulation of epidemics. In: Horsfall, J.G. and Cowling, E.B. (eds) *Plant Disease, An Advanced Treatise: Volume II, How Disease Develops in Populations.* Academic Press, Inc., New York, pp. 203–222.

Waggoner, P.E. (1990) Assembling and using models of epidemics. In: Kranz, J. (ed.) *Epidemics of Plant Diseases: Mathematical Analysis and Modeling. 2nd edn.* Springer-Verlag, Berlin, pp. 230–260.

Wallace, H.R. (1978) Dispersal in time and space: soil pathogens. In: Horsfall, J.G. and Cowling, E.B. (eds) *Plant Disease, An Advanced Treatise: Volume II, How Disease Develops in Populations.* Academic Press, Inc., New York, pp. 181–202.

Xu, X.-M. and Butt, D.J. (1993) PC-based disease warning systems for use by apple growers. *EPPO Bulletin* 23, 595–600.

Xu, X.-M., Butt, D.J. and Van Santen, G. (1995) A dynamic model simulating infection of apple leaves by *Venturia inaequalis*. *Plant Pathology* 44, 865–876.

14

Multi-media Tools for Diagnosing and Managing Pest and Disease Problems

G. Norton

Cooperative Research Centre for Tropical Pest Management, Gehrmann Laboratories, University of Queensland, Brisbane, Qld 4072, Australia
Fax: +61 7365 1855/E-mail: g.norton@ctpm.uq.edu.au

Introduction

The number and range of computer-based aids for the prediction, diagnosis and management of insect pest and disease problems has increased steadily over the past 25 years. As new developments have occurred in information technology (IT), the potential value of computer-based systems in improving pest management has increased accordingly (Scott, Chapter 1, this volume). While considerable progress has been made in the past few years, I would suggest that the value of IT in improving pest management is well short of its potential.

Why is this ? In my view, the reason why computer-based systems have not had more impact is largely because they have been predominantly:

- science or technology driven, rather than being designed to meet specified user needs.
- constructed by biologists with computing ability rather than by specialist programmers.
- developed *de novo*, with the 'software wheel' often being reinvented.
- produced with short-term funding, and lacking resources for up-grading and help services.
- seen as competing with, rather than complementing, existing research and extension effort.

If the practical value of IT in pest management is to be improved, we clearly need to heed the lessons from the past. This paper describes the approach and some of the products developed through a cooperative effort in Australia, involving staff from The University of Queensland, CSIRO Division of Entomology, and two State Departments – Queensland Departments of Primary Industries and Natural Resources. A range of software products are being and have been developed, aimed at contributing to training and decision-making in pest management at the policy, research and farm level. The chapter consists of three sections, dealing with the role of software within a cooperative research environment, providing brief details of some of the products currently available and in development, and concluding with comments on future needs.

The Role of Software in Cooperative Pest Management Activities

Within the Cooperative Research Centre for Tropical Pest Management (CTPM), a range of scientific disciplines have been brought together to tackle important pest management problems in tropical Australia and the Asia-Pacific region. Within this joint venture, IT products complement other key activities, as shown in Fig. 14.1.

The role of software in problem specification and specific research

A feature of the approach developed by the Centre's joint venture activity in pest management is the use of problem specification workshops to facilitate a participatory approach to research, development and implementation. These workshops, which involve key players, such as farmers, consultants, and industry, as well as research and extension scientists, are aimed at specifying the key

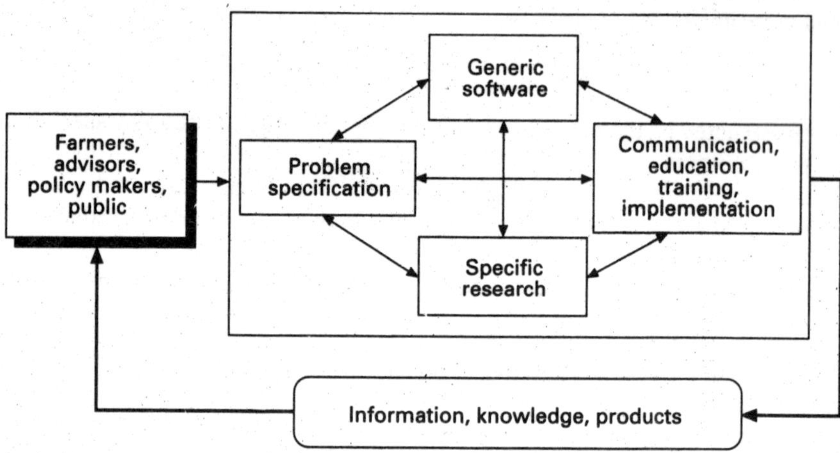

Fig 14.1. The role of software in contributing to improved pest and disease management.

features of the pest management problem and determining the crucial activities that need to occur to achieve improvements (see for instance, Norton and Brough (1995) and Knight *et al.* (1996)).

Various computer modelling tools can help in this problem definition process, as well as providing on-going inputs to the process of determining what key research needs to be done to improve our understanding or practice of pest management. As part of the joint venture, a team of biologists and programmers have developed a suite of generic software products for use in problem specification, and as on-going research tools, including genEsim – a generic population model; CLIMEX – a model that estimates species distribution on the basis of climatic data; and the virtual plant laboratory (further details on these software products are given later in this chapter). These tools are not only being applied to specific problem specification issues and to on-going specific research activity within the joint venture but are also being made available to other potential users through cooperative agreements, commercial sale or consultancies.

The role of software in decision support and training

The second major role that software has to play within our cooperative environment is in decision support and training. Communication, delivery of research results and information, training and decision support constitute the fourth critical activity in achieving improved pest management (Fig. 14.1). To complement the more traditional communication, publication and training activities, a number of software products have been developed that make novel contributions to decision support and training. Software that falls within this category includes: a library of modules that provide a migratory pest (spatial) modelling system, that currently provides information on the movement of *Helicoverpa* (for details see http://fassbinder.ento.ctpm.uq.edu.au./forecast/migration/intro.html); LucID – a generic, interactive taxonomic key; BugMatch – a series of crop based multi-media CD-ROMs providing information to growers; and a series of training modules for grain store managers. Further details of these products, which are sold or developed on contract for industry or donor agencies, are provided below.

Details of the Centre's Software Products

Some of the problems associated with previous attempts to develop software aimed at improving pest management were described earlier. The software products developed as part of the overall joint venture activity have been designed and produced in a way that has attempted to avoid these past problems. The key features of the Centre's software products are:

1. They are produced by a team of programmers, working closely with other scientists and users. A participatory, multi-disciplinary approach to software development, involving programmers, subject specialists, teachers, extension agents, graphic artists and business managers, has been an important factor in achieving current levels of success. In particular, close links with public and private sector extension agents in the development of 'extension software' has provided invaluable contributions to product design and utility.
2. Wherever possible, the software is developed generically, allowing the costs of production to be spread across a wide range of applications. For instance, a number of the Centre's products provide a software player that allows the user to run particular files that the user or content specialists have developed. LucID – the interactive key – consists of a player that allows the user to load and use specific taxonomic or diagnostic keys. LucID also comes with a builder that allows keys to be easily modified or constructed for specific users without requiring any programming skills. Other products which have players and builders include DIAGNOSIS for Crop Protection and the generic population model – genEsim. CLIMEX provides the user with a relatively simple means of entering parameter values to estimate the distribution of a specific species based on climatic data.
3. Training manuals and tutorials are supplied, or are being developed, for educational forms of the software. Since many of the software products are generic, they can be used for a range of purposes. Educational versions have been or are being produced, along with training manuals that provide worked examples and other valuable guidelines for teachers.
4. Monitoring and evaluation of prototype and finished products is becoming standard practice. Social scientists are complementing normal Beta testing and other procedures by providing various means of facilitating feedback on user response to prototypes and finished products, including focussed group sessions and one-to-one discussions.

The remainder of this section describes in more detail the main software products already developed, or in the process of being developed, to train or provide decision support in the diagnosis and management of pest and disease problems. Demonstration versions of a number of these products can be accessed via the World Wide Web, through the Centre's Web pages. The specific address for demonstrations is shown at the end of each product description.

LucID

LucID is a research, educational and decision support tool for interactively identifying biological specimens or diagnosing pest and disease problems. It consists of a Player, that allows the user to load specific keys in the LucID or DELTA format, and to use these keys to interactively identify taxa. LucID incorporates text, images, video, and sound to help the user select those taxonomic characteristics which best describe the specimen of interest, providing an extremely user-friendly means of identification and diagnosis.

The LucID Builder allows teachers, lecturers, taxonomists or decision support developers to build and modify keys to meet the particular requirements of specific users. Keys can be built in various languages and use terminology familiar to the user, allowing the package to be used internationally, across a wide range of abilities. Potential users range from biologists and ecologists, pest and disease management researchers, extension scientists, and farmers, to high school, college, and university students. The LucID Player and Builder were released in 1997. A demonstration version of LucID is available from our web site at http://www.ctpm.uq.edu.au/software/lucid.html

DIAGNOSIS *for crop protection*

DIAGNOSIS for crop protection, a product developed with Massey University, New Zealand, is described in more detail by Stewart (Chapter 33, this volume). It is a user-friendly training tool for tertiary students, crop consultants and others interested in learning skills to correctly diagnose pest and disease problems. The program presents various problem scenarios, complete with graphic displays, video and sound. The user aims to correctly diagnose the problem and make appropriate recommendations. Specific feedback is provided on how to improve diagnostic skills. The DIAGNOSIS Builder allows trainers to create or modify their own scenarios.

DIAGNOSIS for crop protection is sold internationally. An upgrade, to version 2.1, was released in 1997 with many new features. DIAGNOSIS users are able to access free scenarios; a demonstration version can be down-loaded from http://www.diagnosis.co.nz

CLIMEX for Windows

CLIMEX is an interactive computer based system for predicting the potential distribution and relative abundance of species in relation to climate. It provides displays of the distribution of particular species and the effect of climate on various population growth indices and parameters at each site with climatic data. CLIMEX is currently used in over 20 countries to examine the distribution of insects, plants, pathogens and vertebrates for a variety of purposes, including quarantine risk assessment (Sutherst *et al.*, 1989), predicting greenhouse scenario effects (Sutherst, 1991), and providing inputs to the development of classical biological control strategies (Worner *et al.*, 1989).

The Windows version of CLIMEX was launched in August 1995 and includes MetManager and MapManager modules which facilitate the easy manipulation of meteorological data and the display of maps. An educational version of CLIMEX is also available that provides additional items, such as self-paced tutorials and tests, and suggested activities and projects. This product, which has been specifically adapted for use in high schools, has all the features of the full scientific version.

genEsim

genEsim is a highly complex, generic research and educational tool. It provides research scientists with a means of building population models for animal or plant species in an easy to use environment, consisting of menu options requiring the user's response.

genEsim has a modular structure. Each module deals with a specific population process, including development, reproduction, mortality and diapause for each user-defined life cycle stage, such as eggs, seed, flowering stages or adult insects. This package can be used for real-time decision support (for instance as a day-degree model) or to explore the impact of a wide range of conditions, including various control strategies, on pest levels. The educational version of genEsim is expected to be available early in 1998.

Virtual Plants

The CRC for Tropical Pest Management, in collaboration with the University of Calgary, has developed a system for measuring the three-dimensional structural growth of real plants, deriving growth rules from analysis of the data, and using the rules to generate 'Virtual Plants' – that is, computer models of growing plants (Room *et al.*, 1996).

The system can be used to investigate the impact of insects or diseases on plant growth and to explore other agronomic or pest management issues which require a 3D frame of reference, such as the distribution of pesticide droplets in crop canopies relative to distributions of insect pests, beneficial insects, or diseases. Virtual plant models are being developed for cotton, beans, sorghum, citrus, stylo, red cedar and several weeds, and a generic model of insect and plant pathogen behaviour on the 3D structure of plants is being constructed. Further information is available from our web site at http://www.ctpm.uq.edu.au/programs/ipi/ipivp.html

BugMatch for citrus and cotton – on CD

The BugMatch series of crop-based information systems for citrus and cotton have been developed in collaboration with Rhone-Poulenc Rural (Australia) and other agencies. They are designed to help crop consultants and growers with the identification of pests, beneficials and diseases, and to highlight the benefits of integrated pest management. Colour images, sound, text and video clips help with identification and concepts. The reference section provides users with a wide range of information including importance, distribution and monitoring of the main pests, beneficials and diseases found on cotton and citrus in

Australia. The BugMatch series of software is distributed without charge by Rhone-Poulenc Rural Australia Pty Ltd, and illustrates the considerable advantages of such consultancies in delivering software products rapidly to many users.

Pest Management Workbench for stored rice

This program has been developed as part of an Australian Centre for International Agricultural Research (ACIAR) project to provide a suite of training programs for use in the rice industry in Indonesia. The workbench, which has been created in collaboration with CSIRO Division of Entomology and Indonesian colleagues, includes modules from existing software packages – especially LucID, DIAGNOSIS, and a tutorial-building module. The combination of elements within this program provide a powerful teaching and training aid that can be customized to suit individual situations. An unusual feature of this program is the ability to switch from one language to another. Currently, this allows the program to be used in English or Indonesian.

Conclusion

Multi-media and other software tools offer enormous potential for improving training and decision support in pest and disease diagnosis and management. The growing importance of the World Wide Web as a marketing and distribution tool greatly increases these opportunities. However, the extent to which this potential is realized depends as much on the processes involved in the design, development, maintenance and utilization of software products as on the inherent capability of the technology available. The successful range of products described in this paper were the result of a team effort and illustrate: the importance of involving clients and users throughout the whole design and development process; the value of utilizing a range of specialist skills; and the benefit of adopting a generic approach to training and decision support in the diagnosis and management of pest and disease problems.

Acknowledgements

The following people, most formally and others informally involved in the Centre's joint venture activity, are the team responsible for the products described in this paper: Hans Anderson, Bruce Blackshaw, Rick Bottomly, Ethna Brown, Stuart Brown, Merv Cooper, Michele Dale, Martin Dillon, Gary Fitt, Gordon Gordh, Greg Hall, Jim Hanan, Barry Longstaff, Yanni Martin, Daniel Marzano, Gunter Maywald, Robert Merlicek, Steve Richardson, Wayne Rochester, Peter Room, Greg Rutter, Bryce Skarratt, Peter Stevens, Terry

Stewart, Subhash Subaaharan, Bob Sutherst, Matthew Taylor, Kevin Thiele, John Turner, Michael Yare, David Yeates, Tony Young, Myron Zalucki.

References

Knight, J.D., Tatchell, G.M. and Norton, G.A. (1996) A structured workshop approach for problem analysis and solution finding: An example using the problems of barley yellow dwarf virus in the UK. *Agricultural Systems* 52, 113–131.

Norton, G.A. and Brough, E.J. (1995) Cooperation and collaboration: key factors for re-designing pest management. *Pesticide Outlook*. December, 1995, pp. 31–35.

Room, P.M., Hanan, J.S. and Prusinkiewicz, P. (1996) Virtual Plants: new perspectives for ecologists, pathologists and agricultural scientists. *Trends in Plant Science* 1, 33–38.

Sutherst, R.W. (1991) Pest risk analysis and the greenhouse effect. *Review of Agricultural Entomology* 79, 1177–1187.

Sutherst, R.W., Spradbery, J.P. and Maywald, G.F. (1989) The potential geographic distribution of the Old World screw-worm fly, *Chrysomia bezziana*. *Medical and Veterinary Entomology* 3, 273–280.

Worner, S.P., Goldson, S.B. and Frampton, E.R. (1989) Comparative ecoclimatic assessments of *Anaphes diana* (Hymenoptera: Mymaridae) and its intended host, *Sitona discoideus* (Coleoptera: Curculionidae) in New Zealand. *Journal of Economic Entomology* 82, 1085–1090.

15

Information Technology in Applied Plant Pathology – a Decision Support System for Crop Protection

B.J.M. Secher and N.S. Murali

Danish Institute for Plant and Soil Science, Department of Plant Pathology and Pest Management, Lottenborgvej 2, DK 2800 Lyngby, Denmark
Fax: +45 4587 2210/E-mail: dipsly@inet.uni-c.dk

Introduction

In intensive agriculture, the control of pests and diseases has changed from routine treatments to control strategies with rational and field specific components. Decisions on treatment need and dosage are becoming more complex due to the increasing number of factors of quantified importance (Jørgensen, 1994; Paveley *et al.*, 1994). The increasing complexity makes it natural to include decision aids which form a suitable platform for the introduction of new information technologies. As an example, a decision aid can involve integration with detailed weather-based models (Frahm and Volk, 1994), site specific applications (Christensen and Walther, 1995) and calculations based on complex models on the crops, diseases, pests and weeds (Olesen *et al.*, 1994).

A number of such systems for decision support have been developed throughout Europe (Secher and Bouma, 1996). These have materialized in the form of sophisticated computerized systems or relatively simple paper-based rules. A good example is the Dutch system EPIPRE developed in the early 1980s for decisions on disease control in winter wheat (Zadoks, 1981).

In the systems that are most widely used at present, complexity has increased compared to EPIPRE. For example the inclusion of detailed weather data in PRO_PLANT (Frahm and Volk, 1994) and NordPRE (Magnus *et al.*, 1991); use of detailed monitoring techniques in the Bavarian Wheat Model (BWM) (Hofman *et al.*, 1991); and the detailed estimation of appropriate dose rates in the Danish PC-Plant Protection (PC-P) system (Secher, 1991).

 Information Technology, Plant Pathology and Biodiversity (eds P. Bridge, P. Jeffries, D.R. Morse and P.R. Scott)

The information technology (IT) platform for these systems is either a centralized system (NordPRE and BWM) or a stand-alone PC-based system (PRO_PLANT and PC-Plant Protection). At present only regional monitoring and warning systems are being implemented on the Internet (Jensen *et al.*, 1996), but this platform will be integrated in future developments of field-based decision support systems.

In this chapter, the Danish PC-P will be described as an example of an IT-based application of decision support in crop protection. The system is one of the most widely used and has great impact on crop protection recommendations in Denmark. PC-P has been tested in a number of countries and will be implemented in Lithuania in 1997.

PC-Plant Protection

In 1987, the Danish Institute of Plant and Soil Science (DIPS) initiated the development of an integrated information system for plant protection consisting of plant protection recommendation models and information on pesticides, pests, diseases and spraying techniques (Murali, 1991). The project was initiated in parallel with a similar system dealing with weeds and weed control (Baandrup and Ballegård, 1989). The end users of the system were farmers and agricultural advisers.

The objective of the PC-P project was to create a system that could help users in making rational decisions on crop protection based on treatment need and the use of factor-adjusted (reduced) pesticide dosages. The project followed the implementation of the Danish pesticide action plan, which aims to reduce pesticide use by 50% (Jørgensen and Secher, 1996).

The system was released in 1991. The weed module and the pest and disease module have been available since 1993. It is distributed as part of a PC-based farm management system, 'Integrated farm management system', developed by the Danish Agricultural Advisory Centre (DAAC) (Andreasen and Fredenslund, 1993). In December 1996, PC-P was licensed to all local advisers, to 1750 farmers and to agricultural and vocational schools.

Elements of PC-P system

The system contains general information and recommendation models for the control of either pests, diseases or weeds. The elements of the systems are shown in Fig. 15.1. The recommendation model for the control of pests and diseases helps the user to decide on treatment need, choice of pesticide and the actual dose to be used.

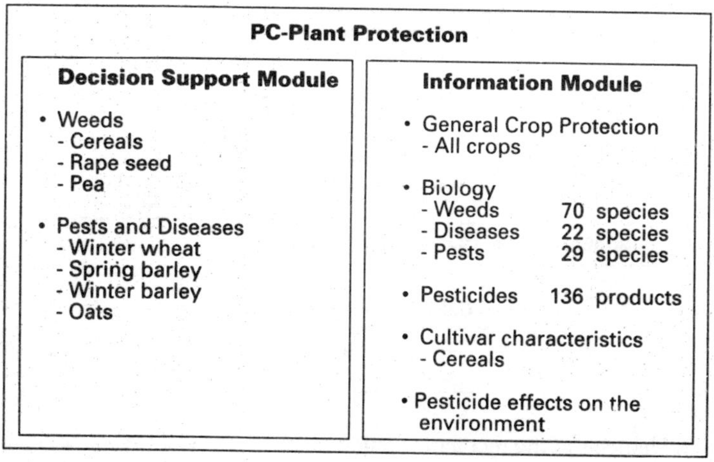

PC-Plant Protection	
Decision Support Module	**Information Module**
• Weeds - Cereals - Rape seed - Pea • Pests and Diseases - Winter wheat - Spring barley - Winter barley - Oats	• General Crop Protection - All crops • Biology - Weeds 70 species - Diseases 22 species - Pests 29 species • Pesticides 136 products • Cultivar characteristics - Cereals • Pesticide effects on the environment

Fig. 15.1. Elements of the PC-Plant Protection information and recommendation system.

Sequence of use

Prior to the season, or when the program is consulted for the first time (Growth Stage 29–30), necessary prerequisites (cultivar, sowing date, etc.) are entered into the system. At every subsequent consultation, the system evaluates which pests or diseases are relevant for control at the actual growth stage, and the user has to enter field assessments together with an updating of the weather records (precipitation and five-day weather forecast) and earlier sprays. Figure 15.2 shows the sequence and main operations of the model.

Field assessments

Field assessments are based on the incidence levels recorded in intervals. In cereals, powdery mildew (*Erysiphe graminis*), rust diseases (*Puccinia* spp.), aphid pests (*Aphidoidea*) and cereal leaf beetles (*Oulema* spp.) require field registration. Models for other leaf diseases such as *Septoria* spp., *Rhynchosporium secalis* and *Pyrenophora teres* are based on precipitation records. To simplify the assessments, the incidence levels are grouped into ranges, for example 0, 1–10, 11–25, 26–50, 51–75 or 76–100% plants affected. Up to Growth Stage 31, incidences are estimated on the whole plant while from Growth Stage 32, incidences are estimated on the upper three leaves on the main shoot or straw with tillers.

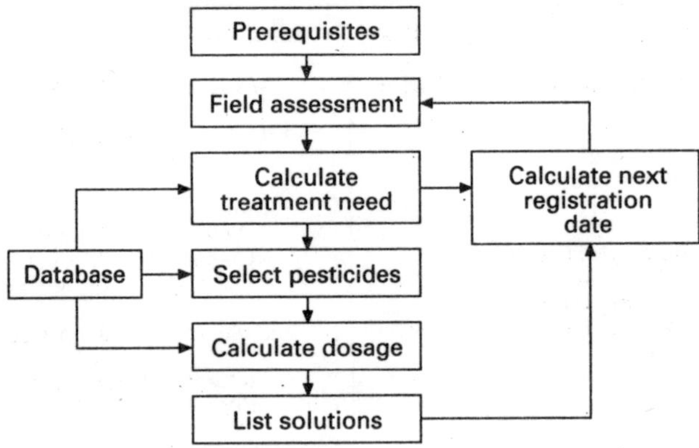

Fig. 15.2. Elements of the recommendation model in PC-Plant Protection and their sequence of operation.

Treatment need

The first step in the model procedure is the calculation of treatment need for the pests or diseases either present or at risk. Decisions are based on the field assessments of the pests or diseases and threshold values related to factors such as the growth stage of the crop, the weather pattern, and the pest or disease to be controlled. For all combinations of prerequisites and growth stage of the crop, a specific threshold value is defined for each pest and disease. A treatment would be triggered if these thresholds are exceeded by the field assessment or if several assessments are close to the threshold value. Thresholds are stored in tables according to the incidence groups and growth stages (Secher, 1991). The various pests or diseases in the system are interlinked so that the presence of other diseases or pests affects the actual threshold.

Dosage calculation

If a treatment is needed, pesticides effective against the pests and diseases are selected. Only pesticides that have been granted official approval by DIPS are recommended. The recommendation on the actual dose of fungicides is calculated on the basis of the relation between growth stage of the crop, level of disease in the crop and specific efficacy of each fungicide on the diseases to be treated (Secher, 1991; Jørgensen and Nielsen, 1994). The recommendation on actual dose of insecticides is calculated on the basis of the relation between growth stage of the crop and the pest to be controlled. The dosage is calculated for each combination of pesticide and the damaging agent and the final recommended dose is the highest of the calculated dosages. The dose adjusting factor for canopy density is shown as a function of growth stage in Fig. 15.3.

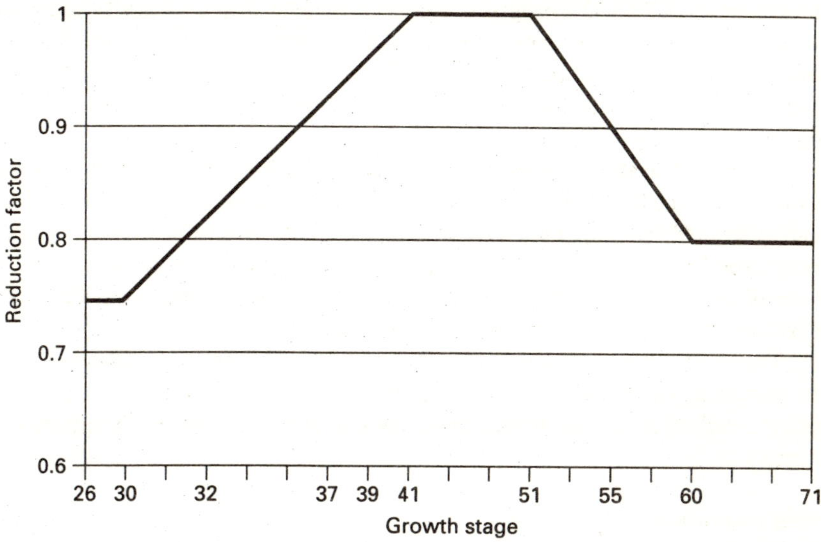

Fig. 15.3. Dosage adjusting factor for canopy density shown as a function of growth stage.

Using these factors, the recommended fungicide dosage can be as low as one-fifth of the approved normal dosage. Similarly, the insecticide dosage can be as low as three-quarters of the approved normal dosage.

Following the consultation, the user is presented with a recommendation. If a treatment is needed, a list of approved pesticides suitable for control of the problem is given. For each solution, the calculated dose is presented, together with the price per hectare and the treatment frequency index (TFI). TFI is defined as the number of approved dosages applied in the season for both insecticides and fungicides or herbicide, calculated as the sum of (dosage used/dosage approved) for all applications made in the season. TFI is one of the parameters expressed in the Danish pesticide action plan to reduce pesticide utilization. Finally, the need for subsequent field registrations is evaluated and, if required, a date for the next field assessment (registration date) is presented, with an option to print a registration form. Additional information on pesticides, pests, diseases and beneficials is available from the system database.

User surveys

PC-P has been evaluated by farmers and advisers, both during the development of the prototype and when the program has been in practical use. The evaluations of the prototype served to adjust the user interface design and performance of the system. In 1995, the program was evaluated by crop husbandry advisers. All advisers in Denmark were asked about the system's usability as an advisory tool in both direct and indirect advisory works. Of the 352 advisers, 80% responded to the questionnaire. Results from this evaluation are presented in Fig. 15.4.

In 1996, the system was evaluated by 682 farmers who had bought the program before the 1995 growth season (Murali *et al.*, 1996). Of the 546 respondents, 89% of the users had actually used the program. Various reasons were given for not using the program; the most frequent was not having time. Some of the results from this user survey are presented in Table 15.1. The overall result was a positive response to the system.

These surveys show that farmers and advisers were very satisfied with the program. The recommendations from the program were well received and its use resulted in a reduction of pesticide use and in financial savings by the farmers.

Field validation

Since 1990, the recommendation model for pests and diseases has been tested in field trials in cooperation with local advisers and coordinated by DIPS and DAAC. Trials were located throughout the country in farmers' fields or in fields at DIPS research stations. Trial plot sizes ranged from 20 to 35 m^2. Those at DIPS had a randomized plot lay-out with four replicates. Those placed with the local advisers had four or five replicates, and were not randomized according to the standards at DAAC.

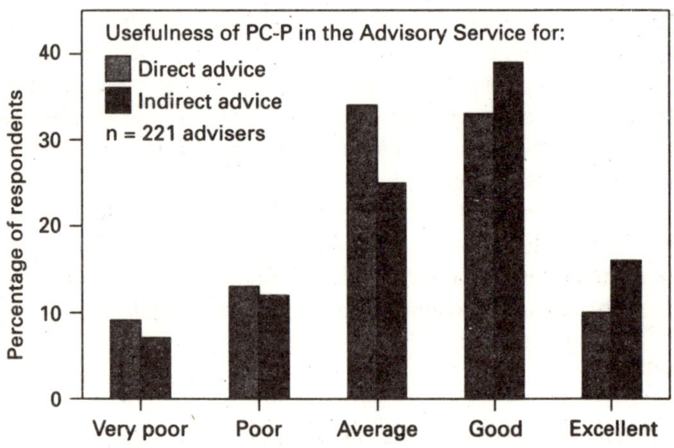

Fig. 15.4. Results from an evaluation of PC-P by Danish crop husbandry advisers.

Table 15.1. Results from a farmer evaluation of the decision support system PC-Plant Protection in 1996.

Question	Response
Has the usage of the system resulted in monetary savings?	5% saved more than 100 DKK ha^{-1} 33% saved 50–100 DKK ha^{-1} 62% did not answer or did not know
Has the usage of the program led to a reduction in pesticide use?	64% answered yes 129 responded with an average 16% reduction
Has the program given good recommendations?	59% responded good or very good. Only 4% did not get satisfactory recommendations
Has it been easy to use the program?	65% answered easy or very easy. Only 3% found the program very difficult to use
Would you recommend the program to other farmers?	58% would recommend or strongly recommend the program. Only 4% would not recommend the program

In each trial, the recommendations of the PC-P were compared with those in an untreated plot and with a standard treatment similar to a spraying programme commonly adopted by farmers without the benefit of decision support. In Denmark, reduced dosages have been widely used and this was reflected in the standard treatments. In spring barley, the standard treatment was two or three applications of a broad spectrum fungicide at 25% or 30% of the approved dose, together with one application of an insecticide. In winter barley, the standard treatment was three applications of a broad spectrum fungicide at 25% or 30% of the approved dose, per application. In 1990 in winter barley, the model was compared with one application of a full dose at Growth Stage 32. In winter wheat, the standard treatment was four applications of a broad spectrum fungicide at 25% or 30% of the approved dose, together with one insecticide.

The field observations necessary to run PC-P were made by technical staff, who communicated by fax with the staff running PC-P at DIPS in Lyngby. The advice was communicated by fax or telephone to the field personnel.

The following equivalents were used to calculate net yields: one approved dose of a broad spectrum fungicide or of an insecticide was equivalent to 0.3 or 0.05 tonnes of grain, respectively. The cost of application was estimated to be equivalent to 0.06 tonnes of grain per hectare.

The five years differed in terms of disease pressure and diseases present. A mild winter in 1989/1990 resulted in widespread overwintering of both powdery mildew (*Erypsiphe graminis*) and yellow rust (*Puccinia striiformis*) so that these diseases were prevalent in 1990. 1991 was characterized by a massive spread of *Septoria* spp. (mainly *tritici*) due to heavy precipitation in June. Disease pressure was low in 1992, and severe disease development occurred at only a few localities. Mildew was the most important disease in 1993 and 1994, in particular with early and severe attacks in spring barley in 1993. Aphids were present in all years.

In both reference plots and the plots treated according to the PC-P recommendation model, diseases and pests were controlled to a satisfactory level although the PC-P treated plots had more disease present late in the season at some localities in some years. Details of the field testing are presented in Secher *et al.* (1995).

The average yields and the extra net yield compared to the untreated plots (net yield response) are presented in Table 15.2 together with the number of treatments and the average TFI.

The field validation has demonstrated that the recommendation model in PC-P has been able to suggest variations in treatment needs between years,

Table 15.2. Average yield, the extra yield and net yield compared to the untreated plots (yield and net yield response), number of treatments and treatment frequency index (TFI) combined for insecticides and fungicides from the trials carried out 1990–1994.

	Average number of treatments	Average total TFI	Yield and yield response (tonnes ha^{-1})	Net yield response (tonnes ha^{-1})
Winter wheat (54 trials)				
Untreated	0	0	6.62	
Reference	4.0	2.08	1.22	0.60
PC-P	2.6	1.28	1.16	0.72
LSD (0.95)			0.27	
Spring barley (54 trials)				
Untreated	0	0	5.09	
Reference	2.6	1.89	0.63	0.17
PC-P	1.5	0.79	0.50	0.26
LSD (0.95)			0.10	
Winter barley (26 trials)				
Untreated	0	0	6.27	
Reference	2.5	0.85	0.56	0.17
PC-P	1.5	0.67	0.58	0.30
LSD (0.95)			0.15	

LSD (0.95) refers to the yield response.

although there has been a tendency to over-treat due to small net yields in some years. The recommended scheme was not able to take into account that the level of disease present in the crops was not the prime cause of decreased yield.

The TFI was, generally, lower in crops controlled by PC-P than in reference crops, although plots in some localities did receive more fungicide treatments than the reference plots in some years. The reference plots in these trials received spray programmes comparable to those used normally by Danish farmers. In comparisons between the recommendations by PC-P and treatments in farmers' fields, the recommended scheme gave an average reduction in pesticide usage of 33% and 24% respectively in winter wheat and spring barley (Secher, 1997a). The level used by Danish farmers is already well below the level used in most other European countries (Secher and Jørgensen, 1995).

Implementation of PC-P

PC-P has been implemented in cooperation with the DAAC, which is responsible for sales and marketing. It has been an advantage that the marketing is closely linked to the advisory service although some advisers have to accept the additional activity of selling and servicing the software. The implementation has included a series of marketing campaigns and the production of specific booklets, folders and courses on the use of the system. These courses are organized by DAAC or by the local advisers selling the system. A large campaign in 1995 followed by a price reduction increased the number of programs used at the farm level to the present level. The price of the program is 1500 DKK (in 1997) and the price of annual updates is 500 DKK. The system is updated at least once before the spring season, but if necessary also in the autumn.

Role in the dissemination process

The development of PC-P has had a beneficial impact on the dialogue between DIPS and DAAC. The process of developing the models has been useful in focusing discussions and research so that gaps in knowledge related to the specific recommendation needs could be filled. In addition, the models have turned out to be a common reference on crop protection measures in both research and advisory work. The implementation of the system has speeded up the process of getting the research findings to the end users since it has proved to be a good tool for placing the concept of need-based crop protection strategy in the context of applications at farm level. It should be noted that the ultimate use of the system at the field level only has had little impact on common recommendations when compared with the role of the system as a reference in newsletters and general publications. The role of the system in the process of disseminating knowledge is shown in Fig. 15.5.

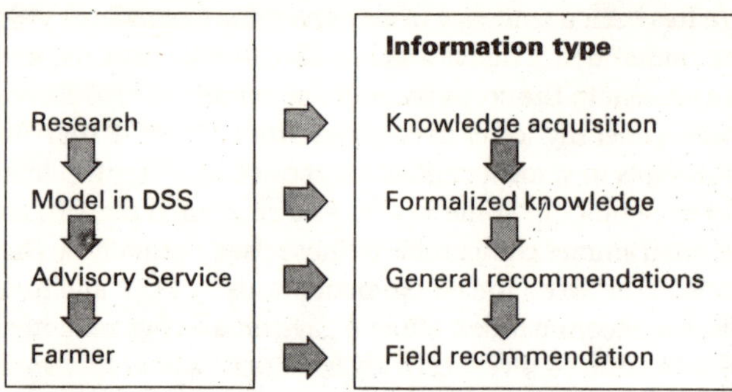

Fig. 15.5. The role of a decision support system in the process of disseminating knowledge – getting the concept to the context.

Future developments

In 1996, the PC-P has been converted from DOS to a Windows 95 operating system. This conversion has been carried out by DAAC in close cooperation with DIPS. As a new feature, the 1997 version will be made as an open system, to facilitate the integration of the program with other farm management systems on the Danish market. The integration will be made so that specific field data can be exchanged automatically with PC-P and recommendations returned to the farm management system. At present, three companies developing farm management systems have expressed interest in integration with the PC-P.

DIPS and DAAC have initiated a project on using the Internet as a means to distribute information. At the Pl@nteinfo Web site (http://www.planteinfo.dk), pest and disease prognoses and warnings are presented on maps which are updated daily during the growing season (Jensen *et al.*, 1996). The Web site is being used on a trial basis, as a platform to collect data from a disease and pest survey carried out during the season. In 1997, an information system for cultivar selection is expected to be launched on the Web (Boesen *et al.*, 1996). It is expected that PC-P will be integrated with information distributed on the Internet, and the Internet can be used to update the program when appropriate. In the long run, it is likely that PC-P will be developed to communicate interactively with the Internet when, for example, weather data or parameters from the cultivar program is needed.

In the present version of the program, some information modules use a simple hypertext facility to present the information in an easily accessible way. In the Windows version of the program, these modules will be developed as a multi-media facility which will integrate illustrations, voice, etc. using HTML.

In the present program, the decisions on disease control are primarily based on field observations. The decisions are therefore based on a scenario that is at least one latent period behind the present. For a number of diseases, risk models are under development that can estimate the risk of development of a particular disease, based on detailed weather data. By integrating the weather-based risk models with the recommendation models in PC-P, it is expected that the robustness of the recommendations can be enhanced. When weather forecasts are distributed electronically, the models can be used in a risk evaluation to prioritize between different possible treatments in various crops (Secher, 1993).

Weather data can originate from a farm-based weather station or from a weather service operated by the local advisory service, or can be downloaded from the Danish Meteorological Institute (DMI) via the Internet. From 1997/98 the Web-based information system from DMI will include the possibility to download weather data from the national weather stations, and optional five-day weather forecasts are expected to be included in these.

The risk models have been validated in field trials since 1995. A Windows 95 version of the weather module will be ready for user evaluation in 1997. If this proves successful, the weather-based models will be implemented in PC-P in 1998.

When detailed weather data are available, it is likely that PC-P will be integrated with crop growth and phenology models (Olesen *et al.*, 1994). The crop growth models are expected to be used to estimate the physiological status of the crop in crop protection decisions. New findings suggest that the physiological response to a disease or pest can vary considerably according to the status of the crop (Paveley *et al.*, 1997). Therefore, a model that takes account of the availability and use of nutrients and keeps track of photosynthetic activity and storage developments in the crop might improve the performance of the disease model. Phenology models can be used to predict the dates for reaching key growth stages for field observations, risk model initiations or treatments.

Use of sensors and GPS

Recent findings show a great potential for a further increase in the efficacy of pesticides if the application is adjusted according to the specific demands of the treated area, i.e. site-specific application of fungicides. A field trial carried out in 1996 has shown a significant 4% yield increase when fungicides were applied in a site-specific way compared with a conventional blanket application (Secher, 1997b).

When this technology is used widely, it will be necessary for PC-P to communicate with the programs used for yield mapping and other systems which deliver the necessary inputs. In addition, it will be essential for PC-P to communicate with software for sprayer control. PC-P will then, with the help of geographic information systems (GIS), link the relevant information and calculate a recommended application map for the field. The exchange of data between the DSS and the sprayer control software can be undertaken via the chipcards that are already used to log yield data. Ultimately, there could be radio communication between farm, tractor computer and weather measuring equipment.

An important factor in the calculation of appropriate doses is an adjustment of the dose according to the specific crop coverage. This can be derived from yields in previous years. However, due to the large annual variations, measurement of the crop coverage in the growing season would be better. This could be achieved with sensors (Christensen *et al.*, 1997) mounted on implements passing through the field and linked to a global positioning system (GPS), thus making a within-season crop density field map. Ultimately, such sensors could be mounted on the sprayer and linked to the tractor computer for final tuning of the applied dosage.

The links between future elements in crop protection decisions are shown in Fig. 15.6. The sequence of elements will form a comprehensive system, and the success of this will depend on the quality of the models involved in the integration of a wide range of information technologies.

Conclusions

The PC-Plant Protection is an example of a DSS with many users and with good market acceptance. The development of the system has proved to be of great value in getting research findings adapted to, and used, in practical farming. This development has also strengthened cooperation between research and extension institutes. Furthermore, this program has shown that DSS are a helpful tool in extension and not a replacement for advisory staff. The system

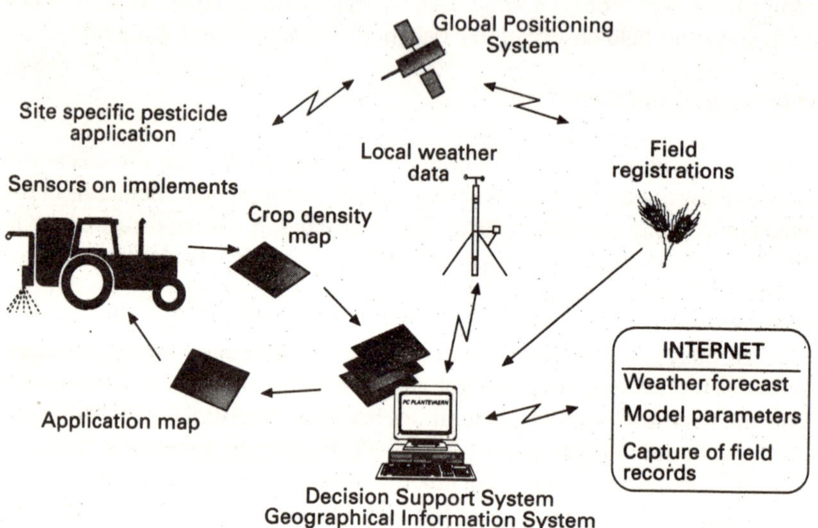

Fig. 15.6. Future elements in the decision support system PC-Plant Protection.

provides an example of a DSS being a suitable platform for the integration of many new information technologies such as the Internet, GIS, GPS and the use of complex weather-based models with various means of communication between different programs. A powerful DSS increases the demands for inputs and new factors can be included in the decision processes. It is important that any inclusion of new technologies is based on a substantial improvement of model performances, and is not done for the demonstration of the new technology itself.

References

Andreassen, F.M. and Fredenslund, E. (1993) The Danish integrated farm management system – sector module for plant production. *EPPO Bulletin* 23, 657–662.

Baandrup, M. and Ballegård, T. (1989) Three years field experience with an advisory computer system applying factor adjusted doses. In: *Proceedings from Brighton Crop Protection Conference – Weeds, 1989*, pp. 555–560.

Boesen, B.D., Bøjer, O.Q., Secher, B.J.M. and Murali, N.S. (1996) A computer based decision support system for selection of varieties. In: Secher, B.J.M. and Frahm, J. (eds) *Proceedings of the Workshop on Decision Support Systems in Crop Protection, Münster. SP-report* 1996 (15), 19–24.

Christensen, S. and Walther, A. M. (1995) Site specific weed management. In: Olesen, S.E. (ed.) *Proceedings of the Seminar on Site Specific Farming, Koldkærgård. SP report* 1995 (20), 151–160.

Christensen, S., Heisel, T., Secher, B.J.M., Jensen, A. and Haahr, V. (1997) Spatial variation of pesticide doses adjusted to varying canopy density in cereals. In. *Proceedings of the first European Conference on Site Specific Farming* (in press).

Frahm, J. and Volk, T. (1994) PRO_PLANT – a computer-based decision support system for cereal disease control. *EPPO Bulletin* 23, 685–693.

Hofmann, G.M., Vereet, J.A. and Habermeyer, J. (1991) Entwicklung und Einführung des 'WEIZENMODELL BAYERN' im Rahmen des Integrierte Pflanzenschutzes. *Gesunde Pflanzen* 10, 333–345.

Jensen, A.L., Thyssen, I., Hansen, J.G., Jensen, T., Secher, B.J.M. and Juhl, O. (1996) An information system for cro protection on World Wide Web. In: Lokhorst, C., Udink ten Cate, A.J. and Dijkhuizen, A.A. (eds) *Proceedings ICCTA 96-Congress on ICT Applications in Agriculture, Wageningen*, pp. 604–609.

Jørgensen, L. N. (1994) Duration of Effect of EBI-fungicides when using reduced dose rates in cereals. In: *Proceedings Brighton Crop Protection Conference – Pest and Diseases, 1994*. pp. 703–710.

Jørgensen, L.N. and Nielsen, B.J. (1994) Control of yellow rust (*Puccinia striiformis*) by ergosterol inhibitors at full and reduced dosages. *Crop Protection*, 13, 325–330.

Jørgensen, L.N. and Secher, B.J.M. (1996) The Danish pesticide action plan – ways of reducing inputs. In: *Proceedings Crop Protection in Northern Britain*, pp. 63–70.

Magnus, H., Munthe, K., Sundheim, E. and Ligaarden, Å. (1991) PC-technology in plant protection warning systems in Norway. In: Secher, B.J.M. and Murali, N.S. (eds) *Proceedings of the Workshop on Computer-based Plant Protection Advisory Systems, Copenhagen. Danish Journal of Plant and Soil Science*, 85(S-2161), 1–6.

Murali, N.S. (1991) An information system for plant protection: I. Development and testing of the system. In: *Proceedings from Colloqium on European Data Bases in Plant Protection, Annales ANPP,* pp. 143–148.

Murali, N.S., Secher, B.J.M., Rydahl, P. and Andreassen, F.M. (1996) Application of information technology in plant protection in Denmark: from vision to reality. In: Lokhorst, C., Udink ten Cate, A.J. and Dijkhuizen, A.A. (eds) *Proceedings ICCTA 96-Congress on ICT Applications in Agriculture, Wageningen,* pp. 146–150.

Olesen, J.E.O., Andreasen, F. and Andreasen, L. (1994) A computer aided integrated crop management system in winter wheat. In: *Proceedings from the conference on Arable farming under CAP reform, Cambridge. Aspects of Applied Biology* 40, 93–96.

Paveley, N.D., Royle, D.J., Cook, R.J., Schoefl, U.A., Morris, D.B., Hims, M.J. and Polley, R.W. (1994) Decision support to rationalise wheat fungicide use. In: *Proceedings Brighton Crop Protection Conference – Pest and Diseases, 1994.* pp. 679–694.

Paveley, N.D., Lockley, K.D., Sylvester-Bradley, R. and Thomas, J. (1997) Determinants of fungicide spray decisions for wheat. *Pesticide Science* (in press).

Secher, B.J.M. (1991) The Danish plant protection recommendation models for cereals. In: Secher, B.J.M. and Murali, N.S. (eds) *Proceedings of the Workshop on Computer-based Plant Protection Advisory Systems, Copenhagen. Danish Journal of Plant and Soil Science* 85(S-2161), 127–134.

Secher, B.J.M. (1993) Weather driven module for calculation of risk indices for disease and pests in a DSS on agricultural crops. In: Secher, B.J.M., Rossi, V. and Battilani, P. (eds) *Proceedings of the Workshop on Computer-based DSS on Crop Protection, Parma. SP report* 1993(7), pp. 155–161.

Secher, B.J.M. (1997a) The impact of research and extension on the development in pesticide uses. In: Anon. *Pesticid Anvendelsen I Dansk Landbrug 1987–1996. SP-report* 1997(11), pp. 41–60 (in Danish).

Secher, B.J.M. (1997b) Site specific control of diseases in winter wheat. In: Western, N.M., Cross, J.V., Lavers, A., Miller, P.C.H. and Robinson, T.H. (eds) *Proceedings from Conference on Optimising Pesticide Applications, Long Ashton. Aspects of Applied Biology* 48, 57–65.

Secher, B.J.M. and Bouma, E. (1996) Survey on European Crop Protection Decision Support Systems. EUNETDSS, European Network for Operational Decision Support Systems in Crop Protection. *SP-report* 1996(16), 95 pp.

Secher, B.J.M. and Jørgensen, L.N. (1995) Current Development in fungicide use – success or failure? In: Hewitt, H.G., Tyson, D., Hollomon, D.W., Smith, J.M., Davies, W.P. and Dixon, K.R. (eds) *Proceedings of the SCI/BCPC Conference. The Vital Role of Fungicides in Cereal Production,* Cirencester. BCPC Monograph, pp. 241–251.

Secher, B.J.M., Jørgensen, L.N., Murali, N.S. and Boll, P. (1995) Field validation of a decision support system for the control of pests and diseases in cereals in Denmark. *Pesticide Science* 45, 195–199

Zadoks, J.C. (1981) EPIPRE: A disease and pest management system for winter wheat developed in the Netherlands. *EPPO Bulletin* 11, 365–369.

From Mainframe to Micro: Information Technology in Plant Breeding

16

A. Marshall

ITpro AG, R-1008.5.11, CH 4002 Basel, Switzerland
Fax: +41 61 697 5600/E-mail: alan.marshall@sn.novartis.com

Introduction

The need to decommission an obsolete mainframe computer in the USA and to replace ageing software in the USA and Europe led us to develop a personal computer (PC) based decentralized software (SW) to support our worldwide breeding operations. The project, christened 'Winbreed', took place between October 1993 and February 1996 with myself as project manager, ably supported by four programmers. This chapter describes a few highlights of this work. The main message of this paper is that modern PCs and SW are powerful and robust enough to support major tasks previously the domain of mainframe and mini computers. After 2 years successful use in the USA, this project has unfortunately now been interrupted by the planned fusion of Ciba and Sandoz. Because of this, not all of the goals could be realized. This chapter was first prepared for the attendees at the British Society for Plant Pathology plus the Systematics Association's December 1996 meeting and is thus aimed at biologists interested in information technology (IT) rather than IT professionals.

Ciba's Seeds Business

- Turnover approximately $200 million.
- 90%+ of business is maize, plus additional activities with sunflowers, soya, sorghum, wheat and barley.
- World leader in genetically modified maize.

 Information Technology, Plant Pathology and Biodiversity (eds P. Bridge, P. Jeffries, D.R. Morse and P.R. Scott)

- Breeding operations in USA, France, Germany, Italy, Brazil, Argentina, Thailand and England.

Historical Background

Our USA operation, consisting of about 12 breeding stations, was supported by an HP 3000 small mainframe computer using SW written in-house in the mid 1980s. This was due for replacement, partly due to obsolescence, partly with a view to saving costs. The 12 breeding stations each used a local PC with permanent connections to this central computer. The PC was used in a terminal mode – the user screen was character based – 80 characters × 25 rows.

In Europe and other non-USA countries, we had been using MS-DOS based SW, originally written by Nagel and Knaack, Kiel, Germany, and later extensively modified for Ciba's specific needs. This was originally designed (early 1980s) to run on IBM PC-XT computers with additional 80MB Tall-Grass external hard disk drives. The SW had been in continual use and evolution since the early 1980s and had become difficult to maintain.

In October 1993, I proposed a project to write a single MS Windows-based application for worldwide use. Because of the urgent need to decommission the HP3000, the initial project priority would be given to the USA with later extension to European and other non-USA countries' needs.

Our Software

Our major SW (excluding 'Financials' and 'Supply Chain') can be classified in several groups, as indicated in Fig. 16.1:

- frontier technology used especially in the marker assisted breeding and biotechnology areas.
- inventory, nursery and breeding trial management SW.
- large plot trials used in product development.
- marketing trials SW.
- electronic mail and general office SW, e.g. document processing, spreadsheets, etc.

The aim of Winbreed was to integrate the inventory, nursery and breeding trial SW which are closely interconnected and also to manage large plot trials since these are technically more-or-less identical to breeding trials (small plot trials). We proposed commencing with the breeding trials component since this would then allow decommissioning of the mainframe computer.

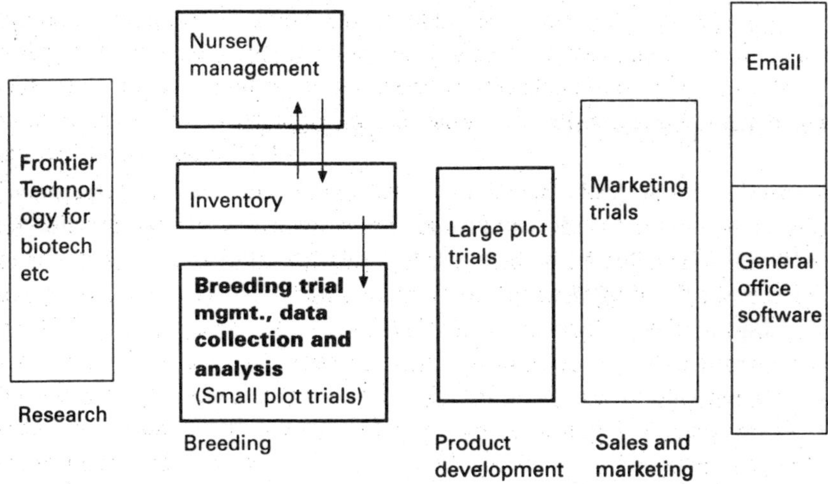

Fig. 16.1. Long term aim: four modules shown with thick border to be covered by one integrated suite of SW.

Terminology

1. Field: the word 'field' is used with two meanings: the muddy agricultural field and the database field. This latter refers to a column in a database table, or if just referring to one record it will be the single cell in the appropriate column. I will try to be clear which is meant.
2. Hybrid or variety: breeding trials may involve, for example, hybrids of maize or varieties of wheat. Arbitrarily I shall use the term hybrid throughout this chapter.

'Winbreed' Project Aims

- A single Windows-based application for worldwide use.
- Initially for hybrid evaluation breeding trials, with later extensions for nursery and inventory management.
- Suitable to cover breeders 'small-plot' trials and product developers 'large-plot' trials.
- Suitable for all crops currently of significance to Ciba.
- Low running costs; locally at stations, centrally at headquarters and for telecommunications.
- No practical limitations on numbers/sizes of items. Our previous SW imposed restrictions due to limitations or compromises dictated by the 1980s HW and

SW. In particular, only about 30 traits were collectable; a maximum of 100 hybrids were allowed in trials and the size of agricultural fields were limited.

Outline Specification

Borland's Paradox for Windows was chosen as the database and programming language, partly because we already had experience of Paradox for DOS. Although the application works on an Intel 486-type computer we are currently using the following:

- Pentium 133Mhz.
- 16–32 MB RAM.
- 1GB hard disk.
- CD-ROM drive.
- Tape or Iomega ZIP™- drive for backup purposes.
- Operating system: Windows 3.1, 3.11, 95 or NT.
- E-mail connection.

The database consists of approximately 30 data tables, plus numerous 'forms' to manipulate this data and display it on the screen, plus several 'reports' to present data on paper. The largest table is of the order of 30MB in size and contains approximately 1.3 million records. The breeding trial evaluation component took three full-time programmers approximately two years to write.

Structure of the Trials Component of Winbreed

Logically the breeding trials component consists of four major modules:
1. Trial design: definition of hybrids to be tested; statistical design choice (Randomized Complete Block, Partially Balanced Lattice, etc.); number of replications; number of different environments; if Split Plot, which treatments. This module also produces early packing labels and paper summaries of the trial design.
2. Field mapping: trials need to be allocated to environments and to particular fields at each environment; several trials need to be mapped into each field thus defining the range and row for each hybrid. Planting labels, field books and field maps need to be printable.
3. Observations and data collection by hand-held computer: much data is collected directly using hand-held computers (Husky FS2). The field plan including pedigrees is downloaded into the hand-held prior to visiting the field. Plant data is collected in the field based on range and row reference. On entry into the PC database this data has to be rearranged to be associated with the corresponding trial. Some occasional data is written in the field book and manually entered into the PC later. Up-to-date field books with current data are printable

from this module. Environmental data concerning weather, soil analysis, fertilizer applications, herbicide and pesticide data is also collectable.
4. Data selection, analysis and reporting: The data selection module serves to compare hybrids under selected conditions. Trials are selectable by year, by trait means, by hybrid, and based on agronomic criteria (e.g. greater than 100 cm rainfall). The analysis module takes the selected trials and calculates means over replications; routine statistical calculations are performed and results presented in tabular or graphical form. ASCII files and Excel files are generated for data export to other programs.

We were surprised by the additional programming effort required to handle 'split plots' and their several 'treatments'. This accounted for approximately 30% of the total programming effort. It was an important design decision to include these from the start, since retro-fitting would be almost impossible; their influence is all-pervasive.

Design Issue 1: Unique PC Identification

We needed to be able to conduct multicentric trials, and to exchange data between breeding stations. This requires having unique trial identities. In the past in the USA we were using a single mainframe computer for all USA breeding stations. As a breeder entered a new trial identification, the computer checked that such a trial ID did not already exist, and complained if it did. In Europe and the Rest of the World (ROW – excluding Europe and USA) we had had little electronic data exchange in the past, and when it did take place, the breeders liaised with one another to select unique IDs. We did not think this approach was acceptable in the 1990s.

The solution we evolved was to give each PC a unique 4 character ID and to use this as part of the trial ID. This solution then lent itself to solving another issue. All breeders were designing their own trials and asking colleagues to plant some of these trials and return the results, so gaining the benefit of additional environments. It was necessary to determine which data on a PC belonged to the PC's own breeder, and which data belonged to a trial being conducted for a colleague. This latter data needed to be sent to the colleague.

When a trial was designed, the unique PC ID was incorporated into the trial ID in a database field called Trial Creating Computer (TCC). After trial designs were transferred between PCs (by E-mail), it was easy to see which trials were for colleagues and which were local trials. If the TCC of a trial was different from the PC's ID then the trial design must have come from another PC (Fig. 16.2).

Data collected from a field of maize (or other crop) must be checked for major mistakes and sometimes adjustments made to take account of local variations. A good example is adjusting for the effects of water holes, i.e. where a certain portion of a field is known to have been waterlogged for days or weeks.

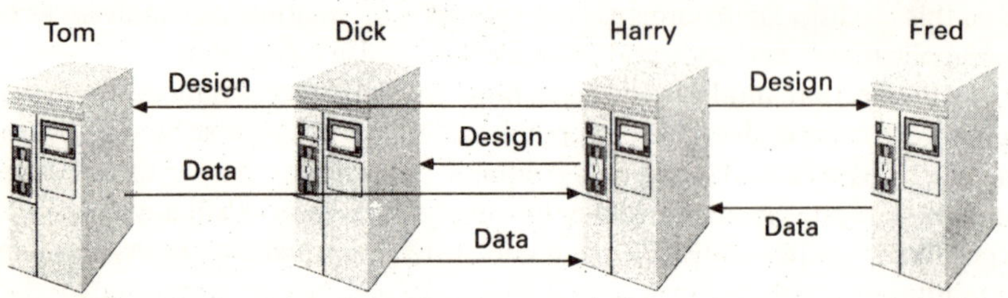

	Tom's PC	Dick's PC	Harry's PC	Fred's PC
Harry's experiment	Mapped into field on this PC = FCC	Mapped into field on this PC = FCC	**Trial designed on this PC = TCC and FCC**	Mapped into field on this PC = FCC
Tom's experiment	**Trial designed on this PC = TCC and FCC**	Mapped into field on this PC = FCC	Mapped into field on this PC = FCC	Mapped into field on this PC = FCC

Fig. 16.2. Using a unique PC identifier allows the originator of trial designs and trial data to be determined.

It makes sense to either discard this data or adjust the yield appropriately. We wished to design the program to only allow the breeder who grew the crop to make such adjustments. The breeder growing the crop may not be the designer of the experiment, but he is the only breeder with first-hand knowledge of the growing conditions and data-collection procedures.

In an experiment grown by several breeders (several environments), the experiment is designed by one breeder on his PC (the TCC) and this design is then sent to several colleagues. Each will map this trial into a field along with his own trials and other trials from colleagues. By associating trial data with the PC on which it is mapped into a field (the Field Creating Computer – FCC), we could then easily control who could modify the data. Only when the FCC associated with the data is the same as the identity of the current PC, can the data be modified.

Design Issue 2: Structure of Trait Table

A limitation of many of our previous programs was that they were restricted in the number of traits they could handle. For example our European MS-DOS SW can handle only 18 traits (breeders have developed tricks to handle more), and our US SW about 35; both are hopelessly inadequate for today's breeding needs.

This limitation arose because we stored the traits 'horizontally' in a database table as shown in Table 16.1 (only four traits shown, for illustration). To

Table 16.1. Our obsolete approach – trait data stored 'horizontally' in a single database table.

Expt no.	Env. no.	Rep. no.	Hybrid no.	Yield (kg/plot)	Moisture (%)	Smut (1–9)	ECB (1–9)	New Col.
A1	4	1	1	15.7	22	5	3	
A1	4	1	2	16.3	21	4	5	
A1	4	1	3	14.7	23	4	4	
A1	4	1	4	15.2	21	6	5	
A1	4	2	1	12.1	24	3	5	

add an additional trait requires a new column and then the programmer must write additional code to access this new column.

With Winbreed we wanted to overcome this limitation – allowing the breeder himself to add new traits, e.g. a disease that suddenly becomes more prevalent in his region and thus worthy of observation. Our solution is to store the data 'vertically' rather than 'horizontally'. Two tables are required as shown in Table 16.2. The first table – trait definition – is rather small. The number of records equals the number of traits (a few hundred to cover our international needs). This table associates a descriptive name and the units of measurement with a unique trait ID.

The second table, trait values, is the real monster – approximately 1.3 million records in a typical breeder's PC. Here are stored all data values relating to plant traits. Each data value is identified by the experiment, environment, replication, hybrid number and trait ID. Thus the first record shows that hybrid 1 had a yield (ID = 001) of 15.7 kg per plot. Adding a new trait is now just a matter of the breeder adding a new line to the definition table, entering a description and the unit of measurement. Although the data values are now stored vertically, they are still displayed, whether on screen or in a paper report, in the horizontal layout. Our application picks out the relevant data from the vertical database table and uses them to fill-up a horizontally designed trait report. The breeder will select which traits and which trials or hybrids he is interested in displaying. We also use this vertical approach to store environmental trait data, e.g. rainfall, sunshine, herbicide applications, soil analysis, etc.

Design Issue 3: Macro Language for Trait Transformation

Trait values as collected in the field are not always in the right form for comparison or for official purposes. For example yield may be collected in kg per plot, but for comparison purposes 'kg per hectare' is more suitable (plot size may vary from trial to trial). For printing in a report, the input trait value (yield – kg per

Table 16.2. Our more elegant design – trait data stored 'vertically' using two database tables.

Trait ID	Description	Units
001	Yield	Kg/Plot
002	Moisture	%
003	Smut	Rating
004	ECB	Rating
005	Root Lodging	Rating
006	Yield	Bushel/Acre
007		
008		

Expt no.	Env. no,	Rep. no.	Hybrid no.	Trait ID	Value
A1	4	1	1	001	15.7
A1	4	1	1	002	22
A1	4	1	1	003	5
A1	4	1	1	004	3
A1	4	1	1	005	7
A1	4	1	2	001	16.3
A1	4	1	2	002	21
A1	4	1	2	003	4
A1	4	1	2	004	5
A1	4	1	2	005	2

Long table: approx. 1.3 million records!!

plot) must be transformed to the appropriate output trait value (yield – kg ha^{-1}). Obviously this transformation is mathematically quite simple, since plot size is obtainable from row length, row width and rows per plot, all of which are recorded in the database.

However, we did not want the programmer to 'hard-code' this equation, since with some traits the transformation equation may suddenly change. A real example: in Argentina, sunflower seed yields are expressed after correction to both a standard moisture and a standard oil content. A few years ago the government changed the standard level and all our hard coded equations contained the wrong conversion 'constant'. Our programmer had to postpone other work, find the original source code, re-understand it and make the necessary modifications.

We therefore developed a macro language allowing the breeder to define an output trait based on input values and constants. For those of you familiar with spreadsheets (e.g. Lotus 1–2-3, MS Excel), our formulas are conceptually similar to the formulas entered in a spreadsheet cell. This has dramatically increased the reporting flexibility of the SW.

Design Issue 4: Historical Data on CD-ROM

Evaluation of more advanced hybrids involves using data collected over 3–4 years. Thus we needed to collect and store this historical data and make it available to all breeders. A large PC was used in our USA Headquarters to collect the data sent in by E-mail. Data from hybrids that were discarded in their first one or two years of testing due to poor performance was not included – such data are essentially only of interest to the local breeder (who designed the cross). Excluding this data reduced the total amount of data by about 50%. Historical data was further streamlined by storing only the means and variance of the several replications, rather than including individual replicate values.

Initially we distributed this data to breeders by magnetic tape or Iomega ZIP diskette, but since have produced our own CD-ROMs. These have a capacity of about 600 MB, more than adequate for 3–5 years data of approximately 50 MB per year. The CD-ROM has the advantage of allowing laptop users and managers to have immediate reliable access to all historical data without the need to first connect to a distant database.

Design Issue 5: Security

Purpose-written SW such as Winbreed represents a significant financial investment (4–20 programmer-years; from one to several million US dollars) and is expected to provide a competitive advantage. To limit the use of our application to authorized users, it is programmed to only run when a HW protection device (dongle) is installed (Hard-Lock from FAST, Germany). This attaches to the parallel port of the PC, between the port and the printer cable – printing from the parallel port is still possible. The data stored in the database is automatically encrypted – a function that is built into Paradox for Windows.

Who's Novartis; Where's Ciba?

On 7 March 1996, it was announced that Ciba-Geigy and Sandoz would merge forming the new Life-Sciences company Novartis. Both Ciba and Sandoz are large, multinational, multidivisional chemical/pharmaceutical companies based in Basle. The legal date of the merger was dependent on approval by the European Commission and the US Fair Trade Commission. These were both available by 17 December 1996 and Novartis was officially formed on 20

December 1996. This paper, presented on 17 December 1996, describes work by Ciba.

Acknowledgements

The following four programmers were involved in this project: Jean Chang (Ciba USA), Ken Holt (early phases; Ciba USA), Wolfgang Nagel (Nagel and Knaack, Kiel, Germany) and Helmut Plakties (Ciba Basle). Numerous colleagues were involved in advising about the design and needs of breeders.

Developing a Model of Expertise for a Taxonomic Expert System

M. Edwards

Computing Laboratory, University of Kent, Canterbury, Kent CT2 7NF, UK and School of Sciences, University of Buckingham, Buckingham MK18 1EG, UK
Email: medwards@nildram.co.uk

Introduction

Expert systems have been recognized as an appropriate tool for assisting with biological identification for a number of years (e.g. Forget *et al.*, 1986; Woolley and Stone, 1987). However, when reviewing approaches to computer-based species identification, Edwards and Morse (1995) showed some reluctance to recommend development of an expert system. This reluctance stemmed from the difficulties of developing a successful expert system rather than any evidence of the superior performance of other computer-based identification techniques. It is now recognized that successful expert system development is greatly facilitated by adopting a modelling approach to the knowledge acquisition phase of expert system development (Neale, 1990; Wielinga *et al.*, 1992). This paper describes how such an approach was used when developing an expert system for water mite identification.

The traditional approach to expert system development is based on rapid prototyping, and this leads to the knowledge acquisition phase of the project being influenced by issues relating to implementation. This approach has been blamed for perpetuating the knowledge acquisition bottleneck (Zaff *et al.*, 1993). This emphasis on the implementation meant that experts were often required to discuss their knowledge in terms of rules and frames regardless of whether or not this approach was natural to the expert or appropriate to the domain, consequently the knowledge acquired could be distorted (Zaff *et al.*, 1993) or the knowledge acquisition phase unsuccessful (Wielinga *et al.*, 1992).

An alternative approach to knowledge acquisition is to view it as a modelling activity. In this context knowledge acquisition is defined as consisting of three activities: the elicitation of the domain specific knowledge (usually from an expert); the interpretation of that knowledge; and the formal representation of that knowledge (Wielinga *et al.*, 1992). This paper concentrates on the elicitation and interpretation of the knowledge. The objective of knowledge acquisition is to develop a model, or model of expertise, which defines the behaviour which the expert system is required to exhibit (Wielinga *et al.*, 1992) and acts as a bridge between the verbal data acquired from the expert and the computer implementation (Neale, 1990). It is expected that the model of expertise will, to some extent, emulate the expert's approach to problem solving. Slatter (1987) discussed the extent to which emulation of the expert's problem solving strategy is both possible and desirable. Neale (1990) stated a number of advantages of adopting a modelling, rather than a prototyping, approach to expert system development:

1. Knowledge acquisition is more efficient and accurate.
2. The clarity and comprehensibility of the project is improved.
3. The expert system has greater potential to give useful explanations: to explain its strategy and justify its rules. If the expert system is not based on a model, the explanation is limited to a simple rule trace which is of limited value to the user.

The remainder of this paper divides into four sections. The first discusses some aspects of taxonomy which need to be taken into account when developing a model of expertise for species identification. The second describes how an appropriate knowledge acquisition technique was selected, and the third section outlines the technique used. The fourth section describes the main features of the model of expertise.

Modelling Taxonomic Knowledge

While biological identification has similarities with other problems of identification and diagnosis (e.g. fault and medical diagnosis), it has some unique characteristics. First, identification takes place in the context of a classification, but that classification is subjective and its definition may vary between biologists and change as new species are discovered. For example, with reference to the water mites, Cook (1974) stated 'The most simple classification scheme would include only four superfamilies, ... Because I am giving major importance to certain morphological characters exhibited by the adults, a total of seven superfamilies are recognised'. The potentially dynamic nature of the classification has implications on the extent to which the expert uses it as the basis of identification, and how a partial identification (identification to a genus or family) is achieved. Second, as the classification is developed to show evolutionary relationships, the characters used in its definition may be of limited value in identification. For example, the characters used in defining the families and superfamilies are shown by the immature rather than adult mites, so they cannot be used identify an adult specimen as a member of a family or superfamily (Cook, 1974). However, the characters shown by the adults are used in defining the species and genera (Cook, 1974), so at these levels of the classification the characters used in identification may correspond to those defining the classification. Arising from these two characteristics are two questions which must be addressed when developing a model of expertise for a taxonomic expert system:

1. To what extent does the expert base identification on the taxonomic classification?
2. If identification is independent of the classification, how can a specimen which cannot be identified as a species be incompletely identified within the classification?

In order to answer these questions it is important that the knowledge acquisition phase of the project uses appropriate tools and techniques to give insight into how the expert handles these issues.

Knowledge Acquisition for Modelling Species Identification

This section considers the selection of a knowledge acquisition technique suitable for eliciting knowledge to define a model of expertise for biological identification. Edwards and Cooley (1993) provide a review of knowledge acquisition techniques from a biological perspective.

The first stage of knowledge acquisition consisted of informal discussions with the expert. These discussions established: the classification of water mites (Fig. 17.1); the main characteristics and terminology of the domain; and the 14 species included in the expert system (Table 17.1). The species were carefully chosen to give a range of identification problems including: species which are both closely related and similar (members of the genus *Albia*); species which are superficially similar, but not closely related (*Teratothyas reticulata* and *Australiobates reticulata*); and species which are very distinctive (*Arrenurus multicornutus*).

During these discussions two points emerged. Firstly, the expert did not perform identification analytically, but by recognition. It was, therefore, important that the knowledge acquisition technique selected elicited the actual characters used by the expert in achieving identification, rather than those he thought were used. Secondly, the expert had an extensive collection of prepared, identified specimens which could be used to provide the scenarios required by a number of knowledge acquisition techniques. In many domains it is difficult, or time consuming, to prepare such scenarios (Schweickert *et al.*, 1987) and the use of such techniques may be constrained by their shortage or absence.

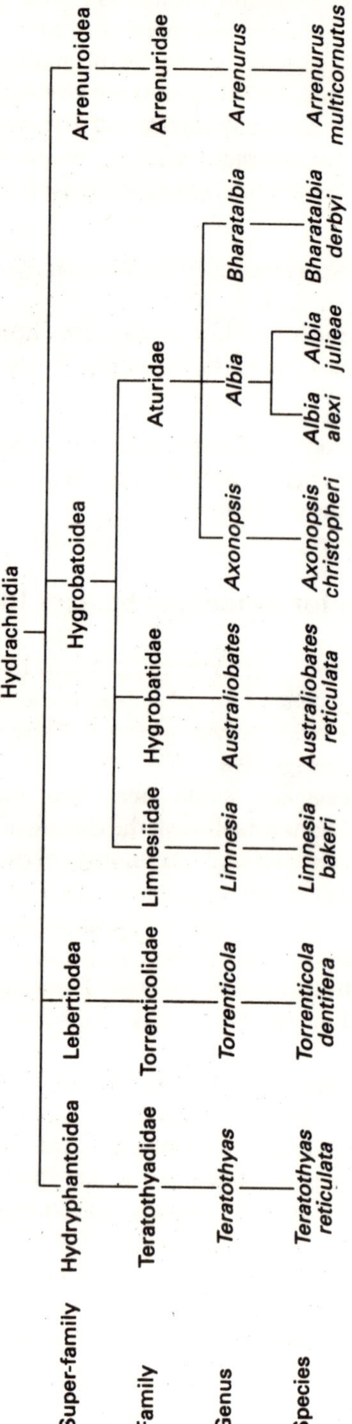

Fig. 17.1. The taxonomic classification of a subset of the water mites used in developing the model of expertise.

Table 17.1. Species investigated during development of the model of expertise.

Limnesia bakeri
Australiobates convexipalpis
Australiobates concavipalpis
Axonopsis christopheri
Bharatalbia derbyi
Australiobates reticulata
Albia alexi
Albia julieae
Albia rectifrons
Albia sulutensis
Teratothyas reticulata
Pseudotorrenticola sharpi
Torrenticola dentifera
Arrenurus multicornutus

From these discussions, and in the light of the questions posed at the end of the previous section, it is clear that the knowledge acquisition technique used should fulfil the following criteria:

1. It should not either explicitly or implicitly encourage the expert to think in terms of the classification.
2. It should enable the actual characters used by the expert to achieve identification to be determined, rather than those he thinks are used.

The first criterion rules out a number of indirect knowledge acquisition techniques. An indirect technique collects information from which inferences can be made about what the expert must have known in order to supply the information (Olson and Rueter, 1987). Indirect techniques largely consist of multidimensional techniques, e.g. multidimensional scaling, repertory grid analysis, and card sort, and are limited in the types of knowledge they reveal. Apart from repertory grids, these techniques tend to only reveal knowledge about relations between objects and concepts (Olson and Rueter, 1987). It was felt that indirect methods would tend to reveal information about the taxonomic classification, and so they would not be appropriate in this case. Similarly the laddering interview technique (Rugg and McGeorge, 1995) is only appropriate when the knowledge engineer suspects there is a hierarchical classification to the domain, and the technique attempts to elicit that classification. Therefore laddering is not an appropriate technique to use for biological identification unless it has already been shown that the expert bases identification on the taxonomic classification.

Introspective knowledge acquisition techniques (Neale, 1988) require the expert to describe how previous problems were solved, or theoretical problems would be solved, so they are not appropriate if the expert solves problems by recognition, as they will only indicate how the expert thinks the problem is solved. Similarly a protocol analysis (Newell and Simon, 1972) relies on the ability of the expert to 'think aloud', verbalizing all his thoughts as the problem is solved, again this technique is not suitable when expertise is based on recognition.

A number of interview techniques, including structured interviews (Schweickert *et al.*, 1987), intermediate reasoning steps and distinguishing goals (Welbank, 1983), rely on the expert describing the relationship between the facts and conclusions in the domain. These techniques are orientated towards the explicit elicitation of rules from the expert, so are unlikely to assist in the development of a model of expertise.

The most promising technique in this case appeared to be the twenty questions knowledge acquisition technique (Welbank, 1983; Schweickert *et al.*, 1987; Neale, 1988), which is also called 'short cut protocol analysis : context focusing' (Wright and Ayton, 1987). The aim of the technique is to determine the information required to reach a conclusion by making the expert explicitly request all the information about the current scenario by questioning the knowledge engineer. During the question–answer dialogue, the answers given by the knowledge engineer are usually restricted, sometimes to yes or no, in order to force the expert to ask detailed, rather than high-level, questions. In order to use the technique, the knowledge engineer must construct the scenarios to be investigated. These scenarios must include all the information likely to be requested and their development is very time consuming (Schweickert *et al.*, 1987), but in biological identification the scenario is the specimen which may be readily available. The twenty questions technique is useful if the knowledge engineer is reasonably knowledgeable about the domain, but it is not without problems. Welbank (1983) stated that the process of asking for information may disrupt the expert's normal line of reasoning. Schweickert *et al.* (1987) found that rule extraction from the resulting transcripts was more dependent on the knowledge engineer than other techniques, and depended on the knowledge engineer's ability to realize when the expert was using information obtained earlier in the discussion. Both Welbank (1983) and Schweickert *et al.* (1987) emphasized the importance of establishing the reasons why the questions were asked as this simplifies interpretation of the dialogue. It was decided to base knowledge acquisition on a modified version of the twenty questions knowledge acquisition technique.

Modified Twenty Questions Interview Technique

This section describes the modified twenty questions interview technique:

1. A mite was selected from a range of specimens. The specimens had previously been identified to species level by the expert – this is referred to as the correct conclusion or identification.

2. The expert did not know the identity of the mite chosen, nor was he allowed to look down the microscope.
3. Identification was achieved by the expert obtaining information about the mite via the knowledge engineer, and took the form of a question–answer dialogue.

The only way that the expert could obtain information about the mite was explicitly through the knowledge engineer which resulted in a protocol describing the identification. For each question asked by the expert the following were recorded:

- The question.
- The reason why the question was asked.
- The answer, which was as brief as possible, either a single word or a short phrase.
- The deduction that the expert made from the answer.

By recording the reasons and deductions relating to the specific questions and answers, it was intended to gather information about the hypotheses investigated and the intermediate conclusions reached by the expert during the identification. Unfortunately, in many of the protocols there were gaps among the reasons and deductions, but this information was obtained retrospectively at the next session. The results of the protocols were arranged in tabular form (Table 17.2) which maintained the temporal order of the questions.

The Model of Expertise

A total of nine protocols were obtained which identified 14 species (some protocols allowed more than one species to be identified). No specialized tools were used in analysing the protocols because of their small size. Interpretation of the protocols was not straightforward as each identification gave a very specific protocol. Although not specifically tested, it is not certain that two protocol analyses identifying the same species would be the same because:

1. The expert would ask different questions from identical initial data. Although Wright and Ayton (1987) stated that the expert should be asked to justify changes in the sequence of questions, this is an unreasonable request if the interviews take place over a number of weeks during which time the expert is working on different areas of the domain.
2. The knowledge engineer would sometimes misinterpret the mite and answer questions incorrectly. In these cases, the identification would proceed in a completely spurious direction until either an incorrect identification was reached or the identification became too implausible.

Table 17.2. Dialogue of the identification of *Limnesia bakeri* obtained using the twenty questions knowledge acquisition technique.

Question	Reason	Answer	Deduction
Are there any dorsal plates?	The answer does not give rise to a taxonomic group but the majority of members of the first 3 superfamilies are soft-bodied — but some have many small plates — as are some members of higher super-families. The mite can be placed in one of two distinct groups	No	Indicates a conventional soft-bodied mite
Is it soft-bodied or sclerotized?	If the mite was heavily sclerotized, it may be possible to 'miss' the dorsal plate. It also enables a check on the amount of membrane present	Soft-bodied	Implies a lower soft-bodied mite (some of which have small plates)
Are there any setae on the dorsal surface?	Setae do not occur on plates. It is a check against a single plate which extends over the entire dorsal surface. Setae are difficult to see on soft-bodied mites	I think so, or there have been	Conclusion — a conventional soft-bodied mite — in particular can exclude all plated mites *Arrenurus*, *Albia*, *Axonopsis* and *Torrenticola*
Is epimera IV triangular in shape? Does it slope upwards towards the middle? Or is it square and flat, bisecting the body?	*Limnesia* (composed largely of soft-bodied mites) is the only genus with a triangular epimera IV, so is easily separated from all other soft-bodied mites	Triangular	*Limnesia*
Are the genital plates separate or fused together? There are two plates – one each side of the genital opening – are they fused together top and bottom, or are they separate? If they are fused, there is normally a large number of setae at the top.	Male *Limnesia* have fused plates, so a fused plate would be further confirmation of *Limnesia*	Plates are separate	Female
Are there 3 or 4 pairs of acetabula?	Number of acetabula can be used to separate the subgenera of *Limnesia*	4 pairs	*Limnesia bakeri*

The different protocols were compared to identify similarities and trends between them in order to determine the model of expertise. Four characteristics of the model of expertise were determined.

Representation of a flexible hierarchy

From the deductions made to the answers, it can be seen that the expert reaches intermediate conclusions, however, these conclusions do not necessarily relate to the taxonomic classification. For example, in the identification of *Limnesia bakeri* (Table 17.2) three conclusions are reached: the mite is a soft-bodied mite, the mite belongs to the genus *Limnesia*, and the mite is *Limnesia bakeri*. The model of expertise must allow the representation of the two intermediate conclusions (the mite is soft-bodied, the mite belongs to the genus *Limnesia*) while recognizing that these intermediate conclusions do not necessarily follow the taxonomic hierarchy and that not all levels of the hierarchy are included. In most cases the first taxonomic group recognized is the genus.

Classes of questions

Different classes of questions are recognized. The reasons associated with the questions were used to indicate both the current hypothesis and the importance of the question. Two classes of questions were recognized, these relate either to the search strategy, or to the strength of evidence associated with a question.

Strategic questions

Three different types of questions were recognized which gave insight into the search strategy adopted by the expert.

General questions. General questions were used to elicit very general information about the identification and they enabled the expert to place the identification in a particular context. It is unlikely that an expert will adopt a very focused problem solving strategy without first acquiring general information about the problem. All identifications began by determining whether the mite was hard or soft-bodied. This is not a recognized taxonomic division, but it divides the mites into higher (hard-bodied) and lower (soft-bodied) mites.

Elimination questions. The expert had a broad search strategy, this showed itself in the inclusion of questions which were asked specifically to eliminate a species, genus or family. These questions were frequently asked after there was an indication that a specimen was not a particular species. For many hard-bodied mites a specific attempt was made to exclude the family Torrenticolidae and the species *Bharatalbia derbyi*, both of which are distinctive hard-bodied mites.

Non-taxonomic questions. Occasionally a question would place a mite in a group which had no taxonomic significance. Non-taxonomic questions are similar to general questions, but they are not necessarily appropriate to all identifications. For example, a reticulate mite belongs to a group consisting of *Teratothyas reticulata* and *Australiobates reticulata* which are not closely related.

Questions concerned with strength of evidence.

Strength of evidence is concerned with the contribution which an item of evidence (in this case an answer to a question) makes to the overall identification. In this context, three different types of questions were identified, confirmatory questions which contributed least to the identification, standard questions, and key questions which were most important in reaching an identification.

Confirmatory questions. Confirmatory questions were easily identified from the protocols, either from the reason for a question or from the deduction made from the answer. For example, when identifying *Limnesia bakeri*, the reason for asking the question 'Are the genital plates separate or fused together?' was given as 'Male *Limnesia* have fused plates, so a fused plate would be further confirmation of *Limnesia*'. Confirmatory questions are of two distinct types:

1. Questions to confirm the current, tentative, conclusion. The expert would often indicate, in the deduction made from an answer, that a mite belonged to a certain genus or species. He would then go on to ask several 'confirmatory' questions before reaching a definite decision. The number of confirmatory questions was increased if preceding questions had been answered hesitantly or inconsistently.
2. Questions to check or confirm the answers to previous questions. These additional questions were only asked if the answer to a question was uncertain. The additional questions were of two types: more detailed questions about the same character or questions about different characters.

A characteristic of both types of confirmatory questions was that if the answer did not support the current conclusion it did not automatically lead to the rejection of the current conclusion. However if all the answers to the confirmatory questions failed to support the conclusion it would be reassessed.

Key and standard questions. Any question which is not confirmatory can be assigned to one of two classes: it is either a key or a standard question. These classes reflect the importance of the questions in the context of the identification, i.e. they indicate different levels of the strength of evidence. Key questions are more important than standard ones (and both are more important than confirmatory questions). The identification of key and standard questions required careful consideration of the protocols and, unless there was clear evidence that a question was of key importance, it was designated as a standard question. It was sometimes possible to identify key questions from the reasons behind the questions. For example, *Limnesia* is the only genus with a triangular epimera IV (Table 17.2). As *Limnesia* can be identified or excluded on the basis of a triangular epimera IV it was designated a key character.

The relationship between strategic and strength of evidence questions

All questions can be identified as having a strength of evidence, and a question may also be identified as a strategic question. For example, the question in the identification of *Limnesia* (Table 17.2) 'Are there any dorsal plates?' was identified as a general question (to determine a hard or soft-bodied mite) and also as having key importance in the identification.

The strength of evidence of a particular question is derived from its context in a particular protocol. It is possible for a question to be identified as a standard question in one protocol and as a confirmatory question in another one. For example, determining the number of acetabula is merely confirmatory to the identification of *Australiobates*, but it is of standard importance in the identification of *Teratothyas reticulata*.

Control strategy

If the expert was having difficulty making an identification and suspected that one or more questions had been answered incorrectly by the knowledge engineer then he would backtrack and repeat questions (often rephrasing the question) in order to clarify the situation. It was felt that this aspect of the expert's behaviour should be included in the model of expertise. Hence, the search strategy in the model of expertise is defined as having two searches. The first search assumes that all the information supplied by the user is correct (apart from information relating to confirmatory questions) and minimizes the number of questions asked of the user either by not asking all confirmatory questions or by assuming information on the basis of answers to previous questions. If no identification is reached then rather than informing the user that no identification is possible, a second search of the knowledge base is made when: all confirmatory questions are asked; no assumptions are made on the basis of previous answers; and when questions may be repeated to check the answer.

Reasoning with uncertainty

Questions were rarely answered with conventional representations of uncertainty (e.g. I don't know, I think so) and never with quantitative responses, but the expert would deduce the amount of confidence that could be placed in an answer by considering the length of time taken to answer a question and the degree of hesitancy in the answer. For this reason, the method of reasoning under uncertainty used by the expert system is not based on that of the expert, and further details can be found in Edwards (1994).

Conclusions

The knowledge acquisition process discussed in this paper has enabled the definition of a domain independent model of expertise as well as the domain specific knowledge relating to water mite identification. The key features of the model may be summarized as follows:

1. The model is not based strictly on the taxonomic classification, but recognizes only those levels of the classification used by the expert (typically genus and species).
2. The recognition of different types of questions, most importantly those which confirm conclusions reached tentatively and questions answered hesitantly.
3. The specific exclusion of possible alternative conclusions.
4. Repetition of questions when there was reason to doubt the original answer.

The expert had a cautious approach to identification, and this led to the model of expertise being called identification by confirmation. The model shows similarities to the criteria tables of Lindberg *et al.* (1982), which have been used in medical diagnosis, as both approaches represent the strength, or importance, of evidence by ranking information. Criteria tables have different types of criteria, e.g. major, minor, required and excluded criteria, which approximate to the different types of questions identified from the protocols.

By adopting a modelling approach to knowledge acquisition, and by making a careful choice of knowledge acquisition technique, it was possible to determine the extent to which the expert used the classification during identification. Interestingly, the expert never reached intermediate conclusions relating to the superfamily and only once relating to the family — these are levels which Cook (1974) stated are defined by immature rather than mature mites.

It is recognized that using the modified twenty questions knowledge acquisition technique has not resulted in a cognitive model of expertise. The expert does not normally achieve identification by such an analytical route, but by recognition. In spite of this, it is argued that the modified twenty questions technique probably provides a good model for the expert system as the expert has to work within the limitations of the knowledge engineer. For example, comparative knowledge is very useful when making an identification, but it is very difficult to elicit when the knowledge engineer does not have the background knowledge on which to base the comparisons, also the knowledge engineer may have difficulty identifying very detailed or obscure characters. The expert admitted to trying to find the easiest route to identification, and used information which would not have been included if writing a conventional key, indicating that he was taking the limitations of the knowledge engineer into account. The twenty questions technique corresponds closely to the 'expert on the end of a 'phone' description of an expert system and, given the interactive nature of expert systems, it is probably not an unreasonable model to use.

Finally, the model of expertise was shown to provide a practical approach to identification, and has been implemented as a research prototype (Edwards, 1994). In addition, there is some evidence that the model can be used as a generic model of expertise as a small-scale expert system has been developed, using the model, to address a problem in plant pathology (Edwards, 1994). The development of generic models is seen as an important area for supporting the development of expert systems (Wielinga *et al.*, 1992) and identification of suitable generic models of expertise for species identification would simplify the development of such expert systems.

Acknowledgements

Thanks are due to the expert, Dr Roy Wiles (University of Buckingham), for his enthusiasm and assistance with the project. Thanks are also due to Dr Roger Cooley (University of Kent) who supervised the research.

References

Cook, D.R. (1974) *Water Mite Genera and Subgenera.* Memoirs of the American Entomological Institute, Number 21. The American Entomological Institute, Michigan, USA.

Edwards, M. (1994) A semi-quantitative approach to representation and reasoning for expert systems. PhD thesis, University of Kent, Canterbury, UK.

Edwards, M. and Cooley, R.E. (1993) Expertise in expert systems: knowledge acquisition for biological expert systems. *Computer Applications in the Biosciences* 9(6), 657–665.

Edwards, M. and Morse, D.R. (1995) The potential for computer-aided identification in biodiversity research. *Trends in Ecology and Evolution* 10(4), 153–158.

Forget, P.M., Lebbe, J., Puig, H., Vignes, R. and Hideux, M. (1986) Micro-computer aided identification: an application to trees from French Guiana. *Botanical Journal of the Linnean Society* 93, 205–223.

Lindberg, D.A.B., Kingsland III, L.C., Roeseler, G.C., Kay, D.R. and Sharp, G.C. (1982) A new knowledge representation for diagnosis in rheumatology. *Proceedings of the first AMIA Congress of Medical Informatics*, pp. 299–303.

Neale, I.M. (1988) First generation expert systems: a review of knowledge acquisition methodologies. *The Knowledge Engineering Review* 3(2), 105–145.

Neale, I.M. (1990) Modelling expertise for KBS development. *Journal of the Operational Research Society* 41(5), 447–458.

Newell, A. and Simon, H.A. (1972) *Human Problem Solving.* Prentice-Hall Inc., New Jersey.

Olson, J.R. and Rueter, H.H. (1987) Extracting expertise from experts: methods for knowledge acquisition. *Expert Systems* 4(3), 152–168.

Rugg, G. and McGeorge, P. (1995) Laddering. *Expert Systems* 12(4), 339–346.

Schweickert, R., Burton, A.M., Taylor, N.K., Corlett, E.N., Shadbolt, N.R. and

Hedgecock, A.P. (1987) Comparing knowledge elicitation techniques: a case study. *Artificial Intelligence Review* 1, 245–253.

Slatter, P.E. (1987) Cognitive emulation in expert system design. *The Knowledge Engineering Review* 2(1), 27–41.

Welbank, M. (1983) *A Review of Knowledge Acquisition Techniques for Expert Systems.* British Telecom Research Laboratories, Martlesham Heath, Ipswich.

Wielinga, B.J., Schreiber, A.Th. and Breuker, J.A. (1992) KADS: a modelling approach to knowledge engineering. *Knowledge Acquisition* 4(1), 5–53.

Woolley, J.B. and Stone, N.D. (1987) Application of artificial intelligence to systematics: SYSTEX — a prototype expert system for species identification. *Systematic Zoology* 36, 248–267.

Wright, G. and Ayton, P. (1987) Eliciting and modelling expert knowledge. *Decision Support Systems* 3, 13–26.

Zaff, B.S., McNeese, M.D. and Snyder, D.E. (1993) Capturing multiple perspectives: a user-centered approach to knowledge and design acquisition. *Knowledge Acquisition* 5(1), 79–116.

18

Information Technology Support for Decision Making – Where from Here?

J.D. Mumford

Centre for Environmental Technology, Imperial College of Science, Technology and Medicine, Silwood Park, Ascot, Berks SL5 7PY, UK
Fax: +44 (0)1344 294308/E-mail: j.mumford@ic.ac.uk

Introduction

A title purporting to cover the future direction of information technology support for crop protection implies both an element of prediction and of preaching. So I will try to conform to such an expectation, and to base the discussion mainly on principles rather than precedents. Throughout this volume readers will find examples of current and planned uses of information technology in the field and I do not propose to review any of these in detail.

The question of where information technology support for crop protection decision making is going can be addressed in three ways: what can happen? what should happen? and what will happen?

What Can Happen?

It is clear that information technology is affecting all aspects of life. Four industries are driving developments in the field, two in the supply of technology and two that dominate demand. The electronics industry, which was originally linked closely to defence, has made the processing power of computers technically possible. The software industry has made that power widely accessible to markets that can and will pay, both serious considerations from the perspective of crop protection. Entertainment and communications are overwhelming influences on the demand for information technology.

We in crop protection are not important players in the information technology revolution and we are certainly not leaders. We can take two views on

 Information Technology, Plant Pathology and Biodiversity (eds P. Bridge, P. Jeffries, D.R. Morse and P.R. Scott)

adapting developments in information technology. On the negative side, we may feel left out of the excitement and challenge of cutting edge innovation. On the other hand, the positive side of this situation is that crop protection can benefit from the almost limitless investment by other industries and adopt proven technology and experience with limited risk. The entertainment and communications industries have the broad appeal and marketing capacity to get both hardware and familiarity with software into large numbers of homes and businesses, including those involved in agriculture. They are also industries that are capable of getting people used to the 'pay as you use' principle that will be essential to finance the development of specialist technology for industries like crop protection.

In general, there are two important areas for improvement in information technology: technical potential and software experience. The technical potential of computing increases continually and the only safe prediction is that it will continue to do so, and fast. There is continuing increase in the capacity to store, process, access and present information. Improved access to information and processing power is an important consideration to useful crop protection decision support because agriculture, and with it crop protection, is a geographically dispersed industry in which decision making has been limited in the past by communication problems. Improved processing, through faster chips, more memory capacity and efficient software applications, makes many operations useful to crop protection decisions feasible at last, such as the combination of spatial modelling of yields across fields and satellite positioning systems for precision spray management.

Experience with software development continues to improve and two types are particularly relevant to crop protection. Educational software, from basic children's programs to those for business users, provides widely tested interfaces and concepts adaptable to crop protection training formats. Other types of software, aimed at financial decision making such as Bayesian risk analysis, also provide starting points for crop protection risk analysis.

What Should Happen?

Before getting too carried away with the growing potential of information technology we should stop and think about three things: how are decisions made? what is wrong with them now? and how can information technology help to improve them?

All rational decisions are made on the basis of perceptions of a problem, the options available for its control and the outcomes expected from using those options. Individuals test these expected outcomes against their objectives and make a choice accordingly. A decision can go 'wrong' for a number of reasons. Perceptions may be unrealistic, the complete range of options may not be available or considered, and the outcomes may be misjudged due to the complexity of the possibilities. People make mistakes and are inconsistent. Individual objectives may be incompatible with wider social goals, such as long term environmental quality. Given the potential for suboptimal decision-making it is fortunate that most systems are fairly robust.

In crop protection, however, we are dealing with a system that may be less robust than many because of both the potential impact of pesticides and the potential losses of basic commodities and incomes due to pests. Two problems resulting from poor decisions cause widespread concern: pesticides are used too much (in relation to many social goals) and they are often not effectively or efficiently used when it is agreed that they are needed to protect crops. Pesticides are overused because, amongst many reasons, pest problems are overestimated, other options are not properly considered and they are seen by users as the best alternative to reducing risk since the future is hard to predict. Pesticides are inefficiently used, again giving only some examples, because of difficulty interpreting field conditions, too little time to process field data on pest development before action may be needed and poor predictive capability in the face of so many environmental and economic variables. On top of these problems there is also a lack of feedback in many pesticide use decisions, for example on environmental impact, which lets people go on making poor decisions without even realizing it.

So decisions are limited. There is a tendency to continue with what 'worked' before, to ignore other or newer options, to change actions only at or after a crisis, to keep management effort to a minimum and to leave much of the responsibility for environmental quality to legislators and consumers.

To solve these problems we could attempt to develop information technology that tries to do it all for you: 'intelligent' computers and prescriptive 'expert' systems but I believe it is more sensible to advocate decision support rather than decision substitution. Table 18.1 presents the forms of support that I think are likely to be appropriate to the crop protection decision problems discussed earlier.

Of the systems suggested in Table 18.1, I consider that the first two, information systems that open up options and improve perceptions, and recording systems that help to formalize decisions and allow better feedback by comparing individual decisions with those of other similar decision makers, offer the most immediate potential for improved crop protection.

What Will Happen?

This final question is obviously the bottom line and clearly difficult to answer with both precision and confidence. Therefore, being risk averse, I will opt for the confident, rather than the precise, route. Sadly I am not as optimistic as I would like to be. I think the pervasiveness and excitement associated with the technology being produced by other industries will mean that pest managers

Table 18.1. Types of information technology support appropriate for crop protection.

Information technology support	Examples
Pointing decision makers in the right direction	Databases of options Dynamic 'what if?' models Crop protection information systems by CD-ROM or Internet
Recording systems	Individual records for quality assurance, interfield and interyear comparison Group comparisons of inputs and outputs Real-time warnings about inappropriate or unusual choices through links to records of other similar decision makers
Data collection	Field and regional crop, pest and meteorological data transfer and storage Mobile telecommunications systems for automatic data transmission
Data management and processing	Interpretation of field data Estimation of returns and probabilities from choices Management of 'precision farming' operations
Training simulations	Dynamic pest management scenarios

will be tempted to explore almost every potential use of information technology in crop protection and they will all be tried, but that much of it will fall by the wayside, unfortunately after confusing and delaying a great deal of useful progress. However, there are definite uses for information technology that can improve crop protection decisions, once systematic analyses have revealed the shortcomings in existing decisions and it has been shown that people are willing to pay to improve them.

Interactive Keys

19

M.J. Dallwitz, T.A. Paine and E.J. Zurcher

CSIRO Division of Entomology, GPO Box 1700, Canberra, ACT 2601, Australia
Fax: +61 2 6246 4000/E-mail: delta@ento.csiro.au

Introduction

Identification is the process of finding the taxon to which a specimen belongs. Several methods are available for aiding this process (e.g. Pankhurst, 1991). The most important are conventional identification keys and interactive keys.

A conventional identification key is a tree with characters at the internal nodes and taxon names at the terminal nodes. Each branch corresponds to a state of the character or characters at the node from which it arises. The user starts at the root of the tree, and follows the branches corresponding to the character states exhibited by the specimen until the taxon name is reached.

Authors of conventional keys try to provide some flexibility for the user by placing alternative characters at each node, but the possibilities for doing this are limited, because the characters must have identical distributions of their states among the taxa remaining in contention at that node. An error by the user in assigning a character state to the specimen inevitably leads to a wrong identification, unless the author has allowed for the possibility of this error by placing the taxon name in the subtree corresponding to the wrongly assigned state, as well as in the subtrees corresponding to states actually exhibited by the taxon. The author's use of this mechanism must also be limited, because each possible error (taxon/character-state combination) treated in this way adds a terminal node to the tree. This increases the size of the printed key (proportional to the number of terminal nodes), and the average number of characters which must be used to obtain an identification (proportional to the logarithm of the number of terminal nodes).

After any identification, it is good practice to check its accuracy by comparing the specimen with a description or illustrations of the taxon, or with other specimens known to belong to the taxon. When a conventional key is being used, the only way to recover from a wrong identification due to an error by the user is to guess where the error was made, return to that node, and try following another branch. If the error is in the key itself (that is, an error was made by the author), recovery is not possible.

An interactive key is an interactive computer program in which the user enters attributes (character-state values) of the specimen. The program eliminates taxa whose attributes do not match those of the specimen. This process is continued until only one taxon remains. The taxon attributes are usually stored as a characters-by-taxa 'matrix'. It is also possible to store the attributes as 'rules', but this kind of program is generally less satisfactory (Dallwitz, 1992).

The criteria presented here are based on experience with the interactive identification program INTKEY (Watson *et al.*, 1989; Dallwitz, 1993; Dallwitz *et al.*, 1993), which uses data prepared in DELTA format (Dallwitz, 1980). They were first published, in summary form, on the Taxacom mailing list (Dallwitz, 1994). All of the features mentioned below are available in INTKEY, unless otherwise stated.

INTKEY was originally developed from an early version of the program ONLINE (Pankhurst and Aitchison, 1975). Many other interactive identification programs are available – see, for example, Edwards and Morse (1995), Dallwitz (1996 onwards), and several chapters in this volume including Schalk (Chapter 20) and White and Sandlant (Chapter 24).

Advantages of Interactive Keys

A well designed interactive key has several advantages over a conventional key.

1. A correct identification can be made in spite of errors by the user or in the data. Taxa are normally eliminated when they differ from the specimen in any way. If it is known or suspected that an error has been made, the program can be instructed to eliminate taxa only if they differ from the specimen in more than one attribute. It is immaterial where the error occurred, and whether it was made by the user or by the author of the data. In INTKEY, this function is controlled by the 'Tolerance' parameter, whose value may be 0 or any positive integer. Taxa are eliminated if they differ from the specimen in more attributes than the current value of 'Tolerance'. The parameter may be set to any permitted value at any time in the identification process, but typically it would be incremented by 1 when an identification has been made and found to be incorrect. The identification process is then continued, exactly as before. If all the taxa are eliminated, the program can increment 'Tolerance' automatically. If a single taxon remains, the program has no way of knowing whether this is the correct identification, and it is up to the user to check the identification, and, if necessary, increment 'Tolerance' manually.

2. Any characters can be used, and their values changed, in any order. Characters which are not available on the specimen, or whose interpretation is not clear to the user, can be avoided (provided that there is sufficient redundancy in the data – see below). The values of any character can be changed at any stage of the identification, or any character deleted from the identification.
3. Numeric characters can be used directly, without being divided into ranges. In conventional keys, numeric characters such as lengths must be divided into ranges before being incorporated in the key. This usually results in loss of information. In an interactive key, the actual range of values exhibited by each taxon can be recorded in the data, and the taxon eliminated if the specimen's value does not fall within this range.
4. The user can express uncertainty by entering more than one state value, or a range of numerical values. A user who is not sure which character-state value applies to the specimen may nevertheless sometimes be confident that some state values do not apply. Entering all the values which may conceivably apply to the specimen eliminates those taxa which never exhibit any of those values.
5. Updating is easy. The key is maintained simply by making corrections and additions to the data matrix. Updating of conventional keys is relatively difficult. Even when the key is generated by computer from a data matrix, major changes to the matrix, particularly the addition of new characters and taxa, can have a large effect on the key structure, which has to be checked and possibly re-optimized.

Desirable Features for Interactive Keys

1. Best characters. The program should be able to advise the user on the most suitable characters for use at any stage of an identification. Because of the very large number of paths which may be taken through an interactive key, the ranking of the characters should be calculated directly from the data matrix for the set of taxa actually remaining at each stage of the identification. It is unsatisfactory to pre-assign rankings for a relatively small number of cases, as, for example, in a rule-based expert system.

The character-ranking algorithm should take into account both the separating power of a character, and its 'reliability' – a subjective measure, usually supplied by the author, of the character's accuracy and/or ease of use. In interactive identification (in contrast to conventional keys), the reliability should be given only a small weight, so that the ranking of characters is mainly influenced by their separating power. This is because the user is in a position to judge how easily he can interpret a character in the particular specimen at hand, but cannot tell how well a character separates the remaining taxa.

The character-ranking algorithm used in INTKEY is the same as that used in the key-generation program, KEY (Dallwitz, 1974). Unlike most such algorithms, it has a theoretical basis and gives sensible results for characters with three or more states, and for numeric characters. The relative weight of the separating power and the reliability can be controlled by both the author and the user.

Ranking of the characters is time-consuming in large data sets, so it is important that the computation is as efficient as possible, and that the user does not have to wait for the ranking to be completed before choosing a character. In INTKEY, the characters are processed in descending order of their reliabilities, and each character is inserted in its correct place in the 'Best' list as the computation for that character is completed. At the start of an identification with a data set containing 8500 taxa, the computation for each character takes about 0.5 sec on a 133 MHz Pentium computer, so that a good selection of characters is available in a few seconds, although the complete calculation (for more than 1000 characters) takes several minutes.

2. Locating errors. The program should be able to locate user and/or data errors which were circumvented by the error-tolerance mechanism. The identification of user errors helps to improve the user's interpretation of characters. Data errors can be reported to the author for correction in later versions.

In INTKEY, errors can be located by using the 'Differences' command to display the differences between the specimen and the remaining taxon.

3. Gaps for integer numeric characters. The recorded values for integer numeric characters should be able to contain gaps, for example, '5 or 10' should be distinguishable from '5 to 10'. For example, the number of stamens can be used to distinguish *Hyacinthaceae* (3 or 6) from *Lowiaceae* (5).

4. Character dependencies. The program should be aware that some values of certain 'controlling' characters make other 'dependent' characters inapplicable. When a controlling character is used in an identification, any characters which thereby become inapplicable should not appear in any lists of characters which are subsequently presented to the user. When a dependent character is used before its controlling character(s), the appropriate values of the latter should be set automatically, and the user informed.

5. Characters for separating a particular taxon. For confirming a tentative identification, it is useful to be able to rank characters according to how well they separate a given taxon from all the rest. Characters highly ranked on this criterion tend to lead to a shorter confirmation than those highly ranked on the ordinary 'best' criterion.

6. Inapplicable/unknown. Inapplicable values, including those not resulting from character dependencies, should be distinguished from unknown values.

7. Text characters. It should be possible to store free-text information about taxa, and to use this information for identification and information retrieval.

8. Expanded ranges for numeric characters. It should be possible, for the purpose of identification, to expand the range of values recorded for a numeric character which has been poorly sampled in a taxon.

This is possible to a limited extent in INTKEY: single recorded numeric values can be transformed into ranges when the original data are converted from DELTA format into the format used by INTKEY. This process should preferably be under the control of the user of the interactive key, so that the size of the range can be varied, and so that the actual-recorded data are available for information retrieval.

9. Flagging of character values which are likely to be assigned to a taxon in error. When recording a taxon description, the author of a data set may record character-state values which are not, in fact, exhibited by the taxon, but may seem to be so, particularly to an inexperienced user. This nullifies the possible effect of the error on identifications, but has the disadvantages that the separating power of the character is reduced, and that the information is misleading in descriptions and for classification. It is therefore desirable that the author should be able to flag such values, and that the user should have the option of ignoring them.

INTKEY does not currently offer this feature, as it is not supported in the DELTA format from which the INTKEY data are derived. The feature is among the enhancements which will be supported in the new DELTA system, now under development (Dallwitz *et al.*, 1993 onwards).

10. Probabilistic identification. The data 'matrix' should have provision for storing the probabilities that a given taxon will exhibit certain character states, and the program should be able to make use of these probabilities in the identification process – see, for example, Willcox and Lapage (1975) and Bryant (Chapter 28, this volume).

The probabilities that the user will make errors in assigning character states to the specimen are often substantial (Stucky, 1984; Fermanian *et al.*, 1989; Morse *et al.*, 1996), and these probabilities need to be taken into account in probabilistic identification, even for deterministic attributes (that is, where the probabilities of the state values are recorded as 1 or 0 in the data matrix). Willcox and Lapage (1975) describe the difficulties of dealing with errors in the context of identification of bacteria, and the problem is likely to be quantitatively worse in general biological identification, because characters are less standardized, and the users of the keys are likely to vary more in their knowledge and experience.

There are other potential impediments to the widespread use of probabilistic identification. Probabilities are not widely available in the literature, and taxonomists might be reluctant to take the trouble to estimate and record them. Similar problems are already evident in the relatively infrequent use of numerical values compared with terms such as 'small', 'large', 'sparse', and 'dense'. Also, there will be greater difficulty for the user in understanding the processes and the results.

Probabilities cannot be recorded in the current DELTA format, except as comments, but they will be recordable in the enhanced DELTA system mentioned above. This will open the way for their use in INTKEY.

11. Restriction to subsets of the characters and taxa. It should be possible to restrict any operation to subsets of the characters and taxa. For example, in identification, the characters could be restricted to those available in an incomplete specimen, and the taxa to those known to occur in the geographical area from which the specimen came.

In INTKEY, it is possible to specify subsets of characters and taxa to which all subsequent operations are restricted, and many commands can make a further restriction applying only to that command. The subsets can be specified either by inclusion or exclusion of characters and taxa.

12. Fixing character values. It is convenient to be able to specify character values which are not to be cleared when a new identification is started. For example, if there is a character whose states specify geographical regions, and several specimens coming from the same region are to be identified, then the appropriate state value need be entered only once.

13. Named subsets of the characters and taxa. There should be a mechanism for attaching names to subsets of the characters and taxa, so that these subsets can be referred to easily. In INTKEY, these names are called 'keywords'. They can be defined both by the author and the user.

14. Glossaries and character notes. Glossaries and character notes should be available. A glossary contains definitions of terms used in the characters. Character notes give details of the interpretation of individual characters, and may include information about the reasons for defining characters in particular ways. Glossaries are not yet implemented in INTKEY.

15. Information retrieval. It should be possible to use the program for information retrieval. This includes finding all taxa which have certain combinations of attributes, and various other operations mentioned below.

16. Descriptions. It should be possible to display descriptions generated from the data used in identification. Some programs can display only text descriptions which have been entered independently of the data matrix.

17. Differences and similarities between taxa. The program should be able to find the differences between members of a set of taxa. There should be no restrictions on the size of the set of taxa.

18. Diagnostic descriptions. The program should be able to find diagnostic descriptions, which distinguish a given taxon from all the other taxa. These provide a quick way of confirming the identity of a specimen. The characters should be chosen from those which have not been used in the current identification, in order to provide an independent confirmation.

INTKEY has a parameter, 'DiagLevel', which specifies the minimum number of characters for which the diagnostic description should differ from all the other taxa. Another parameter, 'DiagType', distinguishes between specimen-diagnostic and taxon-diagnostic descriptions. The latter are allowed to contain characters which may sometimes be inapplicable to specimens belonging to the taxon.

19. Control of value matching. The user should have flexible control over which character values or combinations of values are to be regarded as equal to other values.

In INTKEY, matching in identification and in the Differences and Similarities commands is controlled by the Set Match command, which has the options 'Inapplicables', 'Unknowns', 'Subset', 'Overlap', and 'Exact'. 'Set Match Inapplicables' and 'Set Match Unknowns' specify respectively that 'inapplicable' and 'unknown' values match any value. 'Set Match Subsets' specifies that two sets of values match if one set (usually the values of the specimen) is a subset of the other (for example, '1 or 2' matches '1 or 2 or 4' but not '2' or '3'; '2–5' matches '1–6' but not '4–10'). 'Set Match Overlap' specifies that two sets of values match if they overlap, that is, if they have any values in common (for example, '1 or 2' matches '2 or 3'; '2–5' matches '4–10'). 'Set Match Exact' specifies that two sets of values match only if they are identical. The default setting is 'Overlap Unknowns Inapplicables', which is usually the most appropriate for identification. For information retrieval, the most appropriate setting is usually 'Overlap'.

20. Character-value distributions. It should be possible to display the distribution of character values within a set of taxa.

21. Searching the character list. It should be possible to find text strings in the character list.

22. Searching the taxon names. It should be possible to search the taxon names and synonyms. The INTKEY command 'Find Taxa' searches the taxon names and any text characters which have been designated as containing synonymy or abbreviations.

23. Character illustrations. It should be possible to display illustrations of characters, and to select values from these illustration screens during identification.

24. Taxon illustrations. It should be possible to display illustrations, sounds, and videos of taxa.

25. Flexible display of illustrations. Illustrations of any size should be able to be scaled, scrolled, repositioned, and displayed simultaneously.

26. Text on illustrations. Text should be able to be superimposed on illustrations (instead of being built into them). This facilitates maintenance, ensures that text remains legible when an illustration is scaled, and allows the use of multiple languages with a single illustration.

27. Running without illustrations. It is desirable for a package containing illustrations to be usable without the illustrations. For example, there may not be enough disk space for the illustrations, or network access to the illustrations may be too slow.

28. Import/export DELTA format. The DELTA (DEscription Language for TAxonomy) data format (Dallwitz, 1980; Dallwitz and Paine, 1993 onwards) has been endorsed by the International Taxonomic Databases Working Group as a standard for interchange of descriptive data. It should be possible to create an interactive key from DELTA data, and to export DELTA data from the interactive key (or from the system that generates the key).

29. Links with description writing, key generation, and classification. It should be possible to generate conventional keys and publication-quality descriptions, and to carry out cladistic and phenetic analyses from the same data that are used to construct the identification system. High-quality descriptions require provision for text characters, and for qualifying comments associated with the recorded values of multistate and numeric characters.

30. Command files or macros. There should be a mechanism for storing and repeating a series of operations.

31. Log files. It should be possible to create a file showing the history (input and output) of a session.

32. Online help. The program should have complete, context-sensitive, built-in help.

33. External program text. The program text (commands, help, messages, etc.) should be external to the program, allowing easy creation and use of different language versions.

34. Unlimited field lengths. There should be no arbitrary limits on lengths of taxon names, text of characters and character notes, or text fields in taxon descriptions.

35. Unlimited data size. There should be no arbitrary limits on numbers of taxa, characters, or character states.

36. Reasonable memory requirements. Large data sets should be able to be handled in the amounts of memory commonly available.

37. Fast execution. Execution times for operations which are frequently used in identification should be reasonably small with large data sets.

38. Simple user interface. The user interface should be simple, efficient, and consistent, and should provide an easy transition from use of basic features to use of the full functionality.

INTKEY is suitable for users ranging from primary-school children to professional taxonomists who will find it useful for many aspects of taxonomic work, not just identification (Aiken *et al.*, 1997). There are three modes of operation: simplified, automatic, and manual. The mode is selected by the user from a startup screen. In simplified mode, only options for basic identification are available, and some buttons are omitted from the dialog boxes. If the 'tutorial' option of simplified mode is selected, instructions for basic operations are displayed when the program starts, and hints are displayed in various circumstances during operation of the program. Automatic mode is the same as simplified mode, except that all options are available. The user can work exactly as in simplified mode, and gradually experiment with other options. In manual mode, some operations that are carried out automatically in the other modes must be specifically requested by the user.

The author (or the user) of an INTKEY data set can define a toolbar containing buttons associated with any operation or sequence of operations. As well as providing convenient shortcuts, this facility enables the author to tailor the interface. Options not available through the menus in simplified mode can be provided via the toolbar. It is also possible to suppress most of the items in the menus, so that the interface is completely defined by the toolbar.

Construction of Interactive Keys

To fully realize the advantages of interactive keys, the data must have a considerable amount of redundancy. Otherwise, the user will have limited choice of characters, and the error-tolerance capability will not work well, if at all. In particular, if the information in the data matrix is obtained solely from a conventional key, then the characters which would be used to identify a given specimen with the conventional key will also have to be used to identify the specimen with the interactive key (see also White and Sandlant, Chapter 24, this volume).

INTKEY provides an easy method of checking the data matrix for redundancy: the command 'Diagnose All' is run with a high setting of the DiagLevel parameter, e.g. 5. The program generates diagnostic descriptions for all the taxa, and displays lists of those that are not separable at the requested level. These lists show the actual number of differences between each taxon and the taxon being diagnosed. It is easy to see, from this information, where the data need to be strengthened by filling in missing information or by adding extra characters. When extra characters are added, they need only be recorded for the poorly separated taxa.

Large conventional keys are difficult to construct, so hierarchies of smaller keys are mostly used instead. For example, an identification might be carried out by first using a key to families, then a key to the genera of a family, then a key to the species of a genus. As good interactive key programs can easily cope with many thousands of taxa, it is usually possible to eliminate some of these steps, for example, to construct a single key to all the species of a family. There are three advantages in doing so.

1. The efficiency of the key is not hampered by uneven distribution of species among the higher taxa. For example, consider 8 species belonging to 4 genera which contain 4, 2, 1, and 1 species. Using 2-state characters, the shortest key directly to species contains 3 steps for each species. The shortest key to genera contains 2 steps for each genus; keying these out to species requires 2, 1, 0, and 0 extra steps, respectively, giving an average of 3.25 steps per species.
2. Information loss in going from one key to another is avoided. For example, supposing an identification to genus level uses a character which varies between species of the genus. The possible identity of the specimen has then been restricted to a subset of the species of the genus. However, at the start of the key to the species, all species of the genus are once again in contention.
3. All characters used in a single interactive key are subject to the error-tolerance mechanism, whereas an error made in a higher-level key cannot be corrected once a lower-level key is in use.

One advantage of a hierarchical set of keys is that it can be entered at any level. This facility can easily be incorporated in a single interactive key by including 'characters' giving the higher taxa to which the species belong. This approach has the advantage that, because of error tolerance, a user's error in assigning a specimen to a certain high taxon does not inevitably lead to a wrong identification.

Similarly, including 'characters' for geographical distribution gives most of the benefit of special regional keys, plus the advantage conferred by error tolerance if the specimen is the first record of its taxon in a region.

Strategies for Users of Interactive Keys

The flexibility of interactive keys, which is one of their major advantages, can also lead to poor results if appropriate strategies for their use are not adopted.

The goal is to maximize the probability of obtaining a correct identification. This probability is equal to the product, over all the characters used, of the probabilities of correctly applying each character to the specimen. So the user should aim to choose characters for which there is a low probability of making an error, and which will tend to minimize the number of characters required to complete the identification.

The character reliabilities set by the author will give some guidance in choosing characters with a low error probability, and they influence the order in which the characters are ranked in the 'Best' list in INTKEY. However, the reliability is set for an average user and an average taxon. Even an inexperienced user can improve on this by evaluating how easily he is able to apply the character to the specimen at hand.

Minimization of the number of characters used is achieved by using characters which eliminate as many taxa as possible. For the user who is not familiar with the taxa, this means choosing characters near the top of the list of 'Best' characters. In INTKEY, more precise guidance is provided (except in 'Simplified' mode) by the 'separating power' of the characters. The user who is familiar with the taxa may be able to improve on this by recognizing characters for which the specimen exhibits a state which is unusual in the group of taxa still under consideration, but this is difficult during the later stages of an identification.

In conventional keys, if the user is uncertain which character state applies to the specimen there is usually no option but to guess. This is a poor strategy for use in interactive keys. Rather, the character should be avoided. Alternatively, if the character has more than two states it may be possible to reduce the possibility of error by selecting more than one state.

A strategy often used by people new to interactive keys, and worth mentioning so that it can be avoided, is to use each character, in the order in which they are stored in the program, until a result is obtained. This violates all of the above principles.

Finally, the result of any identification should be regarded as a hypothesis requiring confirmation. INTKEY provides several ways of doing this. The specimen can be compared with illustrations of its putative taxon, with a full description, or with a diagnostic description employing characters not already used in the identification. Alternatively, the error tolerance can be increased, bringing previously eliminated taxa back into contention, and the identification process continued. The 'Separate' command should be used in preference to 'Best' for this purpose, because it ranks characters according to their ability to select a particular, rather than an average, taxon from the rest.

References

Aiken, S.G., Dallwitz, M.J., McJannet, C.L. and Consaul, L.L. (1997) Biodiversity among *Festuca* (Poaceae) in North America: diagnostic evidence from DELTA and clustering programs, and an INTKEY package for interactive, illustrated identification and information retrieval. *Canadian Journal of Botany* 75, 1527–1555.

Dallwitz, M.J. (1974) A flexible computer program for generating identification keys. *Systematic Zoology* 23, 50–57.

Dallwitz, M.J. (1980) A general system for coding taxonomic descriptions. *Taxon* 29, 41–46.

Dallwitz, M.J. (1992) A comparison of matrix-based taxonomic identification systems with rule-based systems. In: Xiong, F.L. (ed.) *Proceedings of IFAC Workshop on Expert Systems in Agriculture.* International Academic Publishers, Beijing, pp. 215–218.

Dallwitz, M.J. (1993) DELTA and INTKEY. In: Fortuner, R. (ed.) *Advances in Computer Methods for Systematic Biology: Artificial Intelligence, Databases, Computer Vision.* The Johns Hopkins University Press, Baltimore, Maryland, pp. 287–296.

Dallwitz, M.J. (1994) Desirable attributes for interactive identification programs. URL http://www.keil.ukans.edu/archive/taxacom.html

Dallwitz, M.J. (1996 onwards) Programs for interactive identification and information retrieval. URL http://www.keil.ukans.edu/delta/

Dallwitz, M.J. and Paine, T.A. (1993 onwards) Definition of the DELTA format. URL http://www.keil.ukans.edu/delta/

Dallwitz, M.J., Paine, T.A. and Zurcher, E.J. (1993) *User's Guide to the DELTA System: a General System for Processing Taxonomic Descriptions.* 4th edn. CSIRO Division of Entomology, Canberra, 136 pp.

Dallwitz, M.J., Paine, T.A. and Zurcher, E.J. (1993 onwards) New features for the DELTA system. URL http://www.keil.ukans.edu/delta/

Edwards, M. and Morse, D.R. (1995) The potential for computer-aided identification in biodiversity research. *Trends in Ecology and Evolution* 10, 153–158.

Fermanian, T.W., Barkworth, M. and Liu, H. (1989) Trained and untrained individual's ability to identify morphological characters of immature grasses. *Agronomy Journal* 81, 918–922.

Morse, D.R., Tardivel, G.M. and Spicer, J.I. (1996) A comparison of the effectiveness of a dichotomous key and a multiaccess key to woodlice. Technical Report 14–96, Computing Laboratory, University of Kent at Canterbury, UK.

Pankhurst, R.J. (1991) *Practical Taxonomic Computing.* Cambridge University Press, Cambridge, 202 pp.

Pankhurst, R.J. and Aitchison, R.R. (1975) An on-line identification program. In:

Pankhurst, R.J. (ed.) *Biological Identification with Computers.* Academic Press, London, pp. 181–185.

Stucky, J.M. (1984) Comparison of 2 methods of identifying weed seedlings. *Weed Science* 32, 598–602.

Watson, L., Gibbs Russell, G.E. and Dallwitz, M.J. (1989) Grass genera of southern Africa: interactive identification and information retrieval from an automated data bank. *South African Journal of Botany* 55, 452–63.

Willcox, W.R. and Lapage, S.P. (1975) Methods used in a program for computer-aided identification of bacteria. In: Pankhurst, R.J. (ed.) *Biological Identification with Computers.* Academic Press, London, pp. 103–119.

20

Archiving Biodiversity: Information Technology Applied to Biodiversity Information Management

P.H. Schalk

ETI, University of Amsterdam, Mauritskade 61, 1092 AD Amsterdam, The Netherlands
Fax: +31 20 525 7238/E-mail: peter@eti.bio.uva.nl

Introduction

The end of this millennium is characterized by an oncoming or perhaps ongoing revolution in the field of information management and fast dissemination by means of the new electronic media. With the large scale introduction of personal computers in the consumer market, developments in information technology have received an enormous impulse. The possibilities of electronic media such as the compact disc and the Internet have literally erased traditional boundaries. Information processing and knowledge dissemination have found a new dimension and influenced science and society in an irreversible way.

The demand for readily accessible biological expertise is growing. In the scientific community and in society there is an increasing need for up-to-date information on biological diversity, stimulated significantly by developments in the context of the Convention on Biological Diversity. In the past, scientific data were mainly for scientists. Now the data and the information derived from them are needed by a far larger group of users for various purposes such as: pure or applied research, industrial processes, nature conservation and monitoring, biodiversity inventories, and policy making. In all these cases, the currency and validity of the information may be vital because important decisions may be based on it.

At the same time the size of the required information sets is growing: we know more, need to update more often, and need to check our data against a bigger variety of existing information. Last but not least, we are more in a hurry than ever before. There is an increasing pressure by society on science to produce the latest data fast and to provide ready access to it; both information providers and users encounter deadlines never heard of before. This calls for innovative instruments to archive, process and handle our information. Information technology (IT) offers new possibilities for this.

This papers deals with some general aspects of the development of these IT tools and the consequences for both creators and users. ETI's LINNAEUS II software and related support and networking activities are described as examples.

Information Technology Tools

Some scientific institutes, or in many cases very dedicated individuals in their employment, have devised IT tools to assists taxonomists, ecologists and policymakers to archive and manipulate biological data electronically. There are IT tools for specimen-based information, such as MUSE (Humphries, 1994), BRAHMS (Filer, 1994), and for species-based information sets, such as LINNAEUS II (Schalk, 1992; Schalk and Los, 1995) and DELTA (Dallwitz and Paine, 1986; Dallwitz *et al.*, Chapter 19, this volume). Some of these systems are 'smart' databases, such as FISHBASE (Froese and Pauly, 1996), computer-assisted identification tools, such as CABIKEY (White and Sandlant, Chapter 24, this volume) and PANKEY (Pankhurst, 1991), biogeographic information systems, such as ERIN (Slater and Noble, 1991), or document retrieving systems, such as ELADA 21 (Simard and Benmouffok, 1994). Some were developed with a broad taxonomic application in mind, such as the DELTA system for plants or LINNAEUS II for all taxa, while others have a far narrower scope but excel in their use for a particular group or application, such as taxon-based information sets, many of which can be found on the Internet, such as the Cichlid databases (Kullander, 1995). These are just some examples; there are many, many more systems available.

To date the exchange of data between these systems has not been standardized and is not always easy or even possible. There is now a trend to formulate standard data exchange formats (e.g. Nexus files for matrices) but no concrete formats have been defined for the various parts of the different IT tools. Perhaps more important than the standardization is the fact that the data are stored in a digital way. This allows combination and polling between systems, despite the fact that porting may need the assistance of a programmer.

The choice of an IT tool depends on the intended use of the data and the needs of the owner. It is important that some level of support is provided and that there is a continuous evolution of the tool to follow the developments in hardware (the processor in computer platforms) and compatibility with the computer's operating systems.

LINNAEUS II Software

Many taxonomists and ecologists now have the desire to store their data in a digital format, to allow easy manipulation of the data and to process large datasets with a view to later use. Furthermore, digital dissemination of the information is becoming more common. This 'electronic publishing' places another set of demands on the IT tools that are available, as it is no longer only the expert dealing with the software, but a potentially wide range of users. For such a publishing application the user-friendliness of the software is very important. Also the use of the IT tool from a CD-ROM or via the Internet sets special boundary conditions for the software that are different from the demands for a hard disk of a single computer.

ETI is a non-governmental organization in operational relations with UNESCO, with the mission to promote the availability, accessibility, quality and easy use of taxonomic, ecological and biodiversity-related information by applying state-of-the-art multipurpose IT tools. ETI has approached the development of an IT tool for biodiversity information from a broad and multi-disciplinary angle and has involved both the data providers and the users in the development process.

ETI developed the LINNAEUS II software package for biodiversity documentation and species identification. It is a multi-functional system that can be used by scientists on their own PCs, in one software shell, to create multimedia biological databases, computer-assisted identification tools, and geographic information systems for mapping distributions and biodiversity. Apart from these basic functional parts, supporting modules can easily be added such as glossaries, literature databases and methodological sections. There is a continuous development, in close contact with users, of new modules that can be added fairly simply, thereby creating optimal applicability and customization for specific purposes, while maintaining a standard approach to information storage and data exchange between different applications of the package. At the same time LINNAEUS II presents a versatile electronic publishing tool, allowing fast dissemination (publication) of large information systems on CD-ROM or the Internet. Figure 20.1 shows the modular structure of the LINNAEUS II software, including some screen dumps of the various sections to illustrate the interactive multimedia aspects. Detailed information on this software is provided on ETI's Internet site (ETI, 1996).

Responsibilities of the Developers

Developing a good IT tool is extremely time consuming and therefore expensive. Furthermore it is a never ending job. The tool may be finished in that it fulfils its requirements at a certain moment, but the environment in which it exists is in

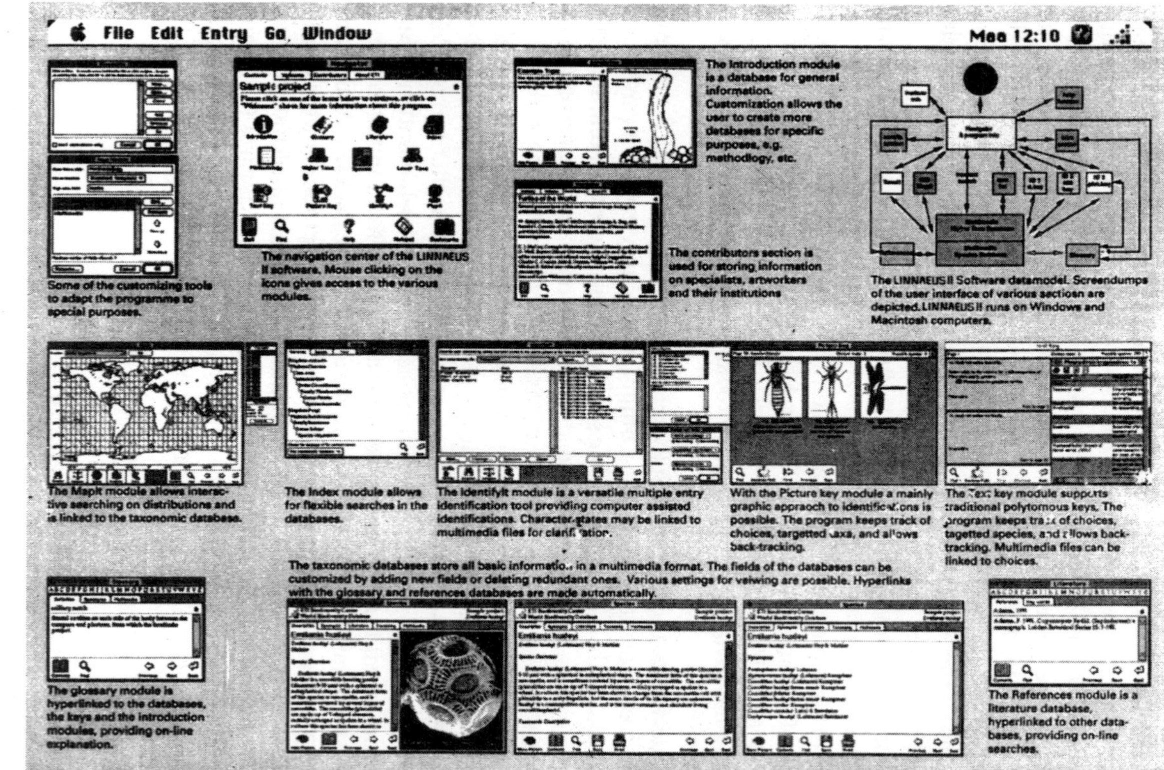

Fig. 20.1. The LINNAEUS II data model and some screen dumps of the user interfaces of the various functional modules it contains.

continuous and fast motion, forcing the developers to update their creations in order to run them on newer machines with faster processors or under changing or different operating systems.

It is important that the developers of these tools, and their employing institutions, realize their responsibilities once they start to distribute, free or for a charge, these instruments to scientists at large. In many cases the users will not have the expertise to understand what makes the tool tick. So once they start to use it they become partly or entirely dependent on the developer, who may not be in a position to offer sufficient user support or the necessary updates.

Up to now, only a few of the tools now available were part of a strategic decision of an institution to create these new instruments as part of their mission. Many of them were derived from the zeal of dedicated scientists, who often put more of their own time into the development than is realized. Often, the host institute does not fully support the IT developer in continuing developments, nor in offering user support. I would like to make a plea for a code of good practice amongst developers, taking into account the ISO standards for software development and taking on responsibilities towards the users. When an IT tool is offered to users outside the developer's own environment, the institutions should take on their responsibilities for support.

Many of the IT tools, some of which may well represent half a life time of hard work for the developer, are poorly known to potential users, due to lack of support of the institution behind the programmers. A central 'library' on an Internet site naming all different electronic instruments available, including a review and comparison between them, would be highly useful to users.

In most cases, the IT tools are being developed by staff of governmental or non-profit organizations. Is this parallel development and competition a waste of effort and a drain on capital resources, in addition to creating confusion among users? In a way it is, but in view of the fact that the IT field and its application is relatively new, these developments and the apparent competition may be considered good. Competition stimulates developers to produce state-of-the-art software products, and it pushes forward the frontiers of the possibilities for hardware and software. Since few tools are developed from a commercial point of view, the sharing of expertise, technical solutions and basic algorithms could be profitable. Perhaps the time has come to form a developers' union?

International Networks of Specialists

As indicated before, there are quite a few software packages available for specific tasks in the field of biological databases, computer-assisted identification tools and geographic information systems. Most of them are under-used, due to a lack of publicity or potential at the developer's institutions to set up user groups.

Because ETI's LINNAEUS II software development is part of its mission, ETI also undertakes to promote the tool and to set up international networks of users. These so called ETI-Partners receive the LINNAEUS II software and a user licence free of charge. Part of the service to the ETI–Partner networks, apart from the obvious technical support in the form of a help desk, is a digitization service and assistance in data entry in order to create complete systematic monographs. ETI also supplies an electronic publishing service (on CD-ROM) for completed information systems under the condition that the materials are peer reviewed prior to publishing in order to maintain quality. Short training courses in the use of LINNAEUS II, and in particular software packages for digitizing pictures and printed text as well as taxonomic standardization, are part of this support service.

The feedback from ETI's networks, which now include well over a thousand specialists, is an important driving force for improvements and new developments in LINNAEUS II. DELTA also has a large user group and CSIRO provides courses and support. The compiled (and compatible) knowledge and potential publication power of these international networks of collaborating scientists, each building their own information system, is new and characterizes the changes brought about by the introduction of IT to science.

A Global Effort: Towards a World Biodiversity Database

In view of the need for vast amounts of validated biological information for the documentation of the earth's biological diversity, and to build a sensible management system or policy on it, international cooperation is imperative.

ETI, with support of the Netherlands Government, started a World Biodiversity Database CD-ROM Series to contribute to making existing knowledge accessible in the context of the Convention on Biological Diversity. The international ETI–Partner networks provide taxonomic monographs including valid names, synonyms, descriptions, illustrations, computer-assisted identification keys and geographic information in a standard LINNAEUS II format. After a peer reviewing procedure these highly detailed reference works are electronically published on CD-ROM. There are now about 15 titles per year (Table 20.1). Each of these CD-ROMs forms a building brick of the World Biodiversity Database. Since all of the titles from this series appear in the same LINNAEUS II format, combining different datasets is relatively easy, allowing for customized datasets centred on a geographic region or higher taxon. In addition, a World Biodiversity Database Support CD-ROM Series started in 1997. Specimen databases, species catalogues, dictionaries, etc. will be published in this series.

Documenting roughly 1.7 million species in the World Biodiversity Database is a big task and all hands are needed to complete it and to deal with the constant changes. The ETI publishing programme is open to all scientists and institutions, including those who have developed their own IT tools and want to make their knowledge electronically available.

Table 20.1. Overview of some ETI CD-ROMs with LINNAEUS II type biological information and identification systems that are currently available or will be produced in the near future. All CD-ROMs are available in Macintosh and Windows formats.

Appeared
- Birds of Europe
- North Australian Sea Cucumbers
- Marine Planarians of the World.
- The BEG Expert System. Arbuscular Mycorrhizal Fungi. Demo CD-ROM
- Marine Lobsters of the World
- Fishes of the NE Atlantic and Mediterranean
- Five Kingdoms. A multimedia guide to the phyla of life
- Marine Mammals of the World
- Birds of Europe, 2nd revised and extended edition
- Etymological Dictionary of Grasses

Productions in 1997
- Pelagic Molluscs of the World
- Turtles of the World
- Bird Remains Identification System
- Turbellarians. A key to families and genera
- Bats of the Indian Subcontinent
- Ostracods of NW Europe
- Lemurs of Madagascar and the Comores
- Cultivation and Farming of Marine Plants
- Copepods: Aetaeidae of the World
- Braconidae of the World. A key to families

Special co-productions
- Preserve & Protect Demonstration CD-ROM (1996)
- Netherlands Indian Ocean Expedition (1995)
- Determinatiesysteem Libellenlarven (1996)
- Identification System Limnologic Fauna (1997)
- North Sea Tidal Data (1996)
- Weed Seedlings Identification Key for The Netherlands (1996)
- Interactieve Flora van Nederland (1997)
- The North Sea CD: A multimedia educational guide (1997)

References

Dallwitz, M.J. and Paine, T.A. (1986) *User's Guide to the DELTA System; a General System for Processing Taxonomic Descriptions.* 3rd edn. Report, Division of Entomology, CSIRO, 106 pp.

ETI (1996) *Expert Center for Taxonomic Identification.* ETI, Amsterdam, Netherlands. World Wide Web page at http://wwweti.eti.bio.uva.nl

Filer, D.L. (1994) *BRAHMS – Botanical Research And Herbarium Management System. A*

Pocket Introduction and Demonstration Guide. Oxford Forestry Institute, Oxford, UK, 28 pp.

Froese, R. and Pauly, D. (eds) (1996) *FishBase 96: Concepts, Design and Data Sources*. ICLARM, Manila, Philippines. 179 pp.

Humphries, J. (1994) *Muse (version 3.4). Tutorial and Reference Manual*. Cornell University, 80 pp. (http://muse.bio.cornell.edu/muse/manual.pdf)

Kullander, S.O. (1995) *Guide to the South American Cichlidae*. Internet site of the Swedish Museum of Natural History. (http://www.nrm.se/ve/pisces/acara/cichpage.html)

Pankhurst, R.J. (1991) *Practical Taxonomic Computing*. Cambridge University Press, Cambridge. 202 pp.

Schalk, P.H. (1992) Computer aided taxonomy. *Binary* 4, 124–126.

Schalk. P.H. and Los, W. (1995) Applications of interactive multimedia software in taxonomy and biodiversity studies. *Global Biodiversity* 7, 12–18.

Simard, R. and Benmouffok, D. (1994) *Elada 21. Agenda 21 Hits the Information Highway*. Ecodecision Fall 1994. Canadian Centre of Teledetection, Ottawa. 3 pp.

Slater, W.R. and Noble, S.J. (eds) (1991) *ERIN Program Brief, June 1991*. ANWPS, Canberra, 45 pp.

Development of Artificial Neural Networks for Identification

L. Boddy[1], C.W. Morris[2] and A. Morgan[1]

[1]School of Pure and Applied Biology, University of Wales, PO Box 915, Cardiff CF1 3TL, UK
Fax: +44 (0)1222 874305/E-mail: boddyl@cardiff.ac.uk
[2]Department of Computer Studies, University of Glamorgan, Pontypridd CF37 1DL, UK

Introduction

Relatively simple and rapid identification techniques for use by biologists not specialist in the taxonomic group of concern are becoming increasingly important. This reflects various demands including: (i) the continuing need for identification but decrease in available systematists; (ii) the need for rapid, 'on the spot' field identification; and (iii) need for identification of large numbers of organisms to provide ecological information. Inevitably these requirements imply the use of modern computer technology. Computerized keys have been produced (e.g. Pankhurst, 1991; Rose, 1993; White and Scott, 1994; other chapters in this volume), but these have many of the same problems inherent in the paper versions, including being sensitive to individual characters, and failing to give a correct identification if characters are missing or assessed incorrectly (Bridge *et al.*, 1994; Morris and Boddy, 1995). Expert systems represent a further possibility (e.g. Woolley and Stone, 1987; Brammer *et al.*, 1991; Morris and Boddy, 1992; Edwards and Morse, 1995). The major problem with these, however, is that it is usually very difficult and time-consuming to extract the relevant rules and then encode these rules in a formal manner.

Artificial neural networks (ANNs) are a further possibility, with great promise, since they can cope with partially contradictory 'fuzzy' data, and while it is not trivial to select and optimize ANNs, their implementation does not need a taxonomic expert beyond the original determinations of example patterns (pathogen species) upon which the system is to be trained (see below). So far they have been used largely only to discriminate a few taxa, for example, six

 Information Technology, Plant Pathology and Biodiversity (eds P. Bridge, P. Jeffries, D.R. Morse and P.R. Scott)

basidiomycetes based on flow cytometric measurements of spores (Morris *et al.*, 1992), five *Penicillium* species based on cultural characteristics (Bridge *et al.*, 1994) and 17 species of *Pestalotiopsis* and the closely related *Monochaetia karstenii* and *Truncatella truncata* from spore morphometric data (Boddy *et al.*, 1994a). That ANNs can be used successfully to discriminate between larger numbers of taxa is shown by recent work using flow cytometric fingerprints to identify over 40 phytoplankton taxa (Boddy *et al.*, 1994b). Current limitations of the technique lie in lack of availability of suitable data sets of characters, which taken together are discriminatory, for training the ANNs rather than in the ANN approach itself.

Artificial Neural Networks: Architecture, Training and Testing

ANNs were initially developed to mimic the brain. Like the brain they are composed of processing elements or nodes, analogous to neurons, each of which has a number of weighted inputs that are combined to give the internal activation level of the node (broadly analogous to dendrites and synapses). The outputs of nodes are passed to the inputs of other nodes. Individual nodes (usually of the order of 100 or less) are connected to form a neural network, as in the brain, but in a very regular pattern, commonly in layers – an input layer, one or more hidden layers and an output layer (Fig. 21.1).

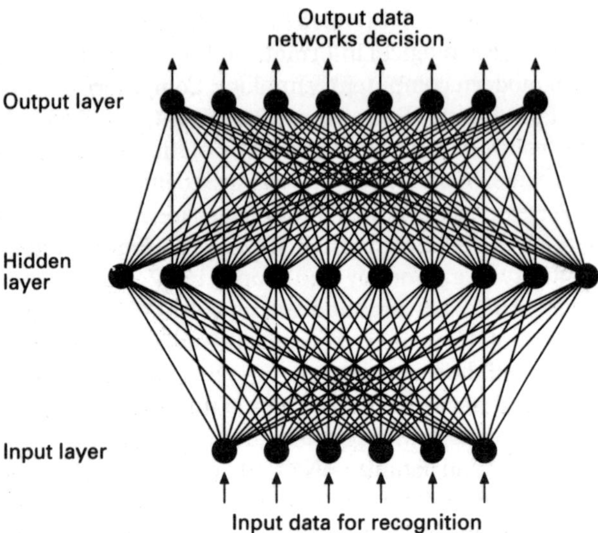

Fig. 21.1. Feedforward architecture common to many ANNs used for identification problems: there is one input node per character, and one output node per taxon upon which the network was trained.

There are a number of different ANN paradigms commonly used in identification problems, including the multilayer perceptron (MLP, also known as a backpropagation network), the learning vector quantization (LVQ) network, the radial basis function (RBF) network, and a powerful variant of the latter, the asymmetric RBF (ARBF) network. These ANNs all differ in the detailed functioning of the hidden and output layer nodes, and the method by which they 'learn' (see below) to represent training data.

Unlike most identification approaches ANNs are not rule-based, but rather 'learn/train' from patterns presented to them. Learning can take two distinct forms, termed supervised and unsupervised. With the former, a number of examples each consisting of a set of known inputs (one per character) and the expected output (i.e. known taxon for that set of characters) are presented to the network, and it learns to recognize the inputs and to classify as the correct output category (taxon) based on these 'known' patterns – hence the term supervised training, since essentially an outside teacher is telling the network the correct classification for a particular pattern of characters. In contrast, with unsupervised training the network is presented with input patterns only and itself 'discovers' similarities between these patterns, and forms classification groups (e.g. Kohonen, 1990; Wilkins *et al.*, 1994).

Supervised training is usually employed for identifying taxa. Essentially, discriminating taxa involves constructing decision boundaries between the example patterns of known taxa in n-dimensional space. With MLP ANNs each hidden layer node (HLN) represents a hyperplanar decision boundary, and the output layer nodes interpolate smoothly to produce arbitrary convex decision boundaries between taxa (Fig. 21.2A). The output node values can be interpreted as Bayesian *a posteriori* probabilities (Richard and Lippmann, 1991), and the node with the highest value indicates the taxon to which the input pattern belongs. When training MLPs an input pattern is presented to the network; values propagate through the network and at the output node are compared with the expected output, and the difference is computed. The connection weights for the output layer are then adjusted slightly to achieve a smaller difference between the actual output and expected output values. The connection weights in the hidden layer are then adjusted similarly. All training patterns are presented sequentially, and the whole process is repeated numerous times (commonly thousands) until the error is acceptably low.

With LVQ ANNs each HLN represents a 'class prototype', and they compete with each other to represent the input pattern. The node whose prototype is nearest to the pattern 'wins' and activates the output node representing that particular taxon, while others are ignored. The location of the class prototypes is adjusted iteratively during training; training patterns are presented randomly and the winning prototype from the correct class moved a little closer, until the distribution of prototypes models the underlying distribution of the training data for each taxon. The decision boundaries are 'piecewise linear', but of arbitrary shape and complexity depending upon the number and location of

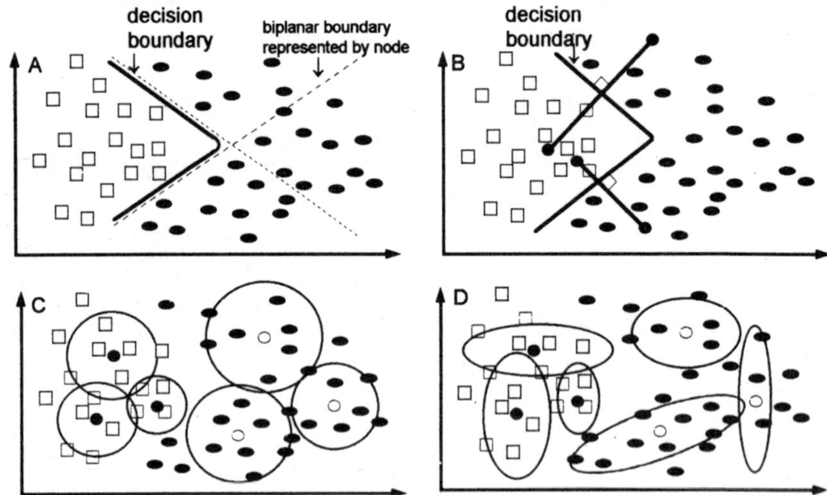

Fig. 21.2. Two-dimensional representation of how decision boundaries are formed by different ANN paradigms. (A) Hyperplanar decision boundaries produced by MLP ANNs. (B) Piecewise linear decision boundaries of arbitrary shape and complexity formed by LVQ ANNs. (C) RBF and (D) ARBF ANNs employ radially symmetric and ellipsoidal kernels respectively to form arbitrarily complex boundaries. Open squares and closed ellipses represent data patterns for two taxa. Circles represent prototypes or kernel centres. See text for further details.

protypes (Fig. 21.2B). Training again involves presentation of the training patterns to the network numerous times.

RBF networks are similar in construction to MLP networks, with the network output depending on the output of all HLNs, though the HLNs function differently to MLPs. Each HLN represents a kernel or basis function, the value of which depends on the distance between the input pattern and the kernel centre. Each kernel only responds strongly to patterns local to it, and there must therefore be sufficient kernels (and hence HLNs) to cover the n-dimensional space in which input patterns may occur. As with LVQ networks, the decision boundaries can be arbitrarily complex depending on the number, size and location of kernels (Fig. 21.2C). The kernels of ARBF networks employ a generalized distance measure termed the Mahalanobis distance metric rather than the Euclidean distance metric employed by RBF networks. This allows non-radially symmetric kernels which are elongated into ellipsoids, which should allow better discrimination of data from arbitrarily complex distributions (Fig. 21.2D). Training involves positioning the kernel centres (e.g. randomly), assigning the nodes among classes (equally) and then repositioning the kernel centres (e.g. using the LVQ algorithm) to ensure that kernel centres approximate the data distributions for each taxon. This is then followed by two systematic passes through the training data set to determine kernel widths and output layer weights.

When trained, network performance should be evaluated by presenting input patterns that the network was not trained upon. The overall percentage of test data patterns that were correctly classified indicates successful training and generalization. How well a network predicts identifications depends on whether the training data covered the range of variation in characters of each species and whether the character set together was sufficiently discriminatory, and depending on ANN type, the number of hidden layers, the number of HLNs and the number of training cycles (i.e. complete presentations of the training data set) amongst others. The last three are best determined by experiment.

Performance for individual taxa can be evaluated by constructing a 'misclassification matrix' (e.g. Table 21.1), which shows the proportion of test data patterns for each taxon that were given each possible identification. When networks have been optimized (in terms of numbers of HLNs, training cycles, etc.) consistent misidentifications of one taxon as another arise from overlap of character distributions. To achieve better discrimination, networks need to be retrained on patterns with different and/or additional characters. For example, an overall successful identification rate of 54% was obtained using optimized RBF networks with 72 HLNs constructed to discriminate 22 species of *Pestalotiopsis* plus *T. truncata* and *M. karstenii* (based on the three characters maximum spore length, maximum spore width and length of median cell). When using an additional six characters (length of each of the five apical appendages and length of basal appendage) there was 74% overall correct identification (Table 21.1). With a further four characters added (branching pattern of appendages, appendage shape (spatulate or not), spore shape (fusoid or clavate) and pigmentation), there was 90% overall success. Obviously, with an increase in characters, the overall misidentifications have decreased. Caution in choice of different/additional characters has, however, to be exercised. Although some of the species consistently misidentified as others in the ANN trained on nine characters were discriminated in the ANN with 13 characters, other new misidentifications arose in the latter, e.g. *P. foedens* was confused with *P. neglecta* on 40% of occasions, and *P. versicolor* was confused with *P. clavispora* on 24% of occasions. This is probably because the additional binary characters used have a stronger weighting within the ANN. This problem could be avoided by using a hierarchy of networks which discriminates taxa at different stages.

The usefulness of characters for discriminating between taxa can be quantified. Firstly, characters can be systematically removed individually or in combination from the training data set and the identification success determined. Secondly, characters can be removed from the test set and replaced with the mean value for all taxa. The former showed, for example, that omitting the character pigmentation resulted in only 56% successful identification, whereas omitting any other character singly resulted in no reduction in success. Using

Table 21.1. Misclassification matrix for a 72 hidden node ARBF network using 25 different patterns per species with nine inputs trained over two complete presentations of the training data set and tested using an 'unseen' test data set. Overall percentage of taxa correctly classified 74%.

	Species	1	2	3	4	5	6	7	8	9	10	11	12	13	14	15	16	17	18	19	20	21	22	23	24
1	*Pestalotiopsis adusta*	**92**		4												4									
2	*P. clavispora*		**52**		4	4			4	4			8		12							4		8	
3	*P. dichaeta*			**84**												8					4				4
4	*P. disseminata*				**48**	8				4		4						24	12						
5	*P. elaeidis*		4		20	**56**														8	4	8			
6	*P. foedens*				4	4	**44**			4	4							16		8	12			4	
7	*P. funerea*							**96**														4			
8	*P. gigas*								**92**													8			
9	*P. glandicola*		8		8	8	4			**52**			12							4		4			
10	*P. guepinii*		8		8	8	4	12		8	**32**							4				12		4	
11	*P. hughesii*											**100**													
12	*P. japonica*				4					16	4		**68**						4	4					
13	*Monochaetia karstenii*													**100**											
14	*P. leprogena*		12							12					**64**		12								
15	*P. mangiferae*	8		4									4			**84**									
16	*P. montellicodes*											4					**96**								
17	*P. neglecta*		8				4			4	4							**64**	12	4					
18	*P. palmarum*			8	4		8						4			4			**64**	8					
19	*P. psidii*		4	4							4		4							**80**					4
20	*P. royenae*		4		4																**92**				
21	*P. theae*		4		4																	**88**		4	
22	*Truncatella truncata*												4	16									**80**		
23	*P. versicolor*		8							4					12							4		**72**	
24	*P. westerdijiki*					4													4	4					**88**

the mean value for all taxa in place of a missing character, had no effect on identification success in about half of the cases, but sometimes resulted in 0% identification, especially if binary characters were involved. Clearly, ANNs often do not cope well if a character is missing from the pattern presented for identification. Thus, if characters are missing from taxa to be identified, identifications should be made using networks trained on a data set with that character omitted.

Choice of ANN Paradigm

It appears to be possible to train MLP, LVQ, RBF and ARBF ANNs to give near equal success in identification, there being little difference in overall percentage successful identification of 19 *Pestalotiopsis* species (A. Morgan, L. Boddy and C.W. Morris, unpublished) or of 12 phytoplankton species (Wilkins *et al.*, 1996) by these methods. Choice of paradigm must therefore be based on other criteria. While MLPs are still the most commonly used ANNs for identification they are not necessarily the most appropriate (Table 21.2). In biological identification, it is essential that taxa are identified correctly. If there is any uncertainty of the identification, estimates of the probability of correct identification must be available, and so should indication of the probability that the identity is something else.

Moreover it is essential that an ANN can indicate when a specimen to be identified is not of a taxon upon which the network has been trained. ANNs will try to assign any pattern to the closest category that it can whatever the pattern. Theoretically, RBF/ARBF ANNs should be able to reject novel taxa as unknown, since decision regions are finite, the HLN kernel output declining to zero with distance from the kernel centre. Rejection of unknown patterns can

Table 21.2. Advantages and disadvantages of different neural techniques.

	MLP[1]	RBF/ ARBF[2]	LVQ[3]
Tried and tested	yes		
Copes well with non-normal data	yes	yes	yes
Outputs can be interpreted as probabilities	yes	yes	
Forms arbitrarily complex decision boundaries		yes	yes
Trains rapidly		yes	
Easy to train; few parameters to adjust	yes		

[1]Multilayer perceptron.

[2]Radial basis function/ asymmetric radial basis function.

[3]Learning vector quantization.

be achieved by placing a threshold on the maximum HLN output, so that a pattern is rejected if the value falls below the threshold (i.e. a pattern is rejected as unknown if it is not sufficiently close to any of the basis function centres to cause any of the HLNs to produce sufficiently large output values) (Morris and Boddy, 1996). Constraints could also be applied to the output nodes such that, for example, there would be rejection as unknown if the winning node did not achieve a sufficiently high value. Both were successful in rejecting novel phytoplankton species (Morris and Boddy, 1996) and *Pestalotiopsis* spores (A. Morgan, L. Boddy and C.W. Morris, unpublished). To achieve complete rejection of individual cells/spores as unknown will, almost inevitably, cause rejections of some knowns as unknown; a balance between what is acceptable for incorrect rejection of knowns and correct reject of unknowns can be achieved by varying the threshold. Thus for *Pestalotiopsis* a rejection threshold anywhere between 0.4 and 0.8 on the HLN with the largest output might be a good compromise, giving 80 to nearly 100% rejection of novel taxa while wrongly rejecting 10–20% of known taxa (Fig. 21.3). Where many spores are involved, if a large number of them are rejected as novel or attributed a certain identification, then there is a much higher probability that this is correct than if identifications are being made on just one spore. In practice, where samples of spores from a single sporocarp are being used for identification, complete rejection of novel taxa, and much higher probability of correct identification of the species can often be achieved.

With MLPs the decision regions are often infinite in extent, so they are likely to be less good at rejection of novel taxa. With *Pestalotiopsis*, 70% rejection of novel taxa with 10% rejection of known taxa was achieved (A. Morgan, L. Boddy and C.W. Morris, unpublished). Further, MLPs take a long time to train if there are a large number of taxa, and sometimes do not train satisfactorily even though RBFs can (Morris and Boddy, 1996). Overall RBF/ARBF ANNs are currently probably the best option for biological identification problems.

Current Status and the Future: Pros and Cons

It is now quite evident that rapid reliable identification can be achieved by training ANNs on suitable data sets, and indeed these methods are at least as successful as multivariate statistical approaches (Wilkins *et al.*, 1996; A. Morgan, L. Boddy and C.W. Morris, unpublished). However, still relatively few taxa have been used for discrimination. Scaling up may not be trivial, but a variety of options are available. For example, hierarchies of networks could be employed, in a similar way to using a key first for families, then for genera and then for species, though there is not necessarily any *a priori* reason for sticking to traditional biological hierarchies, especially when the identification characteristics are not those conventionally employed by taxonomists, e.g. flow cytometric signatures. Also, when it is only necessary to know whether or not the organism

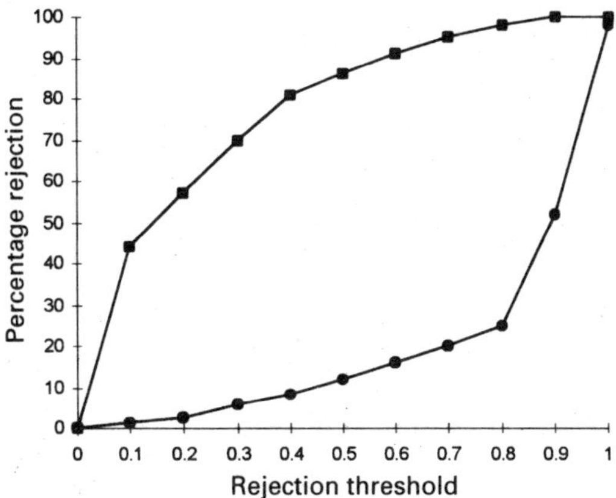

Fig. 21.3. Overall percentage rejection of 'known' taxa (i.e. the taxa upon which the network was trained; circles) and of 'unknown' taxa (i.e. taxa which were not used for training the network; squares) by an ARBF ANN, when employing a threshold on the largest hidden node output value. Twelve known and 12 unknown taxa were used.

in question belongs to a specific taxon, a network could be trained with two outputs – one for the taxon and one for everything that is not that taxon; ideally the training set would contain all the species with which that organism could be confused.

The major advantage of ANNs lies in the flexibility of characters that can be employed, and the rapidity with which they can predict identity. While all the examples presented here have concerned spore morphology data which might feature in any traditional key, ANNs can cope with many and varied types of data. For example, basidiospores (Morris *et al.*, 1992) and microalgae (e.g. Boddy *et al.*, 1994b) have been discriminated by ANN analysis of flow cytometry data. Bacteria have been discriminated from pyrolysis mass spectra (Goodacre *et al.*, 1994), whilst bacteria and fungi (Bridge *et al.*, 1994; A. Morgan, L. Boddy and C.W. Morris, unpublished) have been identified from cultural and biochemical characteristics, algae by image analysis (e.g. Thiel *et al.*, 1995), and bacterial genera by neural network analysis of aligned rRNA sequence data (e.g. Wu and Shivakumar, 1994). With these new data analysis procedures, there is no absolute necessity for using traditional morphological criteria for identification. Since trained neural networks make predictions rapidly, they could be used for automatic identification in real time in combination with automatic analytical equipment, e.g. flow cytometry, mass spectrometry, image analysers.

ANNs should not, however, be seen as a panacea for all identification problems. If there is no pattern or consistent relationship within the data they cannot be expected to discern one, and will not train successfully, although if a human can find a relationship then an ANN will usually be able to detect one. More importantly, ANNs can often find patterns when humans cannot!

A major limitation lies in the lack of suitable data sets for training. Since ANNs learn by example they must be presented with sufficient patterns to cover biological variation, with sufficient others left to test the trained network. Most taxonomic data are stored in the form of keys, which are a summary of significant characteristics rather than individual data patterns. Also, the information in keys refers to what are perceived as good taxonomic characters, yet there could be many more, which could certainly discriminate taxa well, when taken together.

One disadvantage of ANNs compared with dichotomous keys is that, very often, in the latter an identity is reached before all characters have been examined. With ANNs most of the characters upon which a network has been trained are needed. A way around this would be to use sets of networks employing different characters. Further, when confusion arises between two or more taxa, an expert system could indicate which additional characters would be most likely to discriminate between these taxa.

Perhaps the largest drawback for taxonomists is that ANNs do not exhibit transparent logic (i.e. it is not directly apparent how conclusions have been arrived at). However, ANNs can be interrogated to some extent. The importance of different characters in discriminating taxa can be investigated by systematically removing each in turn and retraining the network, and determine percentage successful identification (as illustrated above). Also, weightings can be examined to see which relationships are most significant, though this has not yet been done with biological identification ANNs.

In conclusion, ANNs are exciting systems with huge potential as identification and classification tools, though as yet we have only just begun to tap the potential.

References

Boddy, L., Gimblett, A.M., Morris, C.W. and Mordue, E.J.M. (1994a) Neural network analysis of fungal spore morphometric data for identification of species in the genus *Pestalotiopsis*. In: Dagli, C.H., Fernandez, B.R., Ghosh, J. and Kumara, R.T.S. (eds) *Intelligent Engineering Systems Through Artificial Neural Networks*. Vol. 4. ASME Press, New York, pp. 605–612.

Boddy, L., Morris, C.W., Wilkins, M.F., Tarran, G.A. and Burkill, P.H. (1994b) Neural network analysis of flow cytometric data for 40 marine phytoplankton species. *Cytometry* 15, 283–293.

Brammer, R.J., Bryant, T.N. and May, J.H.R. (1991) Investigation of an expert systems approach to bacterial identification. *Computer Application in Biological Sciences* 7, 461–469.

Bridge, P.D., Boddy, L. and Morris, C.W. (1994) Information resources for pest identification

– an overview of computer-aided approaches. In: Hawksworth, D.L. (ed.) *Identification and Characterization of Pest Organisms*. CAB International, Wallingford, UK, pp. 153–167.

Edwards, M. and Morse, D.R. (1995) The potential for computer-aided identification in biodiversity research. *Trends in Ecology and Evolution* 10, 153–158.

Goodacre, R., Neal, M.R., Kell, D.B., Greenham, L.W., Noble, W.C. and Harvey, R.G. (1994) Rapid identification using pyrolysis mass spectrometry and artificial neural networks of *Propionibacterium acnes* isolated from dogs. *Journal of Applied Bacteriology* 13, 157–160.

Kohonen, T. (1990) The self-organizing map. *Proceedings of the IEEE* 78, 1464–1480.

Morris, C.W. and Boddy, L. (1992) Intelligent computing in microbiology. *Society for General Microbiology Quarterly*, 70–72.

Morris, C.W. and Boddy, L. (1995) Artificial neural networks in identification and systematics of eukaryotic microorganisms. *Binary* 7, 70–76.

Morris, C.W. and Boddy, L. (1996) Classification as unknown by RBF networks: discriminating phytoplankton taxa from flow cytometry data. In: Dagli, C.H., Akay, M., Chen, C.L.P., Fernandez, B.R. and Ghosh, J. (eds) *Intelligent Engineering Systems Through Artificial Neural Networks*. Vol. 6. ASME Press, New York, pp. 629–634.

Morris, C.W., Boddy, L. and Allman, R. (1992) Identification of basidiomycete spores by neural network analysis of flow cytometry data. *Mycological Research* 96, 697–701.

Pankhurst, R. J. (1991) *Practical Taxonomic Computing*. Cambridge University Press, Cambridge, xi + 202 pp.

Richard, M.D. and Lippman, R.P. (1991) Neural network classifiers estimate Bayesian a posteriori probabilities. *Neural Computing* 3, 461–483.

Rose, D.A. (1993) ROTTERS – An expert key for the identification of wood-rotting fungi in culture. *Binary* 5, 9–12.

Thiel, S.U., Wiltshire, R.J. and Davies, C.J. (1995) Automated object recognition of blue green algae for measuring water quality – a preliminary study. *Water Research* 29, 2398–2404.

White, I.M. and Scott, P.R. (1994) Computer information resources for pest identification: a review. In: Hawksworth, D.L. (ed.) *Identification and Characterization of Pest Organisms*. CAB International, Wallingford, UK, pp. 129–137.

Wilkins, M.F., Boddy, L. and Morris, C.W. (1994) Kohonen maps and learning vector quantization neural networks for analysis of multivariate biological data. *Binary* 6, 64–72.

Wilkins, M.F., Boddy, L., Morris, C.W. and Jonker, R. (1996) A comparison of some neural and non-neural methods for identification of phytoplankton from flow cytometry data. *CABIOS* 12, 9–18.

Woolley, J.B. and Stone, N.D. (1987) Application of artificial intelligence to systematics: SYSTEX – a prototype expert system for species identification. *Systematic Zoology* 36, 248–267.

Wu, C. and Shivakumar, C. (1994) Back-propagation and counter-propagation neural networks for phylogenetic classification of ribosomal RNA sequences. *Nucleic Acids Research* 22, 4201–4299.

22

Mixing Elements from Different Identification Systems

P. Bridge

International Mycological Institute, Bakeham Lane, Egham, Surrey TW20 9TY, UK
Fax: +44 (0)1784 470909/E-mail: p.bridge@cabi.org

Introduction

The identification process, whether computer-assisted or manual, must involve a final decision making step. This final step is arrived at either through a series of earlier decisions, as in keys and expert systems, or as a result of a numerical value in probability and identification score-based schemes. Many manual schemes such as some dichotomous keys cannot fail to give a final identification, whereas numerical systems generally employ some form of critical cut-off value in order to validate the final identification. Increasingly, computer-assisted identification systems will use combinations of elements from keys, probability scores or profile matching to make a final decision on the validity or likelihood of the final identification. This process may not only validate the identification but may also be able to provide insights into the reasons for a poor or unsatisfactory identification.

Any identification system will always be placed under limitations imposed by the quality of the reference data, particularly in biology where the reference data must be sufficiently comprehensive in order to reflect the natural variablity within a taxon. All identification systems are subject to this shortcoming, and numerical systems that rely on a critical cut-off level can be particularly sensitive. Different identification coefficients will react differently to aberrant or atypical data, and combining the results from more than one coefficient can result in a more 'robust' scheme.

Results obtained from different numerical coefficients can be correlated to give further information on the validity of an identification. The relationship of the numerical schemes to both profile matching and key character approaches will also be considered. The combination of these different elements can be used to provide a single robust identification program which will be demonstrated with a frequency matrix of taxonomic characters for a defined group of fungi.

The purpose of identification is to compare an unknown against reference items and, if possible, to match the unknown with one or more of these. In biological systems, the most common application of identification is the matching of newly isolated organisms with existing genera and species.

In most computer-aided biological identification systems, the identification is arrived at through a series of steps which involve the application of existing knowledge, with or without numerical calculations. The final result may then be given as one of four outcomes; the unknown may be correctly identified, the unknown may be correctly identified but considered atypical, the unknown may be incorrectly identified, or the unknown may fail to identify (Bridge *et al.*, 1994).

Different numerical identification strategies all contain some method of determining which category an unknown will be placed in, and this is often based on either an identification score exceeding a threshold value or some function derived from the identification coefficient. In an ideal system all unknowns would be either identified correctly or fail to identify and results would be clear-cut; however, in biological systems variability in identification features, or their interpretation, can lead to unknown organisms being identified as atypical. This then brings about the need for a threshold value that can be used to reliably differentiate between an atypical correct identification and an incorrect identification.

The performance of any identification coefficient is dependent on the type and quality of the reference data. If the reference taxa are well defined and very distinct from each other, then atypical isolates are both less likely to occur and unlikely to lead to misidentification. If, alternatively, the reference data consists of a number of very similar poorly defined groups, then the identification performance will be reduced and atypical isolates are more likely to be misidentified (Sneath and Sokal, 1973). In reality, biological systems often consist of a mixture of groups that vary in both the level of their definition and their similarity to each other, and so reference databases generally contain a mixture of elements from both extremes. This variation is generally included in the data by representing the reference data as a frequency matrix. Computer-assisted numerical identification systems should, as a prerequisite, be expected to correctly identify entirely typical representatives of the reference data. However, the identification of atypical isolates, and the detection of misidentifications, will vary between systems. This contribution will consider how threshold values and comparisons between different coefficients can be used to support the identification of atypical organisms from percentage frequency databases.

Numerical Values

The simplest numerical value for an identification measure is the number or proportion of discrepant characters. This has been widely used in profile matching systems, where a specific number of mismatches is permitted in making an identification (e.g. Barnett *et al.*, 1990). The significance of any individual mismatch will depend upon the frequency of the character in the reference data set, for example a discrepancy of a negative test result in an unknown organism when compared to a value of 99% positive for that character in a reference database is more significant than a discrepancy where the unknown is negative and the reference value is 51% positive. It can therefore be useful to set identification cut-offs related to the frequency values and their significance when determining discrepant characters.

Although a considerable number of numerical coefficients have been suggested for computer-aided identification systems with frequency matrices (see Sneath and Sokal 1973; Sneath, 1979), in practice three types have been most commonly used. These are the normalized probability measures based on Bayes' theorem (Willcox *et al.*, 1973), the modal likelihood coefficient (Dybowski and Franklin, 1968), and distance measures derived from the taxonomic or Euclidean distance (Sneath and Sokal, 1973; Sneath, 1979). Each of these will behave differently with different reference datasets and as a result different cut-off or threshold values of the coefficients have been suggested at each application. The range of results around these threshold values, and comparisons made between two or more coefficients, can be used to validate or dismiss individual identifications.

Probability-based measures

Probability-based measures are used to provide a measure of the likelihood that an unknown belongs to any given taxon in the reference dataset. They are based on Bayes' theorem and give an absolute value which will decrease as the number of characters considered increases. This is due to the practice of representing all-positive or all-negative values in the frequency matrix as 0.99 and 0.01, reflecting a limitation of Bayes's theorem that if any individual value is zero, the final probability will also be zero and so an organism aberrant for that character could never be correctly identified. This problem does not apply if Laplace's law of succession is followed for sampled data, and values approaching 0.01 or 0.99 would be expected for samples of around 100 records (Willcox *et al.*, 1973).

Bayes' theorem as adapted by Willcox *et al.* (1973) can be represented as:

$$P(t_1 \mid R) = \frac{P(t_1)\,P(R \mid t_1)}{\Sigma_1 P(t_1) P(R \mid t_1)} \tag{22.1}$$

Where $P(t_i | R)$ is the probability that an organism giving the test result R is a member of taxon t_i, and where $P(t_i)$ is a prior probability for that taxon, and $P(R | t_i)$ is the probability that a member of the taxon will give test results R.

Bayes' theorem in its full form allows for the inclusion of prior probabilities, and so can be adjusted to take into account taxa that are more likely to show variation. Again in practice the difficulty in calculating prior probabilities has meant that they have not been widely used (Willcox *et al.*, 1980), although the commercially produced API microbial identification kits include prior probabilities in their calculation of an identification score (bioMérieux, 1993).

The decrease in the absolute probability with the number of characters leads to very small likelihood values from large databases and so it is usual to represent the final identification score as a normalized score, that is the value for the most likely taxon divided by the sum of the likelihoods to all taxa. When this likelihood coefficient was first introduced in bacteriological studies, a threshold level of 0.999 was suggested for definite identification (Willcox *et al.*, 1973). Since then a number of different threshold scores have been proposed between 0.999 and 0.85 for different data sets (Williams *et al.*, 1983; Priest and Alexander, 1988; Bridge *et al.*, 1992). These thresholds have been suggested after considerable screening of new and old isolates and testing of each identification system, and it seems that in practical circumstances the final threshold value will depend on the number of taxa in the database, the variability of the characters included, and the underlying similarity between taxa. This last feature is a very important consideration in the use of normalized probabilities, because if two or more taxa are very similar then the scores of an unknown to these taxa may be very similar and so the normalized value will be considerably reduced. Another feature of normalized probabilities is that if an unknown is very unlikely to belong to one taxon, but is significantly even less likely to belong to any other, then the resulting normalized value can in fact become large.

Modal likelihood fraction

The modal likelihood fraction was proposed by Dybowski and Franklin (1968) and has been used in both the API microbial systems (Boeufgras *et al.*, 1987) and the PENIMAT fungal identification system (Bridge *et al.*, 1992). Essentially, this coefficient is a fraction derived from the likelihood that an unknown belongs to the taxon, divided by the likelihood that would be obtained for the most representative member of the taxon. Therefore, if the unknown is very typical, then the MLF will approach 1, whereas atypical organisms will give increasingly lower scores. In a frequency matrix where 0.99/0.01 have been used, a single complete discrepancy will reduce the MLF by a factor of approximately 0.01, and a truly variable reference character will affect the score by a factor of approximately 0.5. As a result, in a large dataset the MLF value can rapidly become very small if there are a significant number of discrepant or variable characters.

Taxonomic distance

The taxonomic distance is a measure of the distance of an unknown organism from the centre of a taxon as defined as a hypersphere in multidimensional Euclidean space, that is where each dimension is representative of a character and is added at right angles to the overall model (Sneath and Sokal, 1973; Gyllenberg and Niemelä, 1975). In this representation the members of a taxon are a cluster of points, defined by the reference tests, and the identification is made by determining how far an unknown is from the centre of each of these groups when placed in the same representation. Most distance measures assume Euclidean space and so are based on Pythagoras' theorem to give the overall Euclidean distance as:

$$D= [\Sigma_{1-n}(x_1 - x'_1)^2+(x_2 - x'_2)^2 \ldots\ldots(x_n - x'_n)^2]^{1/2} \qquad (22.2)$$

This value is then generally scaled according to the number of tests to give an average or taxonomic distance

$$d = \sqrt{}D^2/n \qquad (22.3)$$

where n = number of characters. One problem with distance coefficients is in their comparison between groups. The reference group is defined in hyperspace on the basis of character values, and a group where the character values for individual members vary will therefore be described by a larger hypersphere than a group that contains very little variation. Distance measures can be scaled by considering the information already contained in the reference data, and one way in which this has been achieved is to calculate the standard error score of the distance for the unknown in comparison with the reference data (Sneath, 1979). Similarly, character correlations may lead to elongation or other distortion of the hypersphere, and these, if they occur, can have a significant effect on the interpretation of distance values.

Comparison of coefficients

Comparisons of the performance of different identification coefficients have been published by several authors (see Bridge *et al.*, 1994). All agree on general trends in coefficient values for good identifications, and all have considered the relative merits of the different coefficients.

Fungal Data

The performance of three identification coefficients, a normalized likelihood measure, the modal likelihood fraction, and a scaled form of the average distance were evaluated with two different fungal data sets consisting of percent positive frequency data. One data set comprised 57 physiological, biochemical and morphological characters for 37 groups of terverticillate penicillia, published in the package PENIMAT (Bridge *et al.*, 1989, 1992), and the second data set comprised 24 biochemical characters for 63 species of yeasts, published as part of the API32C identification package (bioMérieux, 1993). The two reference data sets therefore consisted of between 1500 and 1900 entries, one with an excess of characters to groups (penicillia) and one with an excess of groups to characters (yeasts). Both data sets contain variable values, that is frequency values other than 99 and 1, the relative proportions of these differed and were 0.32 of entries in the penicillia data set and 0.38 of entries in the yeast data set.

The two data sets were 'anticlustered' to check the independence of characters. The anticlustering involved determining similarities between characters on the basis of the reference organisms' response and then clustering the resulting matrix. Independent characters would be expected to appear ungrouped, whereas grouped or clustered characters could indicate linkage or dependence between characters (P.H.A. Sneath and M.J. Sackin, personal communication). In the penicillia data, two characters (both secondary metabolites) appeared linked, whereas with the yeast data, two of the carbon assimilation characters appeared linked (Figs 22.1 and 22.2). Both data sets were tested with up to 32 known appropriate reference organisms, and the absolute and normalized probability, the modal likelihood fraction, the taxonomic distance and the number of discrepant characters were recorded.

Discrepant characters

Cut-off values for discrepant tests were set at 20% and 80%. Therefore, a positive response for the unknown was only considered discrepant if the property was shown by less than 80% of the reference taxon, and a negative response was only considered discrepant if the feature was recorded for more than 20% of the reference group. Under these conditions, all of the unknowns identified showed at least one discrepant character. With the yeast data set the number of discrepant characters ranged from 1 to 4, whereas with the penicillia data set the number of discrepant characters ranged from 3 to 11. These results however represented similar ranges for the two data sets, being 4–17% of the number of characters in the yeast set, and 5–19% of the characters in the penicillia set.

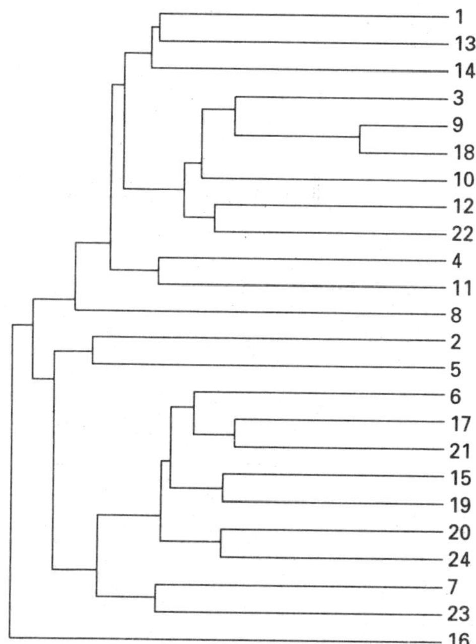

Fig. 22.1. Dendrogram of the 24 characters contained in the yeast data set. The anticlustering was performed by calculating the differences between characters with the Average Distance Coefficient, and clustering these by UPGMA average linkage.

Probability measures

The absolute probability values obtained for the identifications were very low. They ranged from 0.0508 to 0.0010 for the yeast data, and from 4.5×10^{-6} to 1.58×10^{-18} for the penicillia data. This difference is due entirely to the different numbers of characters in the two data sets, as the absolute value decreases as the number of characters increases. The wide range of values obtained for each data set preclude the use of an absolute value as a threshold score for identification.

The normalized values were, however, very similar for the two data sets and the two sets of unknowns. The normalized values ranged from 1 to 0.694 for the yeast data and from 1 to 0.618 for the penicillia data. Twenty two per cent of the identifications with the yeast data gave normalized scores of below 0.9, as compared to 33% with the penicillia data. The lower identification scores were generally obtained against taxa which showed a higher proportion of variable values and the lowest scores for the yeast data and the penicillia data were to groups where the proportions of variable characters were 0.58 and 0.47

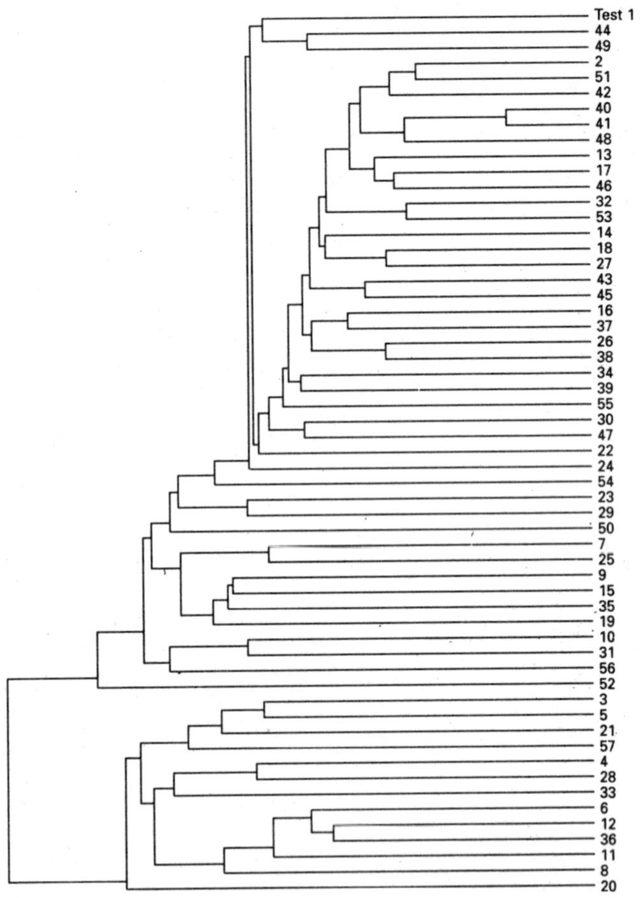

Fig. 22.2. Dendrogram of the 57 characters contained in the penicillia data set. The anticlustering was performed by calculating the differences between characters with the Average Distance Coefficient, and clustering these by UPGMA average linkage.

respectively. These contrast to the overall proportions of variable characters of 0.38 and 0.32 for the complete data sets.

Modal likelihood fraction

The modal likelihood fraction values varied considerably both within and between the two data sets. MLF scores for the yeast data ranged from 0.897 to 1×10^{-4} and from 7.5×10^{-3} to 2.7×10^{-15} for the penicillia data. As with the absolute probability values, the wide range of values obtained and the discrepancy between data sets would preclude the use of the MLF as a threshold level.

Taxonomic distance

The taxonomic distance values were calculated as average values, scaled for the number of characters. The distance values ranged from 0.194 to 0.318 for the yeast data, and from 0.2728 to 0.8138 for the penicillia data. These values would not be suitable for threshold values for the same reasons as the MLF and absolute probability values.

Mixing elements

Sneath (1979) demonstrated some of the effects of different identification coefficients and measures, and warned that spurious values may arise under different conditions with different measures. In practice, the yeast data set of this study is generally used with two cut-offs, an identification score based on the normalized likelihood, and a corrected form of the MLF; in addition, the ratio of the highest probability to the next highest is also used to aid in interpretation (bioMérieux, 1993). The penicillia data set has similarly been published with an identification system based on the normalized probability, the MLF and the proportion of discrepant tests (Bridge *et al.*, 1992). However, one result of the small trial performed here is that the MLF value is very sensitive to the data set used, and a correction to allow for the number of variable tests included is required. Such a correction factor is included in the yeast API system (Boeufgras *et al.*, 1987) but not in the penicillia system and so a direct comparison of performance cannot be made here. The MLF value appears to provide a reasonable measure of the number of tests against the identification in larger data sets. Figure 22.3 shows a graph of the number of discrepant tests for each identification plotted against the log MLF value for that identification. It is possible to fit a curve between the two sets of values, but this is a line of best fit and the deviation from it can be large for some individual values. The MLF would be expected to decrease by a factor of 0.01 for each discrepant character if all reference values were all positive or all negative. However, the inclusion of variable values in the data set prevents this occurring. The proportion of variable values in each reference group will therefore have a significant effect on this score.

The normalized probability was described by Sneath (1979) as related to taxonomic distance, the final value being the proportion of the sum of the antilogarithms of the distances from the centroids of the reference groups. However, the value obtained for the normalized probability presupposes that the unknown belongs to one of the reference groups (Willcox *et al.*, 1973; Willcox *et al.*, 1980). The direct calculation of the taxonomic distance is not subject to this condition, because if the unknown is not represented in the data set then all of the distances will be large (Sneath and Sokal, 1973). However, the

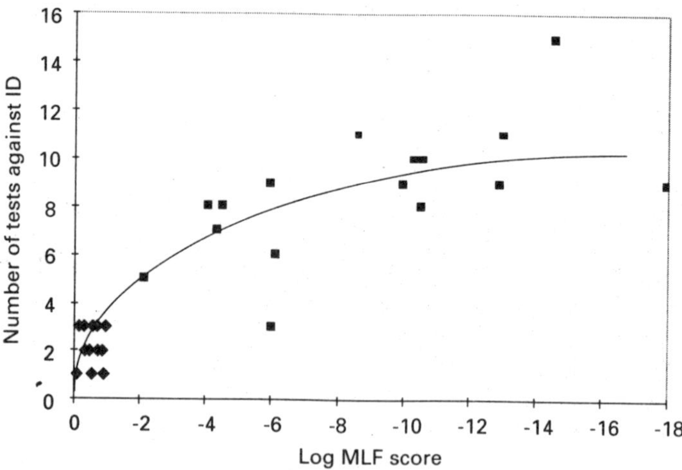

Fig. 22.3. Graph of the number of discrepant tests for each identification against the log MLF value for that identification.

distribution of the members of the reference group in hyperspace will not be consistent for every reference group and so the radii, and possibly the shape of reference groups, will vary. Sneath suggested using the standard error score of the distance, and assumed each reference group was distributed as a Gaussian hyperspherical swarm (Sneath, 1979). The warning that distance values may be affected by linked or correlated characters does not appear to be of major concern with these data sets as the anticlustering showed only one, possibly correlated, pair of characters in each data set. The problem that is encountered, however, when determining the distribution of the organisms in a reference group is that of sample size, and how representative this is of the true distribution. As is often the case, the penicillia data used here consist of a number of commonly and less commonly isolated groups, and this is reflected in the numbers of representatives available to construct reference data. However, no information is available on the numbers included in the yeast reference groups.

An alternative approach is to derive an estimate of the size of the reference group from the number of variable characters, that is taking the assumption that a reference group containing a large number of variable characters will occupy a larger area of hyperspace. The true taxonomic distance value can then be scaled from the proportion of variable characters. The calculation of the taxonomic distance depends upon the sum of the differences in character values, which for a truly variable character would be 0.5. In the calculation of distance this would become 0.5^2 for each variable character. However, this difference may be towards or against an identification and so the element can be reduced by half to represent this. This then gives a modified version of the taxonomic distance of

$$d(scaled) = \sqrt{((D^2-((0.5^2)/2)*v)/n)} \quad (22.4)$$

where v = number of characters described as variable for that reference group. This adjustment appears to work well with the two data sets used here. The 20% and 80% cut-off values that were used in determining discrepant characters were used to identify the 'truly' variable characters. The adjustment to the calculation resulted in a greater reduction to the distance value for groups based on larger numbers of variable characters. As a result, the range of distance values for the penicillia data was found to be 0.2193 to 0.463 after adjustment, whereas the adjusted values for the yeast data varied from 0.185 to 0.258. Although these ranges are substantially different, they do allow an estimate to be made of the relative group sizes in hyperspace, particularly in respect of the different sizes of the data sets.

Conclusions

It would appear from the data sets analysed here that there is no single coefficient or combination of coefficients that can be used to provide a universal 'threshold' value for identification. Although the overall trend for each approach and coefficient is the same for each data set, the suitable threshold values will be different for each. This is particularly significant in that, in most cases, the user of a computer-aided identification system is generally not the person who has constructed it, so users will often require a clear interpretation to be made for them. Sneath suggested some general trends and conditions for a series of identification measures with a single data set. His findings, coupled with those obtained here, can be incorporated into a computer-based system to provide a level of general interpretation of identifications (Sneath, 1979). It is clear from the two fungal data sets used here that the actual threshold values will differ with the data set used. As a result it is necessary to test new data sets to determine appropriate threshold values for inclusion in these qualifications and interpretations.

1. The limited trial here suggests that a reasonable score of better than about 0.7 on the normalized probability may be sufficient for fungal data, provided that this is supported by an appropriate score with one of the other coefficients, such as a scaled distance.

2. There is, as mentioned earlier, a relationship between distance measures and probabilities. However, the distance measures can be used as a check against spurious identifications for unknowns that represent groups not included in the data set. In these conditions, it is possible to obtain a spuriously high normalized value, but this would be countered by a large distance measure. These circumstances can be included in the identification programming to provide a qualifying condition for the identification.

3. The proportion of very variable values in the reference data can also affect the quality of the identification, as was seen with the scaling of the distance measure and to a lesser extent with the MLF value. The values of these coefficients can be used to provide further qualifications.

4. The increase in tests against an identification will rapidly decrease the MLF value and so a good score on the normalized value, an acceptable distance value but a low MLF value may indicate a correct identification of an atypical unknown. This circumstance can be compared with **2** to provide a suggestion as to whether the final identification is spurious, as compared to an atypical unknown.

References

Barnett, J.A., Payne, R.W. and Yarrow, D. (1990) *Yeast Identification PC Program*. Barnett, Norwich.

bioMérieux (1993) *ID32C Analytical Profile Index*. BioMérieux, Marcy-l'Etoile.

Boeufgras, J.M., Balzer, J.L., Allard, F. and Diaz, I. (1987) *A New Computer Program for Routine Interpretation of API Systems.* 2nd Conference on Taxonomy and Automatic Identification of Bacteria, Prague.

Bridge, P.D., Hawksworth, D.L., Kozakiewicz, Z., Onions, A.H.S., Paterson, R.R.M. and Sackin, M.J. (1989) A reappraisal of terverticillate penicillia using biochemical, physiological and morphological features. II. Identification. *Journal of General Microbiology* 135, 2941–2966.

Bridge, P.D., Kozakiewicz, Z. and Paterson, R.R.M. (1992) PENIMAT: a computer assisted identification scheme for terverticillate *Penicillium* isolates. *Mycological Papers* 165, 1–59.

Bridge, P.D., Boddy, L. and Morris, C.W. (1994) Information resources for pest identification: an overview of computer-aided approaches. In: Hawksworth, D.L. (ed.) *The Identification and Characterization of Pest Organisms.* CAB International, Wallingford, pp. 153–167.

Dybowski, W. and Franklin, D.A. (1968) Conditional probability and identification of bacteria: a pilot study. *Journal of General Microbiology* 54, 215–229.

Gyllenberg, H.G. and Niemelä, T.K. (1975) New approaches to automatic identification of micro-organisms. In: Pankhurst, R.J. (ed.) *Biological Identification with Computers.* Systematics Association Special Volume No.7, Academic Press. London, pp. 121–136.

Priest, F.G. and Alexander, B. (1988) A frequency matrix for probabilistic identification of some bacilli. *Journal of General Microbiology* 134, 3011–3018.

Sneath, P.H.A. (1979) BASIC program for identification of an unknown with presence–absence data against an identification matrix of percent positive characters. *Computers and Geosciences* 5, 195–213.

Sneath, P.H.A. and Sokal, R.R. (1973) *Numerical Taxonomy*. W.H. Freeman, San Francisco, 573 pp.

Willcox, W.R., Lapage,S.P., Bascomb, S. and Curtis, M.A. (1973) Identification of bacteria by computer: theory and programming. *Journal of General Microbiology* 77, 317–330.

Willcox, W.R., Lapage, S.P. and Holmes, B. (1980) A review of numerical methods in bacterial identification. *Antonie van Leeuwenhoek* 46, 233–299.

Williams, S.T., Goodfellow, M., Wellington, E.M.H., Vickers, J.C., Alderson, G., Sneath, P.H.A., Sackin, M.J. and Mortimer, A.M. (1983) A probability matrix for identification of some streptomycetes. *Journal of General Microbiology* 129, 1815–1830.

23

The Role of the User in Computer-based Species Identification

G.M. Tardivel and D.R. Morse

Computing Laboratory, University of Kent at Canterbury, Canterbury, Kent CT2 7NF, UK
Fax: +44 (0)1227 762811/E-mail: g.m.tardivel@ukc.ac.uk

Introduction

Species identification, particularly by experts, is a difficult and often frustrating task. As well as the dichotomous key familiar to most ecologists, taxonomists have developed a range of tools and techniques aimed at easing the task of identification. Increasingly, these tools have used computers either in the production of the key or in the key itself (see Edwards and Morse (1995), Pankhurst (1991) and other papers in this volume for reviews).

While the theory and implementation of computer-based species identification tools is well developed in a number of areas, much less is known about how these tools perform in practice (Edwards and Morse, 1995). Exceptions to this include Stucky (1984) and Wright *et al.* (1995) who described evaluation experiments where volunteers were asked to identify species using two identification tools. Both studies compared a novel identification tool (a hypertext key in the case of Wright *et al.* and a polyclave in Stucky's) to a dichotomous key. They both found that the novel identification tool was preferred over the dichotomous key and that there was no difference in the frequency of correct identifications. Most worrying is that both studies reported a high frequency of misidentifications, reaching 30% in Stucky's (1984) study. Whilst this figure lumps all forms of misidentification, it is high and suggests that there is much work to be done in making the identification process easier and more accurate.

In addition to computer-based implementations of dichotomous keys, the multi-access key is a popular form of computer-based identification tool (Pankhurst, 1991; Legg 1992a, b; Dallwitz *et al.*, Chapter 19, this volume;

White and Sandlant, Chapter 24, this volume). Multi-access keys potentially have at least two advantages over a dichotomous key:

1. The most obvious characters will tend to be selected first and the user is more likely to be able to identify such characters correctly.
2. The user is not required to 'guess' the states of uncertain characters.

The chief disadvantage of allowing the user to control the identification is that, while they may select characters that are easily observed, the characters may not make a significant contribution to the identification by discriminating between the remaining species.

The aim of this investigation was to conduct an empirical evaluation of the efficiency, accuracy and usability of two computer-based implementations of a key, one a dichotomous implementation, the other a multi-access, comparing them to a paper counterpart. Preliminary work had shown that volunteers were no less effective when using the computer-based version of the key than they were when using the paper version of the key, although they were slower in achieving an identification (Wright *et al.*, 1995). We included a multi-access key in the evaluation experiment because we were interested in comparing it to the paper key in order to see if the perceived advantages of multi-access keys manifested themselves in an improved identification accuracy. In some ways, the hypertext key can be viewed as a control in these experiments as it is computer-based but used the same modus operandi as the paper key whereas the multi-access key was both computer-based and had a different underlying theory to the paper key.

The Keys

Three keys were used in the evaluation experiment. They were all based on Hopkin's (1991) *A Key to the Woodlice of Britain and Ireland* which is one of the Field Studies Council's AIDGAP keys. There are several reasons why an AIDGAP key and Hopkin's key in particular were chosen for the evaluation experiment:

1. Hopkin's key had been developed recently and the text of the key was available in machine-readable form.
2. Woodlice are one of the more 'user-friendly' groups of invertebrates. They are fairly hardy, can be cultured in the laboratory and several species are very abundant. Indeed, the 'big-five' woodlice species are known as such because of their abundance and wide distribution in the UK (Hopkin, 1991).
3. The AIDGAP (Aids for the Identification of Difficult Groups of Animals and Plants) project aims to identify those groups for which the difficulty in identification is due to the absence of a simple and accurate key rather than being due to insuperable taxonomic problems, and to produce simple, well written aids to identification (Tilling, 1987).

Three versions of this key were developed:

1. The original key (Hopkin, 1991) published as an AIDGAP guide.
2. A computer-based version of the key, developed using the PC Guide hypertext system (Office Workstations Limited). This key contained the same text, diagrams and photographs as Hopkin's key. The main difference between this version and the first was in the manner of presentation: computer-based rather than paper.
3. A multi-access key developed using Legg's (1992a) key. This package was chosen rather than some of the other commercially available multi-access keys (Edwards and Morse, 1995) because it is one of the cheaper packages available, it has modest hardware requirements and it is an open system. That is, users of the package are encouraged to develop their own keys and programs are provided to help them do this. The implementation of the multi-access key was text-based with additional graphic information supplied in ancillary paper-based annexes.

The woodlice multi-access key differed from most other multi-access keys in one important respect, namely that the character × species matrix had many more empty (that is absent or unknown) characters than is usual or desirable in a multi-access key. These unknown elements arose because of the way in which the matrix was developed, from the information contained in Hopkin's (1991) key which was not augmented in any way. Investigations by White and Sandlant (Chapter 24, this volume) suggests that some dichotomous keys contain less than 20% of the possible information in a data matrix.

Had the matrix been augmented from other sources then the three keys would not have contained comparable amounts of taxonomic information. It is true that the multi-access key is sub-optimal because of all the unknowns in the matrix. However, if all these gaps had been filled in from other sources, it could no longer be compared to the other keys because it would then differ in at least two respects: (i) the different user interface and manner of operation; and (ii) the amount of taxonomic information it contained.

Methods

The evaluation experiment was integrated into the University of Sheffield's Zoology course. It formed one of four practicals being taken by second year students following the Animal Diversity module. The practicals were run on a 'round-robin' basis with students being assigned to one of eight groups. One group participated in the evaluation experiment in each practical class.

Each student performed the same task with one of the three keys, to which they were randomly assigned. They were first asked to identify a specimen of *Ligia oceanica* to familiarize themselves with the key they were using. This was intended to lessen the unfamiliarity of the two computer-based keys, which was one reason

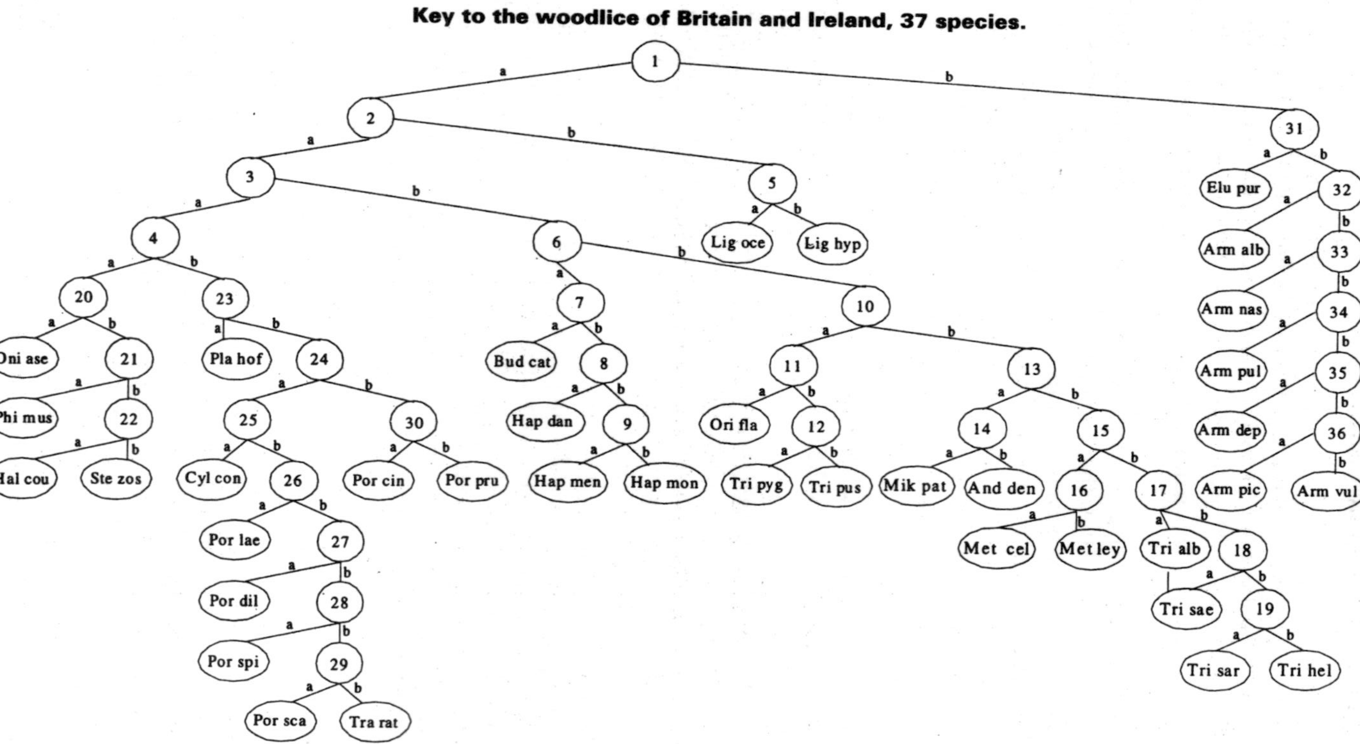

Fig. 23.1. A diagram of the decision tree used in Hopkin's (1991) key. The circles represent the couplets with the couplet numbers as indicated. The two branches of each couplet are labelled 'a' and 'b' on the diagram. Species names have been abbreviated and are shown in the ellipses at the point at which they 'key out' in the decision tree.

why the hypertext key had been found to be slower than the paper key in earlier trials (Wright *et al.*, 1995). Then, the students were asked to identify an unknown live woodlouse and to note how long the identification took. They were also asked to fill in a short questionnaire detailing their prior experience of identifying woodlice and how they had found various aspects of using the key. We also asked them to indicate how confident they were in their identification.

Three species of woodlice were chosen: *Armadillidium vulgare*, *Philoscia muscorum* and *Porcellio scaber*. The former two species 'keyed out' at approximately the same level in the identification tree for the hypertext and paper keys (Fig. 23.1), while the latter species keyed out at a much lower level, requiring many more choices to be made before it could be identified correctly.

As three species of woodlice were chosen, and there were three keys, each trial involved nine students. We were extremely fortunate in that 72 students took the Animal Diversity course. There were eight practical classes giving us nine students per session, allowing us to run one complete trial per practical class.

Results

All 72 students achieved an identification, with only one student having had prior experience of identifying woodlice. Table 23.1 shows that 53 students achieved a correct identification, corresponding to an average accuracy of identification of 53/72 = 74%. Table 23.1 also shows that two species, *A. vulgare* (79% correct) and *P. muscorum* (83% correct), were identified much more readily than *P. scaber* (58% correct) although this difference was not statistically significant ($\chi^2 = 4.43$ with 2 degrees of freedom, $P > 0.10$). This could be due to the distinctiveness of the two former species compared to the latter, or to the fact that with the paper and hypertext keys many more choices had to be made before a correct identification was achieved for the latter species compared to the former (Table 23.1 and Fig. 23.1).

Turning to the different identification media, Table 23.2 shows that the two computer-based methods were slower than the paper key although the hypertext key was only 19% slower (2 minutes and 39 seconds). Table 23.2 also shows that the multi-access key was slightly more accurate than the paper key and the hypertext key was the least accurate. Again, these differences are not statistically significant ($\chi^2 = 1.00$ with 2 degrees of freedom, $P > 0.50$). Overall, the accuracy of identification across all three keys was 66%. Most intriguing about the data in Table 23.2 was how long it took to achieve a correct or incorrect identification. For each of the three media, it took longer to achieve a correct identification than it did an incorrect identification, by at least 3 minutes 18 seconds.

The time taken by each individual to achieve an identification (summarized in Tables 23.1 and 23.2) can be analysed in a two-way analysis of variance (ANOVA), with the eight practicals being treated as replicates in this analysis. Table 23.3 and Fig. 23.2 summarize the results. In Table 23.3 it can be seen

Table 23.1. Frequency of correct identifications broken down by species but pooling across the methods used to identify the specimens. The number of couplets column shows how many choices have to be made before a correct identification is achieved in the paper and hypertext keys (see Fig. 23. 2). Also shown are the average times to achieve an identification and the difference between the time taken to achieve a correct and an incorrect identification.

Species	Number of couplets	Average time taken			Identification		Total
		Correct	Incorrect	Difference	Correct	Incorrect	
Armadillidium vulgare	7	20:00	14:43	5:03	19	5	24
Philoscia muscorum	6	14:55	10:19	4:36	20	4	24
Porcellio scaber	11	22:22	15:52	6:30	14	10	24
Total					53	19	72

Times are given in minutes: seconds.

Table 23.2. Frequency of correct identifications broken down by the method used to identify the specimen. Also shown are the average times to achieve an identification and the difference between the time taken to achieve a correct and an incorrect identification.

Method	Average time taken			Identification		Total
	Correct	Incorrect	Difference	Correct	Incorrect	
Hypertext	18:01	14:43	3:18	16	8	24
Multi-access	22:28	18:21	4:07	19	5	24
Paper	15:31	10:33	4:58	18	6	24
Total				53	19	72

Times are given in minutes: seconds.

Table 23.3. Two-way analysis of variance on the time taken for each individual to obtain an identification.

Source of variation	Degrees of freedom	SS	MS	F
Species	2	425.2	212.6	3.61 *
Key	2	661.5	330.7	5.61 **
Interaction	4	618.4	154.6	2.62
Error	63	3708.9	58.9	
Total	71	5413.9		

The variance ratio (*F*) for the Species term is significant at the 5% level (*) and that for the Key term is significant at the 1% level (**).

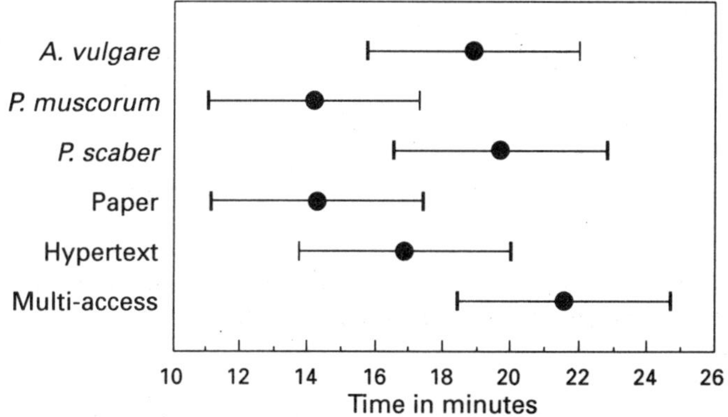

Fig. 23.2. The mean (of 24 observations) and 95% confidence interval of the time taken by each individual to identify a given specimen of *Armadillidium vulgare*, *Philoscia muscorum* and *Porcellio scaber* using either the original paper key, the hypertext version or a multi-access key. The datum for each individual appears in both the top and bottom halves of the graph as each individual can be classified according to which species they identified and which key they used.

that both the species the students were asked to identify and the key they used to identify it had a significant effect on the time it took them to perform the identification. Figure 23.2 shows that the statistically significant effect of species was due to *P. muscorum* which was significantly quicker to identify than the other two species. Figure 23.2 also shows that the multi-access key was significantly slower than the other two keys. The main purpose of this experiment was to compare the hypertext and paper keys. These were not found to be statistically significantly different from each other in the time it took the students to achieve an identification.

Table 23.4. Student assessments of how confident they are in their identification scoring each student's assessment according to whether their identification was by species, pooling across the methods used to identify the specimens, and right or wrong.

		Confidence in identification			
Species	Identification	Definitely correct	Probably correct	Don't know	Total
Armadillidium vulgare	Right	7	11	1	19
	Wrong	2	2	1	5
	Total	9	13	2	24
Philoscia muscorum	Right	8	12	0	20
	Wrong	1	1	2	4
	Total	9	13	2	24
Porcellio scaber	Right	3	10	1	14
	Wrong	0	5	4	9
	Total	3	15	5	23
Total		21	41	9	71

Table 23.5. Student assessments of how confident they are in their identification according to the method used to identify the specimen. See the legend to Table 23.4 for further notes on the data.

		Confidence in identification			
Media	Identification	Definitely correct	Probably correct	Don't know	Total
Hypertext	Right	8	8	0	16
	Wrong	1	2	4	7
	Total	9	10	4	23
Multi-access	Right	6	12	1	19
	Wrong	0	3	2	5
	Total	6	15	3	24
Paper	Right	4	13	1	18
	Wrong	2	3	1	6
	Total	6	16	2	24
Total		21	41	9	71

Students were asked to enter the confidence they had in their identification in one of five categories: 'Definitely correct', 'Probably correct', 'Don't know', 'Probably incorrect' and 'Definitely incorrect'. One student did not fill in this part of the questionnaire, hence the totals in Tables 23.4 and 23.5 are 71 and not 72. None of the students thought their identification was either 'Probably' or 'Definitely' incorrect so these categories were not used.

Looking at the students' confidence in their identification by species (Table 23.4) it can be seen that the students are more confident in their identification of the two species where they obtained a higher frequency of correct identifications (*A. vulgare* and *P. muscorum*). They were less confident in identifying *P. scaber* (3 Definitely and 15 Probably correct, compared to 9 Definitely and 13 Probably correct with the other two species) which had many more incorrect identifications.

How confident were the students in their identification with respect to the different media? They had equal confidence in the identifications obtained using the paper and multi-access keys, although with the paper key there were more 'Definitely correct' identifications which were in fact wrong (Table 23.5). There was a wider spread of confidence in the hypertext key, although it appeared that students were in general more confident of their identification with the hypertext key than the other two keys (9 Definitely and 10 Probably correct, compared to 6 Definitely and 15 or 16 Probably correct). However, it was noted above that the hypertext key was the least accurate of the three keys, although it was only a little worse than the other two keys.

Finally, Table 23.6 summarizes the students' impressions of the keys. Many of their comments concerned usability and navigation issues. In particular, students commented on the frequently encountered problem of finding their way round the paper key. Easing this task was one of the original motivations for developing hypertext keys (Wright *et al.*, 1995). In general, students found both computer-based keys easy to use but the diagrams and colour plates in both keys were criticized, either because they weren't there (the multi-access key) or because of the poor quality of some of the images and text (the hypertext key).

Discussion

The evaluation experiment was integrated into the University of Sheffield's second year Zoology course. Other workers have found such an approach to be an efficient and effective means of obtaining a large number of volunteers to evaluate computer-based material (see Boyle *et al.*, 1993; Hutchings *et al.*, 1993, 1994; Viau and Larivée, 1993; Watkins *et al.*, 1995 for examples). In contrast to these studies, we were not evaluating a piece of courseware per se, rather we were evaluating a multimedia tool which was designed to perform a well defined task with a number of easily measurable outcomes. In addition, we used a designed experiment, with students being assigned different 'treatments' at

Table 23.6. A summary of the features of the keys which students liked, those they disliked and their suggestions for how they would improve the key they used.

	Likes	Dislikes	Improvements
Hypertext	Navigation Availability of information Fun, easy	Scanned diagrams Labelling on diagrams Only one diagram available at a time	Improve diagrams Increase size of glossary
Multi-access	Easy to use Easy navigation Ability to skip characters Next best character Probabilities	Swapping between screen and book Abbreviations of character descriptions Difficult to remove characters	More built-in help Diagrams on screen
Paper	Availability of information Ability to see overall structure of the key	Moving between different parts of the key (e.g. glossary, colour plates, etc.) Numbers to follow paths through key	Improve movement between different parts of the key

random. Both these factors considerably increase the power of the experiment and the ease with which the results can be interpreted.

An overall frequency of correct identifications of 74% is high when it is considered that only one student had had any prior experience of identifying woodlice. This compares favourably with Stucky (1984) who found a misidentification frequency as high as 30% (corresponding to 70% correct identifications) in a dichotomous key and polyclave to weed seedlings. It is also an improvement on an earlier study (Tardivel and Morse, unpublished) where Year 12 and Year 13 pupils were asked to identify woodlice using the paper and hypertext keys. In that experiment, the frequency of correct identifications fell as low as 60%, although those subjects had little experience of using keys and virtually no experience of identifying woodlice. How many errors experts (either in the taxonomic group or in the use of the key) would make is not known.

This study confirms the findings of a previous experiment (Wright *et al.*, 1995) where it was found that the hypertext key was both slower and less accurate than the paper key. The difference in times between the two keys was less in this experiment because of the familiarization phase which each student undertook before commencing the identification proper. This familiarization phase has been found to be an important feature of other hypermedia evaluation studies (Hutchings *et al.*, 1993, 1994). The higher misidentification frequency could be due to students becoming 'mouse-button happy' and selecting one or other couplet even when they are not sure which one is correct. Edwards and Morse (1995) proposed that in such a situation users of the paper key would be more likely to backtrack or start again than in the hypertext key. The observation that students who obtained the correct identification took about four minutes longer than those who did not, regardless of the media they used to identify the specimen, is intriguing. It could be that they were more careful during the identification and hence they were slower than their counterparts who did not achieve a correct identification. Alternatively, they could have spent the four minutes checking the species description and photographs, or both. Only close observation of people when they are identifying specimens and experiments like this will confirm the effect if it exists and reveal the difference between the two groups of people.

This is probably the first experiment in which volunteers were asked to estimate the confidence which they place in their identification. In retrospect, the scale on which volunteers were asked to judge their confidence was too crude. This group of students appeared unlikely to admit that they thought their identification was 'Probably incorrect'. It is more likely that they would try again until they achieved an identification in which they had some confidence. While more data are needed to confirm this conclusion, the students were reasonable judges of whether their identification was correct or not. However, only nine people admitted that they did not know, of which seven were incorrect and two had the correct identification (Table 23.3). An overall nineteen students were incorrect in their identification (Table 23.1).

Another way of looking at the experiment is that it was a comparative evaluation of three different user interfaces to the same taxonomic information. In general, the quantitative differences between the three keys are small, although in some cases they may well be important differences, such as the accuracy of the identification and the confidence which the students had in their identification. On the other hand, the students' subjective impressions of the keys (summarized in Table 23.5), could be more important in determining the future development of taxonomic keys. For example, it is encouraging that students found both computer-based keys easy to use and, in the case of the hypertext key, fun! There is clearly room for improvement in all three keys as the list of suggested improvements in Table 23.5 shows. In the paper key, the students' dislikes and their suggested improvements stem from the linear nature of paper documents (Nielsen, 1995). Hundreds of years of dichotomous key development (Pankhurst, 1991) has not yet overcome these restrictions. In contrast, it was the quality rather than the accessibility of the ancillary information which was commented on in the two computer-based keys.

Whilst legibility, speed of reading and speed of display of screen text have improved, the reading of on-screen text is still more problematic than that of paper (McNight *et al.*, 1989). Improvements to graphic user interfaces in the near future may further improve the display of on-screen text. However, in general, diagnostic hypertexts present text in small units, which are less likely to cause users problems than large chunks of text.

This study suggests that users find diagnostic hypertexts of dichotomous form easier to use than their paper counterparts. Presenting three or more choices rather than two might increase the efficiency of large scale diagnostic hypertexts, though possibly with a reduction in accuracy. It is possible that a simple choice between three or more similar elements is more effective than a tedious and artificial nesting of choices which separate off one unit at each node. Such multiple choices would be easier to navigate in a hypertext than in a linear document.

The time taken to become familiar with the media tested was not measured. This might be an important factor when comparing the different media. Students were given a broadly similar introduction to each. All had had experience of paper-based keys and all had used a computer, mouse and Microsoft Windows-based software. There were no apparent problems associated with learning to use the hypertext package. It is worth noting that the use of (paper-based) keys is part of the 'A' Level biology curricula. Gaining this skill is regarded as something that requires teaching and repetition. Comments from students suggest that it would be at least as easy to learn to use the hypertext package as the paper medium. As the students saw navigation as easier in the hypertext package, there is no reason to believe that retention of the skill between uses would be a problem.

The hypertext produced for this study was created to have, as nearly as possible, the same functionality as its paper counterpart. How much further the hypertext key, and other diagnostic hypertexts like them, can be improved is not clear, but it is clear that dichotomous keys restricted to a paper medium have just about reached the limits of their evolution. There are various features which could be included in hypertext implementations which cannot be included in paper versions. Such features include history mechanisms, calculation of ratios, range checking and links to supporting information. Future evaluations of enhanced hypertext identification keys will assess the usefulness of these and other features.

Acknowledgements

The British Ecological Society funded the project. Steven Hopkin and Steve Tilling gave permission to convert the woodlice key to hypertext form. John Spicer gave us permission to work with his students during his practical classes. Last, but by no means least, we are very grateful to the 1994/95 cohort of Zoology students at the University of Sheffield who participated in the experiment.

References

Boyle, T., Gray, J., Wendl, B. and Davies, M. (1994) Taking the plunge with CLEM – the design and evaluation of a large-scale CAL system. *Computers and Education* 22(1–2), 19–26.

Edwards, M. and Morse, D.R. (1995) The potential for computer-aided identification in biodiversity research. *Trends in Ecology and Evolution* 10(4), 153–158.

Hopkin, S.P. (1991) A key to the woodlice of Britain and Ireland. *Field Studies*, 7, 599–650.

Hutchings, G.A., Hall, W. and Colbourn, C.J. (1993) Patterns of students interactions with a hypermedia system. *Interacting with Computers* 5(3), 295–313.

Hutchings, G.A., Hall, W. and Thorogood, P. (1994) Experiences with hypermedia in undergraduate education. *Computers and Education* 22(1–2), 39–44.

Legg, C.J. (1992a) Random-access identification guides for a microcomputer. *Field Studies* 8(1), 1–30.

Legg, C.J. (1992b) Random-access guide to sedges of the British Isles using a microcomputer. *Field Studies* 8(1), 31–57.

McKnight, C., Dillon, A. and Richardson, J. (1989) Problems in hyperland? A human factors perspective. *Hypermedia* 1(2), 167–178.

Neilsen, J. (1995) *Multimedia and Hypertext: The Internet and Beyond.* Academic Press, London.

Pankhurst, R.J. (1991) *Practical Taxonomic Computing.* Cambridge University Press, Cambridge, xi + 202 pp.

Rouse, S.H. and Rouse, W.B. (1980) Computer based manuals for procedural information. *IEEE Transactions on Systems, Man and Cybernetics*, SMC-10, 506–510.

Stucky, J.M. (1984) Comparison of 2 methods of identifying weed seedlings. *Weed Science* 32, 598–602.

Tilling, S.M. (1987) Education and taxonomy – the role of the Field Studies Council and AIDGAP. *Biological Journal of the Linnean Society* 32, 87–96.

Viau, R. and Larivee, J. (1993) Learning tools with hypertext – an experiment. *Computers and Education* 20, 11–16.

Watkins, J., Davies, J., Calverley, G. and Cartwright, T. (1995) Evaluation of a physics multimedia resource. *Computers and Education* 24(2), 83–88.

Wright, J.F., Morse, D.R. and Tardivel, G.M. (1995) An investigation into the use of hypertext as a user-interface to taxonomic keys. *Computer Applications in the BioSciences* 11(1), 19–27.

Computerized Insect Identification: a Comparison of Differing Approaches and Problems

I.M. White and G.R. Sandlant

International Institute of Entomology, 56 Queen's Gate, London SW7 5JR, UK

Introduction

Most keys are printed and dichotomous, and aside from the obvious advantage that a book requires no electricity or costly electronics, they have little to commend them when a comparison is made with computerized keys. However, despite the fact that computerized key production methods have now existed for several years (e.g. Dallwitz and Paine, 1986) there are still very few computerized keys. One reason may be that the emphasis has been placed on the production of multiple-entry keys, which are perceived as being very expensive and difficult to produce. This slow take-up for the technology is to be regretted when one considers that identification is the foundation-stone of plant quarantine legislation, environmentally friendly pest management techniques and biodiversity studies. Clearly with numbers of taxonomists dwindling it makes sense to encourage these users of taxonomic services to do some routine or initial identification work themselves, and therefore we should be using computers to provide them with user friendly tools for the task. In order to speed the production of electronic keys the authors have in some cases opted for a simple computerized dichotomous key in favour of using a multiple-entry key approach. The purpose of this paper is to compare the relative merits of multiple-entry keys with a simple transferral of dichotomous keys on to the computer, and to illustrate these with a few entomological examples based on work being carried out at the CABI International Institute of Entomology (IIE). Such a comparison should be of value when proposing new projects and therefore needing to decide which approach to take.

A simple dichotomous key program typically presents each couplet as an illustrated screen, with each half couplet taking half the screen on which the user can click the mouse to give an answer. Such a program may also support additional text and a linked glossary. IIE's software of this type is called TAXAKEY and was developed by G.G. Kibby and has recently been applied to crop pest aphids (Blackman *et al.*, in press), orders of arthropods, families of Lepidoptera (adults and larvae), Heteroptera and Homoptera (CABI, 1997). The structure of these dichotomous keys may be visualized as a tree, with couplet 1 as the root and each taxon as a terminal point. This type of key was called a hypertext key by Wright *et al.* (1995).

Conversely, multiple-entry (multi-access, synoptic or polyclave) keys have no tree structure to represent their data. Instead they simply have a matrix (or table) in which each taxon is a row and each character a column, and the cells of the matrix hold the state of each character for each taxon. The user (or the software) selects a character which should be presented as a picture with similar links to text and glossary as those described above. The user then indicates which state(s) apply and taxa which never have any of those states are eliminated. Further questions are answered and taxa eliminated until a single taxon remains, and that is taken to be the identification. Several multiple-entry key programs have now been developed, e.g. INTKEY (Dallwitz and Paine, 1986), ONLINE (Pankhurst, 1991) and LucID (CRCTPM, 1996). Other examples are listed by Dallwitz (1995). Our own software is called CABIKEY (White and Scott, 1994) and has recently been applied to species of Dacini fruit flies (White and Hancock, 1997) and European thrips (Mound and Moritz, 1996), mosquito genera (adults and larvae) (Harbach and Sandlant, 1997) and beetle families (Booth *et al.*, 1994).

Information Content

To compare dichotomous and multiple-entry keys we need first to consider how each of them represents the taxonomic information space. The total information space describing a group of taxa may be thought of as a matrix of taxa by characters, e.g. if 100 taxa were described with respect to 100 characters, then the information space describing them holds 10,000 pieces of information [this is simplified by ignoring the fact that some characters may have more states than others]. A multiple-entry key can express the whole of that information space since it holds a matrix of taxa by characters, although in practice most multiple-entry keys include some characters that are not scored for all taxa since they are either non-applicable or unknown. For example, White and Hancock (1997) scored data on 513 taxonomic units (507 species but with six treated separately for each sex) with respect to 169 characters in a study of Dacini (Diptera, Tephritidae) fruit flies, and they only completed 58% of the matrix (total matrix 86,697 cells) due to 32% of points being non-applicable and the remaining 10% unknown (usually due to only one sex being known).

Although multiple-entry keys have the potential to describe the entire information space, that is far from being the case for dichotomous keys. In a dichotomous key each couplet infers the state of a character for all the taxa that branch from it and its daughter couplets. Thus it is possible to attempt to fill a taxa by characters matrix with information extracted from a dichotomous key. When this was attempted for 33 genera of mosquitoes (data from Mattingly, 1971), which included data on 90 characters, only 17% of the total 2970 cells in the matrix could be filled. Regrettably, there is no simple way of calculating how much of the information space any given dichotomous key describes. However, we can consider the two extremes of key structure and derive a simple formula for each of them. At the extremes, a key may be asymmetric (each couplet gives rise to one taxon and another couplet) or symmetric (each couplet leads to two more couplets to a certain depth and then all couplets lead to two taxa). Consider the trees depicted in Fig. 24.1.

In both structures couplet 1 gives us information on the state of a character for all four taxa. In the symmetric structure couplets 2 and 3 each provide information on the state of a character for just two taxa, so in total we can extract eight pieces of the information space from the symmetric tree. Conversely, in the asymmetric key, couplet 2 gives us data on the state of a character for three taxa, and then couplet 3 yields data on the state of a character for two taxa, so in total we can extract nine pieces of the information space. If we had constructed a dichotomous key for the 513 taxa of fruit flies mentioned earlier, then we would use 131,840 pieces of information to construct a key that was completely asymmetric (an impossible feat needing at least 512 characters as there would be 512 couplets), or just 4619 pieces of information if the key was perfectly symmetric (also an impossible feat) (Table 24.1). The formulae for calculating the amount of information contained in a perfectly symmetric or perfectly asymmetric key, assuming just one character is used per couplet is shown below. i = information points. t = taxa.

$$i_{(symmetric)} = (\lg_2 t)\, t \qquad i_{(asymmetric)} = t\,(t + 1)/2 - 1$$

In practice, such a key would be somewhere between these extremes and need perhaps 10,000–20,000 pieces of information, which with the 169 characters used by White and Hancock (1997) would be just 12–23% of the information available. Dichotomous keys are therefore extremely wasteful of information, and will be so even when two or three characters are used in some couplets. Furthermore, it is not realistic to attempt to turn a dichotomous key into a multiple-entry key unless there are also very full and consistent descriptions available of all taxa, in which case it may be possible to draft a multiple-entry key matrix.

Table 24.1. Amount of the taxonomic information space included in a perfectly asymmetric or a perfectly symmetric dichotomous key in which each couplet uses a single character.

Taxa	Symmetric key	Asymmetric key
10	33	54
50	282	1274
100	664	5049
150	1084	11,324
200	1529	20,099
250	1991	31,374
300	2469	45,149
350	2958	61,424
400	3458	80,199
450	3966	101,474
500	4483	125,249

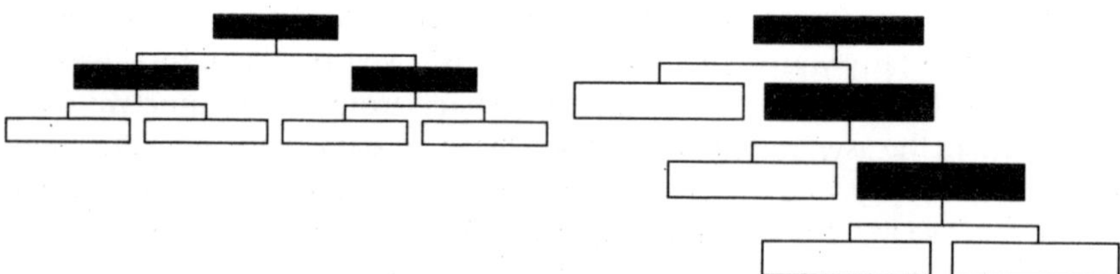

Fig. 24.1. Two extremes of dichotomous structure linking four taxa; black boxes represent couplets and open boxes taxa. The left tree is symmetric and the right asymmetric.

Why Is So Much Information Useful?

There are several reasons, including the fact that multiple-entry keys work by process of elimination from a matrix describing the taxonomic information space, and are therefore amenable to special configuration. For example, the keys to dacine fruit flies (White and Hancock, 1997) and mosquito genera (Harbach and Sandlant, 1997) both allow users to restrict the identification process to one or more zoogeographic regions. Furthermore, the fruit fly key can be configured to only consider a selection of 51 pest species if the flies were reared from a commercially grown fruit, or to consider a selection of 407 well described species, or to consider all 507 species. In this respect the dacine multiple-entry key is far more flexible than the separate hard copy publications presenting dichotomous keys for the Australasian and Oceanian Dacini (Drew, 1989), or for the fruit flies of Thailand and the Philippines (Hardy, 1973, 1974), or the many isolated papers covering other areas of Asia. Such a multiplicity of scattered publications made fruit fly identification impossible for entomologists having to identify a range of organisms found in plant quarantine, as they were often uncertain of the true origin of the specimens they had before them.

Another major attribute of the multiple-entry key system is the power it gives users to avoid answering particular questions and still stand a very good chance of being able to reach an unambiguous identification. Users may wish to avoid questions because body parts are missing or costly to analyse, or they may simply find them too difficult.

The wealth of data known to a multiple-entry key program can only be harnessed as a powerful error checking tool provided that most characters that are applicable to all taxa are scored for all taxa. Some multiple-entry key programs are able to construct a diagnosis which lists only a minimal set of characters and their states which separate a given taxon from all others in the key. This can be done by looking for the character which separates the reference taxon from the greatest number of other taxa and then regarding those taxa as 'eliminated'. The process is continued by finding the character that separates the greatest number of remaining taxa, and so forth until all other taxa are eliminated (if the matrix is imperfect in any way the system will run out of characters before eliminating all taxa). This diagnosis tends to list characters that were not used in the initial identification. Such a diagnosis may not always be the shortest that could be found (Pankhurst, 1991) but it will be a good list of characters that should be checked in order to verify an identification. Furthermore, the characters listed in such a diagnosis often differ markedly from those used to make an identification. This is because the separation score (or number of Pankhurst, 1991) system used by CABIKEY, INTKEY and ONLINE, tends to favour those characters that have a very even distribution of states across taxa. Conversely, the diagnosis is calculated by searching for characters with a highly skewed distribution across taxa. For example, White and Hancock (1997) was used to identify a species known as the melon fly, *Bactrocera cucurbitae* (Coquillett), and eleven questions were answered to reach an identification [the ID process assumed the specimen was of uncertain geographic origin]. The diagnosis was then calculated and found to include five questions, only two of which were considered during the identification process. In a subjective test carried out by one of us (IMW), students (New Zealand Ministry of Agriculture entomologists) were given the task of identifying a wide range of pest fruit flies using printed keys (primarily Drew, 1989; Hardy, 1973, 1974) and a pre-release version of the White and Hancock (1997) system. Use of the dichotomous (printed) keys resulted in many undetected misidentifications whereas those few misidentifications that were made with the multiple-entry key were detected by the students themselves as a result of the built-in error checking facilities.

During the construction of that key to Indo-Australasian Dacini, about 30 new species were discovered in the collections of The Natural History Museum. Each of these species was run through the multiple-entry key, which only included already published species. Although most ran to a name (i.e. a misidentification), when the diagnosis was calculated as a check on identification, most specimens failed to fit the diagnosis. The only exceptions were a few species belonging to a very difficult complex of closely related species [the Oriental fruit fly or *Bactrocera dorsalis* (Hendel) species complex] that were separated by characters that had not been needed to separate the presently described fauna. In summary, the representation of the entire information space in a multiple-entry key means that the key has 'knowledge' of every permutation of character states that has ever been described in the group. A 'new' species with a different permutation of character states may run to a 'name' but will probably not fit the diagnosis, and so the fact that a misidentification has been made becomes easily detectable. The same can seldom be said of a dichotomous key, unless the author goes to a great deal of trouble to supplement it with diagnostic notes on each taxon or the user goes to the trouble of checking an entire description.

Misidentifications can not only result from user error and the discovery of 'new' species, but also from data error (either a coding mistake or inadequate available specimens to see the full range of variation). Mistakes may also be caused by aberrant individuals (allowing for all potential rare variants could render a key unworkable). Again, practical experience has indicated that when a misidentification is made the diagnosis will show up the error. However, multiple-entry keys also offer some powerful techniques to help users recover from having made an error. CABIKEY, in common with some other multiple-entry key programs, allows the user to find out which other taxa are similar to the questions answered or to the taxon that they have 'run' to. Such software also presents an option to replace (un-answer) or re-answer any question (character). Users can therefore examine the diagnosis of other similar species to see if any fit or they can 'undo' their answers to any questions they were uncertain of. When questions are replaced (undone) in a multiple-entry key, the software re-calculates which taxa still match the answers to the remaining answered questions. The user can either try re-answering the questions at fault or ignore them in favour of other questions. Conversely, the user of a dichotomous key must go all the way back to a problem couplet and start again the other way from that point.

The other great advantage that the multiple-entry key gains through its dynamic use of the entire taxonomic information space is in user efficiency. In a dichotomous key, the user has to follow a fixed route to any given taxon and the more symmetric the key, the more efficient it is (Pankhurst, 1991). However, in a multiple-entry key, characters can be selected using a separation number (Pankhurst, 1991) which changes according to which taxa remain un-eliminated. A comparison of the relative efficiency of dichotomous and multiple-entry keys was carried out. White and Hancock (1997) included 292 species of Dacini from the Australasian and Oceanian Regions (discounting species from Hawaii and other north Pacific areas). Drew (1989) presented a series of dichotomous keys covering 290 species from the same areas. This fauna includes 20 *Bactrocera* species that were regarded as pests by White and Elson-Harris (1994). Those 20 species were each 'run' through Drew (1989) and through White and Hancock's (1997) multiple-entry key. It was found that using the dichotomous keys the user had to answer between 7 and 21 couplets, and in total 244 couplets were answered to identify the 20 species (mean = 12). Conversely, with the multiple-entry key 8–14 questions had to be answered, and in total only 179 questions were needed to identify the 20 species (mean = 9) [in each case geographic origin was regarded as unknown and characters were selected strictly by highest separation score]. In summary, the dichotomous key required 36% more effort to identify the 20 species than the multiple-entry key. Amusingly, it was the inaptly named *B. distincta* (Malloch) which required the greatest number of steps with either key. The results are displayed graphically in Fig. 24.2.

Discussion

The relative advantages and disadvantages of the dichotomous and multiple-entry key approach are listed in Table 24.2. As expected, the multiple-entry key approach wins by a large margin and, since construction of a dichotomous key involves an intimate knowledge of the entire taxonomic information space, one

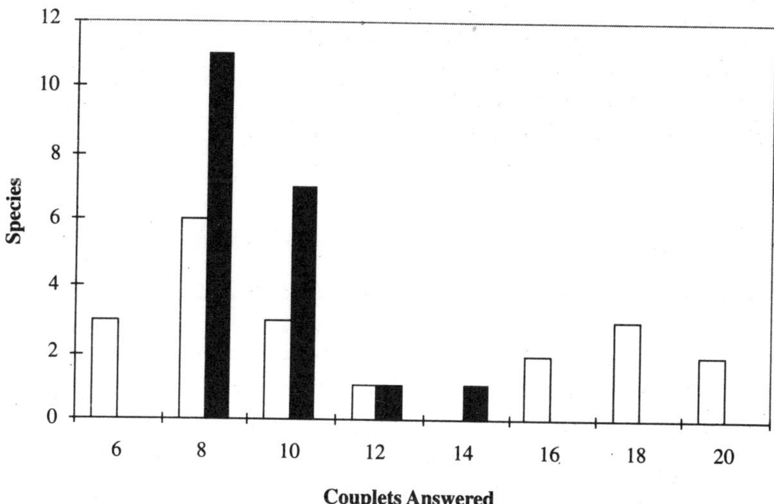

Fig. 24.2. Number of Australasian-Oceanian *Bactrocera* pest species requiring the user to answer any given number of couplets in a dichotomous key or questions in a multiple-entry key to achieve an identification. □ Dichotomous; ■ multi-entry.

is tempted to think that it is a criminal waste not to express that information in the form of a multiple-entry key. In this respect we do not agree with Wright *et al.* (1995) who said it was cheaper to produce hypertext (dichotomous) keys than multiple-entry keys; it is only cheaper if the substantial task of studying the taxonomic information space has already been carried out and expressed as a key by somebody else.

Given that we have now analysed the relative shortcoming of the dichotomous approach, were we right to have used it for our keys to aphids, arthropod orders and some family keys? When asked to produce order and family level keys for the CABI Crop Protection Compendium (CABI, 1997) we were faced with a limited budget and limited time. Furthermore, the dichotomous key does have one advantage (Table 24.2) in that it can be produced from an existing printed key. We concluded that if an existing printed key exists, works well and the added flexibility offered by a multiple-entry key is not demanded, then a computerized form of that dichotomous key is to be preferred (subject to permission). The crop pest aphid key was also produced in this way because it was based on a book (Blackman and Eastop, 1984) which had lots of short keys, each to the aphids of a single crop, and the electronic version followed the same pattern. A similar conclusion was reached in an earlier review (Edwards and Morse, 1995).

Clearly the multiple-entry key approach was the right one for the dacine fruit flies (White and Hancock, 1997), given that existing keys, at least to the Asian fauna (e.g. Hardy 1973, 1974), all started with a subgeneric key based on male secondary sexual characters, so the females of one of the most devastating groups of pests in the region could not be identified except by a few experts who knew how to jump-start the key. This situation was made all the more ridiculous as the females often provide the best diagnostic characters between species of dacine fruit flies. However, the multiple-entry key to 60 economically important beetle families (Booth *et al.*, 1994) proved a costly and very

Table 24.2. Summary of the advantages and disadvantages of dichotomous and multiple-entry keys as discussed in this paper.

	Dichotomous	Multi-entry
Characters presented as pictures	Yes	Yes
Links to additional information	Yes	Yes
Information content	Low	High
May be produced from existing key	Yes	No
Users may avoid a question	No	Yes
Standard of error detection	Bad	Good
User efficiency	Bad	Good
Total advantages	3	6

time consuming exercise taking about three man days per family, compared to the fruit flies which took little over half a man-day per species. The most likely reason for this high cost of a family level key is that to produce a key to a supra-specific level of taxa requires the analysis of variation across many included taxa, although the authors could hardly be expected to test it for all of the one-third of a million known beetles, choosing instead to ensure it should work for the roughly 300 pest species discussed by Booth *et al.* (1990).

Experience gained from producing multiple-entry keys to species (fruit flies), genera (mosquitoes) and families (beetles) also suggests that it is increasingly difficult to apply the multiple-entry key approach with increasing taxonomic level. Dichotomous keys often contain couplets which say if it has something then it must lack something else, i.e. complex permutations of characters embodied in a single question. In theory, these should be avoided in a multiple-entry key as each character is coded as a separate question and non-matching taxa are removed bit by bit as each of the questions is answered. In practice, the authors of the beetle and mosquito keys found this impossible, as supra-specific taxa often lack a simple homogeneous definition. To overcome these problems, supra-specific taxa could sometimes be subdivided to make more homogeneous groupings, but that did not always work and the authors then resorted to combining awkward permutations of characters into single questions. On completion, the beetle family key contained 17 questions that combined more than one character, the mosquito larval key four, mosquito adults three, and fruit flies none, i.e. the higher the taxonomic level the more combination questions we used.

One aspect of multiple-entry keys that is not appreciated by those who have not tried constructing one is their value as research tools, the most obvious use being that the data matrix can be used with little modification for cladistic studies. During the construction of the dacine fruit fly key (White and Hancock, 1997) numerous new synonymies and incorrect subgeneric placements were discovered. Many of these would have been missed if we had simply constructed a dichotomous key, particularly if we had made it follow the traditional subgeneric structure, since in some cases the synonyms were between a species known only from males and one known only from females and formerly placed in different subgenera.

Building a matrix in order to run a multiple-entry key (or for cladistic analysis) also enforces good discipline on the taxonomist, since the characters have to be interpreted uniformly across all taxa. However, that may prove difficult in some groups. In addition, multiple-entry keys work best where a high proportion of characters are essentially yes/no in form and are therefore very easily applied to insect chaetotaxy (bristle characters) but not so easily applied where most characters are descriptors of shape or pattern. In fact, lepidopterists do not even like using dichotomous keys to identify butterflies and moths, preferring instead to browse hundreds of pictures.

In conclusion, when there is a demand for a user friendly identification tool, the multiple-entry key is still to be preferred for its power to be configurable, to allow user control in character choice, for its good error checking and recovery aids, and for its value to the author as a research tool. However, if a good dichotomous key already exists and these other features are not specifically required, then it makes sense to simply computerize the dichotomous key. Caution should also be exercised in trying to apply multiple-entry keys to higher group taxa, especially if the groups lack homogeneous definitions and many of the characters are hard to score in a practical manner across all taxa.

Acknowledgements

We are grateful to our Director, Professor V.K. Brown, Dr R.E. Harbach (The Natural History Museum), and our colleagues N. Arkas, Dr M.L. Cox and G.G. Kibby, for their help with the manuscript.

References

Blackman, R.L. and Eastop, V.F. (1984) *Aphids on the World's Crops: An Identification Guide.* Wiley, Chichester. vii + 466 pp.

Blackman, R.L., Eastop, V.F. and Kibby, G.G. (in press) *TAXAKEY to Aphids on the World's Crops.* Windows and Macintosh CD-ROM. CAB International, Wallingford, UK.

Booth, R.G., Cox, M.L. and Madge, R.B. (1990) Coleoptera. *IIE Guides to Insects of Importance to Man.* 3, vi + 1–384.

Booth, R.G., Cox, M.L. and Madge, R.B. (1994) *CABIKEY to Major Beetle Families.* DOS Floppy disk. CAB International, Wallingford, UK.

CABI (1997) *Crop Protection Compendium: Module 1.* Multimedia CD-ROM. CAB International, Wallingford, UK.

CRCTPM (1996) *LucID: Identification Tool for Teachers, Taxonomists and Ecologists.* Cooperative Research Centre for Tropical Pest Management, Brisbane, Australia. World Wide Web page at http://www.ctpm.uq.edu.au/software/lucid.html

Dallwitz, M.J. and Paine, T.A. (1986) *User's Guide to the DELTA System; A General System for Processing Taxonomic Descriptions.* 3rd edn. Report, Division of Entomology, CSIRO, 106 pp.

Dallwitz, M.J. (1995) *Programs for Interactive Identification and Information Retrieval.* BIOSIS, York, UK. World Wide Web page at http://www.york.biosis.org/zrdocs/zoolinfo/int_keys.htm

Drew, R.A.I. (1989) The tropical fruit flies (Diptera: Tephritidae: Dacinae) of the Australasian and Oceanian regions. *Memoirs of the Queensland Museum* 26, 1–521.

Edwards, M. and Morse, D.R. (1995) The potential for computer-aided identification in biodiversity research. *Trends in Ecology and Evolution* 10, 153–158.

Harbach, R.E. and Sandlant, G.R. (1997) *CABIKEY to the Mosquito Genera of the World.* CAB International, Wallingford. Windows CD-ROM.

Hardy, D.E. (1973) The fruit flies (Tephritidae – Diptera) of Thailand and bordering countries. *Pacific Insects Monograph* 31, 1–353.

Hardy, D.E. (1974) The fruit flies of the Philippines (Diptera: Tephritidae). *Pacific Insects Monograph* 32, 1–266.

Mattingly, P.F. (1971) Contributions to the mosquito fauna of Southeast Asia. XII. Illustrated keys to the genera of mosquitoes (Diptera, Culicidae). *Contributions of the American Entomological Institute (Ann Arbor)* 7(4), 1–84.

Mound, L.A. and Moritz, G. (1996) *CABIKEY for Common Thysanoptera of Europe*. DOS Floppy disk. CAB International, Wallingford, UK.

Pankhurst, R.J. (1991) *Practical Taxonomic Computing*. Cambridge University Press, Cambridge, xi + 202 pp.

White, I.M. and Elson-Harris, M.M. (1994) *Fruit Flies of Economic Significance; Their Identification and Bionomics*. Reprint with addendum. CAB International, Wallingford, UK. 601 pp.

White, I.M. and Hancock, D.L. (1997) *CABIKEY to the Dacini (Diptera, Tephritidae) of the Asia-Pacific-Australasian Regions*. Windows CD-ROM. CAB International, Wallingford, UK.

White, I.M. and Scott, P.R. (1994) Computerised information resources for pest identification: a review. In: Hawksworth, D.L. (ed.) *The Identification and Characterisation of Pest Organisms*. CAB International, Wallingford, UK, pp. 129–137.

Wright, J.F., Morse, D.R. and Tardivel, G.M. (1995) An investigation into the use of hypertext as a user interface to taxonomic keys. *CABIOS* 11, 19–27.

25

Automated Analysis of Insect Sounds using Time-encoded Signals and Expert Systems – a New Method for Species Identification

E.D. Chesmore, O.P. Femminella and M.D. Swarbrick

Environmental Electronics Research Group, Department of Electronic Engineering, University of Hull, Hull HU6 7RX, UK
Fax: +44 (0)1482 466664/E-mail: e.d.chesmore@e-eng.hull.ac.uk

Introduction

The application of acoustics in entomology can be divided into the following categories:

1. Analysis of communications and sound production mechanism. Traditional analysis methods include oscillograms and frequency response to determine the physical and physiological basis for sound generation in many groups of insect such as Orthoptera, Hemiptera, Coleoptera and Lepidoptera. Modern high speed computers and signal processors allow more rapid and accurate analysis of sounds.
2. Detection of pests in stored produce. Much research has been devoted to the detection of pests in, for example, grain by the acoustic emission caused by movement or eating (Hagstrum *et al.*, 1990; Shuman *et al.*, 1993). Work on mating signals of grain pests has also been carried out (Trematerra and Pavan, 1995).
3. Automated identification of species. Research into automated species identification has been reported for birds and amphibia using machine learning techniques (Mills, 1995; Taylor *et al.*, 1996). However, little work has been carried out in this area for insects.

The topic of this Chapter falls into the third category and describes a novel method for automatically identifying insect species which also lends itself to the first two categories.

The work described here is the result of a final (fourth) year MEng undergraduate project carried out in the 1995/96 academic year in the Electronic Engineering Department at the University of Hull (Femminella, 1996). The project's remit was to investigate the application of digital signal processing and artificial intelligence to the automated identification of Orthoptera (grasshoppers and crickets) although the techniques developed are equally applicable to other insect orders and phyla.

The project's scope can be summarized as follows:

1. The development of a user-friendly PC-based software package written in 'C' under Windows™ called ISAR (Insect Sound Analysis and Recognition).
2. Various digital signal processing (DSP) tools were used to analyse signals and derive parameters including resonant frequencies, syllable duration, silence detection and tooth impact rates.
3. A multiple expert system known as a 'blackboard system' was used to further process the results for hierarchical analysis (e.g. song type) and recognition.

The project occupied one semester (12 weeks) and ISAR is currently in prototype form but shows considerable promise. The remainder of the paper describes the operation of the system in detail and discusses future directions and application to other acoustic signals.

Sound Production and Morphology in the Orthoptera

It is beyond the scope of this paper to fully describe the mechanisms involved in sound production in insects. However, the mechanisms of song production in the Orthoptera will be briefly discussed together with the resulting acoustic structure as this is fundamental to the identification process. Many books and papers have been written on Orthoptera sound production; some useful references are Chapman (1982); Stephen and Hartley (1995). Sound production mechanisms in the Orthoptera are frictional and consist of two components, the file and scraper. The file is a row of teeth, line of hairs or a number of pegs that are rubbed against the scraper which may be a single ridge or hard projection such as a raised wing vein. These pairs of components are found on different parts of the legs, tegmina and abdomen which are rubbed together to produce sound. In Orthoptera, elytral and elytrofemoral methods are widespread, the former being found in Gryllidae and Tettigoniidae where modified veins act as file and scraper on the elytra. One elytron overlaps the other and both are moved across each other producing a vibration whose characteristics are dependent on the rate of movement, the number of file elements and the resonant characteristics of the elytra. In the elytrofemoral mechanism the file is a raised ridge on the surface of the hind femur and the scraper is a vein of the tegmen raised into a sharp ridge. The up-down movement of the femur held close to the body causes the pegs to strike the scraper. The size, number and spacing (density) of pegs vary according to genera and species and can provide species-specific characters.

Acoustic characteristics

In the Orthoptera, the gryllids produce the purest sounds of any insect; frequencies produced vary between species and are generally in the range of 2–8 kHz at intensities up to 70 dB. The signal is strongly resonant and the frequency spectrum is therefore narrow and simple. In contrast, the tettigoniids produce complex wideband spectra, often with ultrasonic components over 100 kHz. The wide spectrum is due to the large number of pulses produced by each elytral movement. In acridids, one movement of the stridulatory mechanism may produce more than one pulse of sound and both hind legs are often used at the same time resulting in complex sounds with frequencies between 2 and 40 kHz. Figures 25.1 to 25.3 show the time and frequency domain signals for a

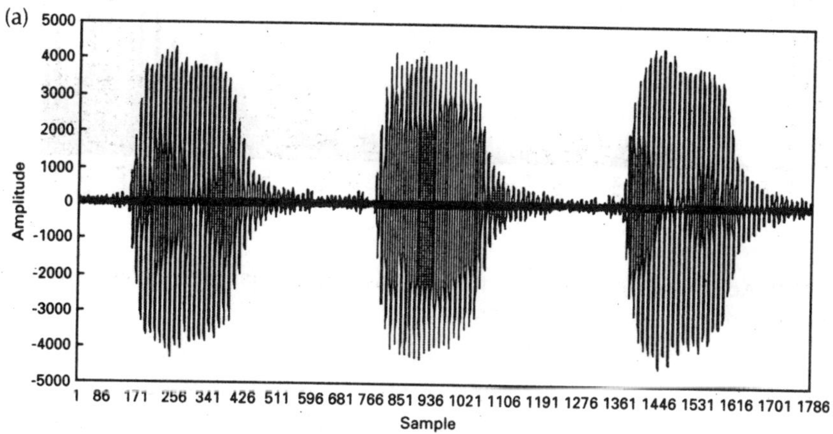

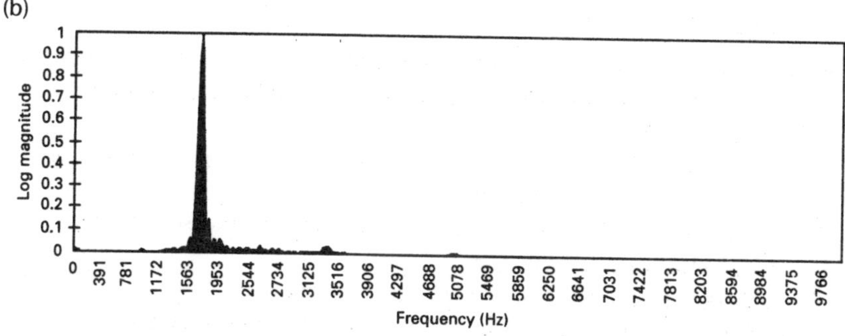

Fig. 25.1. Acoustic characteristics of the mole cricket *Gryllotalpa gryllotalpa* L. (a) Time domain plot; (b) spectrogram.

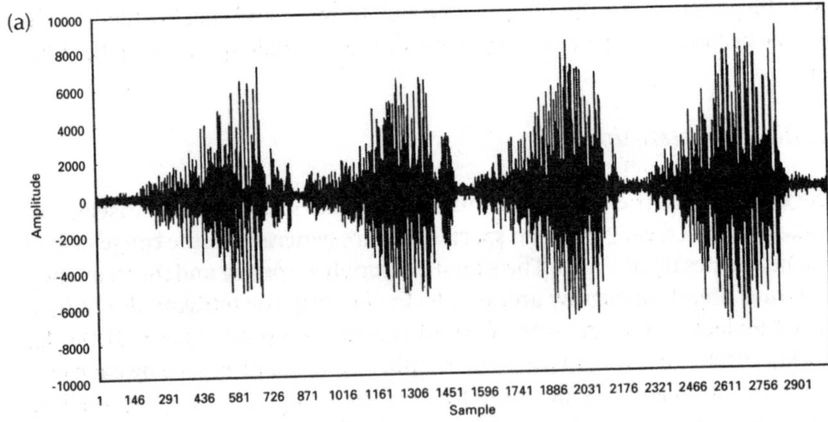

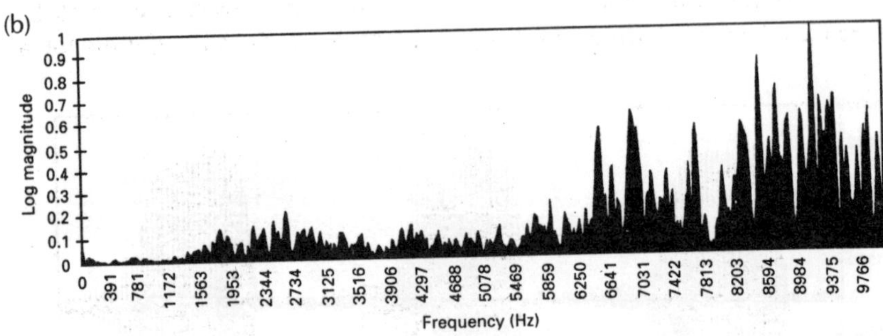

Fig. 25.2. Acoustic characteristics of the grey bush cricket *Platycleis albopunctata* G. (a) Time domain plot; (b) spectrogram.

representative species in each group. The sound structure can be described by the following terms:

1. Pulse. Each tooth strike produces an underdamped acoustic transient which may have several oscillations before decaying. The gap between tooth impacts varies between species.

2. Syllable. One upward or downward stroke of the scraper over the file is termed a syllable and is composed of a number of pulses.

3. Trill. Multiple consecutive syllables (no inter-syllable pause) which can be produced for some considerable time.

4. Chirp. Syllables occurring in repeated short sequences.

5. Song. The song structure varies considerably between species and group. Songs can also have variable structures to represent courtship, rivalry and copulation.

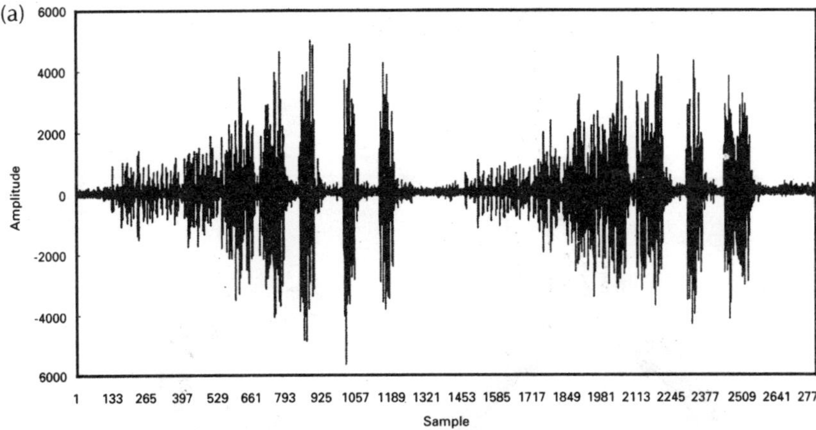

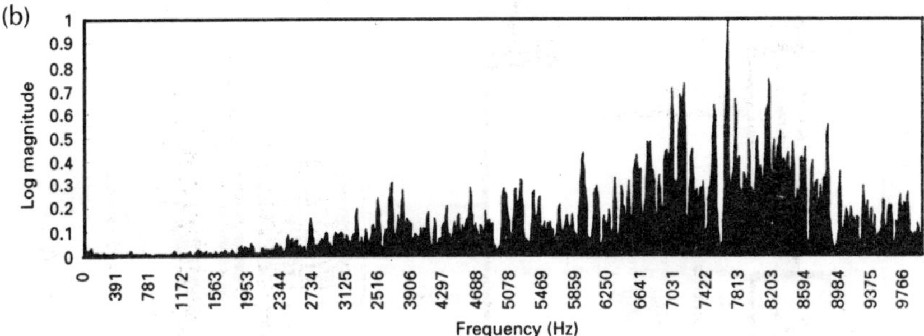

Fig. 25.3. Acoustic characteristics of the lesser mottled grasshopper *Stenobothrus stigmaticus* R. (a) Time domain plot; (b) spectrogram.

It is possible to extract some or all of these features by examining the time and frequency domain components of the acoustic signal. Traditionally, this has been carried out manually, however, the availability of high speed computing can provide a high degree of automation. Once the features have been extracted, it is then possible to attempt recognition.

Insect Sound Analysis and Recognition (ISAR) System

Figure 25.4 shows a block diagram of the ISAR system which comprises a PC (486 or better) with an analogue to digital (A–D) converter interface which may be as simple as a soundblaster stereo card. The system operates under Windows and was written in ‘C’ for Windows and has been designed for modularity and ease of use, making full use of the Windows interface. Other work involving acoustic signal analysis at the University of Hull makes use of a digital signal

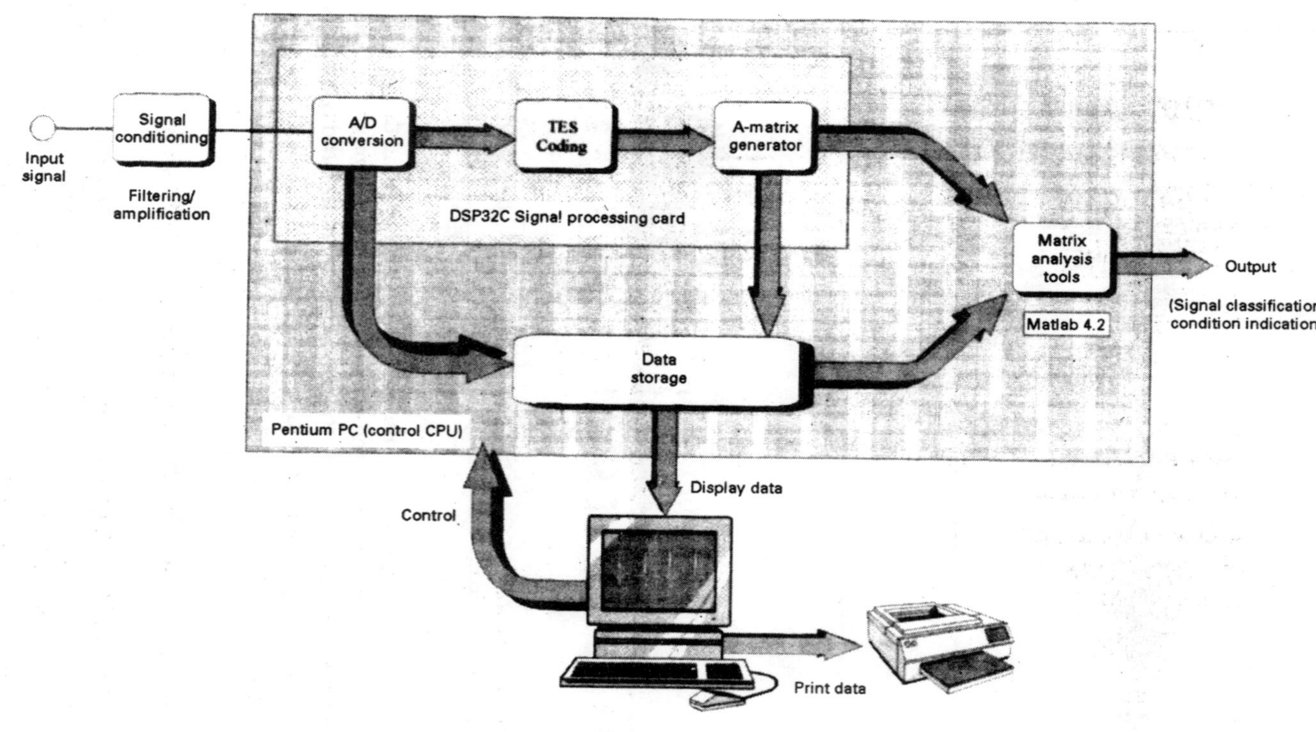

Fig. 25.4. Block diagram of the Insect Sound Analysis System.

processing card (DSP32C) to carry out some of the more numerically intensive work.

In simple terms, ISAR performs the following operations:

- extraction of information on the sound production mechanism such as number of tooth impacts, resonances, etc.
- higher level parameter extraction including many of the features described in 'Acoustic characteristics'.
- recognition of species and intra-species call type.

The sampling rate currently used is 20 kHz giving a theoretical maximum signal of 10 kHz which in reality is closer to 5 kHz due to the presence of non-ideal anti-aliasing filters at the input of the A–D converter. This restriction should be borne in mind when considering the spectrograms.

The signal processing techniques employed within ISAR fall into three categories – general functions commonly used for signal processing, time-encoded signals and artificial intelligence techniques. These are described in the following sections.

General signal processing functions

Signal processing functions implemented in ISAR are listed below:

1. Short-time average zero-crossing rate. This gives an indication of the frequency content, particularly for narrow-band low noise signals.
2. Average energy. The short time average energy of the signal is calculated as the square of the amplitude averaged over an interval. This is used to determine, for example, gaps or pauses between song elements.
3. Amplitude probability density function (APDF). The APDF gives an estimate of the range of signal amplitudes which has use in statistical descriptions (Scalabrin *et al.*, 1996).
4. Short-time autocorrelation. The autocorrelation function provides information on the presence or absence of periodicity in the signal. Although implemented, this function has not been used.
5. Power spectral density (PSD). The PSD (derived from a fast Fourier transform) is a well known and widely used technique for determining the frequency content of a signal.

Time-encoded signals

Time-encoded signal (TES) processing was developed in the 1970s by King (King and Gosling, 1978) as a purely time domain approach to the compression of speech for digital transmission. It has subsequently been used in a number of applications including acoustic condition monitoring of machinery (Lucking and Chesmore, 1994; Lucking *et al.*, 1994). TES characterizes any bandlimited signal between successive zero-crossings (termed epochs). Each epoch is described in terms of its duration in samples (D) and shape (S) usually taken as the number of minima as indicated in Fig. 25.5 which shows a 21 sample epoch with two minima (D = 21, S = 2). Signal energy may also be employed as a descriptor. The number of possible D–S combinations (or symbols) is termed the natural alphabet which can often be non-linearly mapped onto a smaller symbol set resulting in signal compression. In the original speech application, the coded symbols were transmitted and used to regenerate the speech signal at the receiver thus providing digital speech transmission at substantially reduced data rates.

TES can be described as the concatenation of a signal's D–S symbols and one analysis method is to examine the occurrence of pairs of symbols over time to give a histogram, A, which describes the number or proportion of symbols i and j occurring in succession, i.e. the number of times i is followed by j by a lag l. A two-dimensional histogram, the A-matrix, can be formed, expressed mathematically as:

$$a_{ij} = \frac{1}{(N-1)} \sum_{n=l+1}^{n=N} x_{ij}(n) \tag{25.1}$$

where l = lag, $x_{ij}(n) = 1$ if t(n) = i and t(n-l) = j (0 otherwise), and t(n) = nth TES symbol.

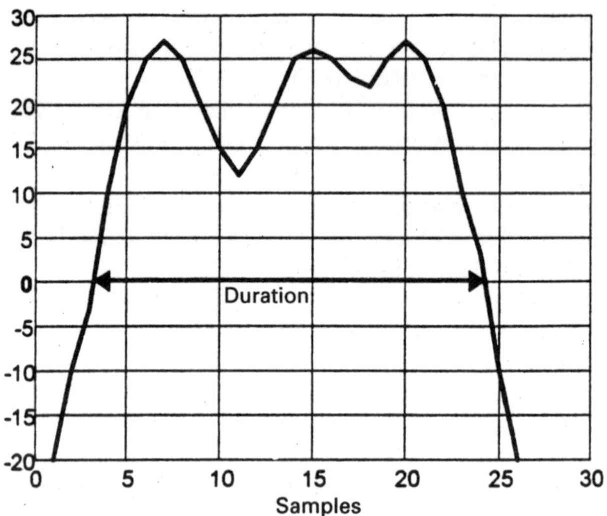

Fig. 25.5. Definition of an epoch in TES.

This fixed size histogram with time-invariant dimensions is the feature set usually used for classification purposes. The entry at position (i,j) represents the number (percentage) of occurrences of the TES symbol pair i and j where j is delayed relative to the first (in epochs). In this application, a lag of one epoch is used; multiple lags may also be employed giving rise to multi-dimensional matrices. The A-matrix is independent of any gain factors if the input signal has no dc component and it is insensitive to relative energies of different segments of the signal.

In this application, epochs are defined in terms of their duration and energy (i.e. D–S symbols), examples for Orthoptera are given in Figs 25.6, 25.7 and 25.8 and their significance explained in 'Preliminary Results'.

Artificial intelligence

In this application, artificial intelligence is used to integrate the data obtained by the various signal processing operations. There are two main approaches – expert systems and artificial neural networks; both have been used successfully in automated identification systems from fish species using ultrasound (Scalabrin *et al.*, 1996; Simmonds *et al.*, 1996) to bird song (Mills, 1995) and frog calling song (Taylor *et al.*, 1996). ISAR uses the expert system approach although work at the University of Hull has shown that neural networks are also feasible.

In ISAR, a multiple expert system termed a blackboard system is employed as this is more suited to problems with a mixture of heuristic and algorithmic

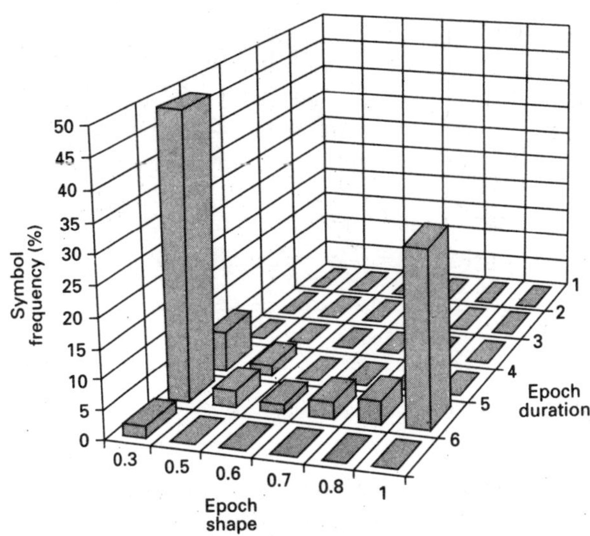

Fig. 25.6. A-matrix of *Gryllotalpa gryllotalpa* L.

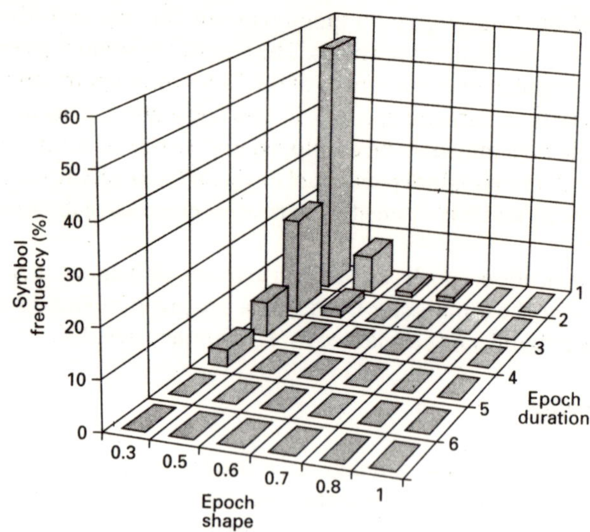

Fig. 25.7. A-matrix of *Platycleis albopunctata* G.

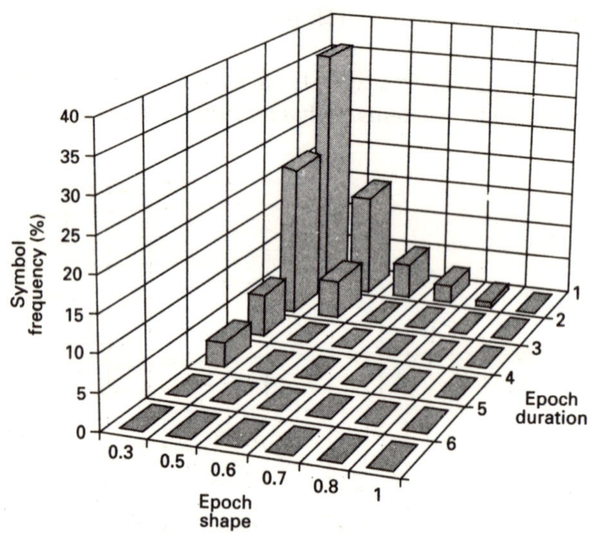

Fig. 25.8. A-matrix of *Stenobothrus stigmaticus* R.

operations. The blackboard system consists of a number of independent knowledge sources (experts) accessing a blackboard with levels corresponding to a hierarchical abstraction of the signal into its constituent units such as syllable, chirp/trill and phase/song type. Hypotheses about these song units are posted to the blackboard at the appropriate level for examination by other knowledge sources. For example, the lowest level of the blackboard contains the acoustically segmented signal which operates at the syllable level. The inclusion of a blackboard gives two advantages:

1. The ability to include non-algorithmic procedures in the form of heuristics or 'rules-of-thumb' and to mix these with algorithms such as PSD calculation.
2. It is relatively simple to include new knowledge sources (KS) as they are designed to be independent, allowing the system to be updated and reconfigured for new applications without changing the core control structures.

Features used in the determination of the segmentation are amplitude, zero-crossing rate and fundamental frequency with the main parameter being zero-crossing rate. The blackboard structure is given in Fig. 25.9, the levels and knowledge sources are defined as:

1. Level 0 (Parameters). Here, the song is partitioned into syllables by the SYL SEG syllable segmentation knowledge source which uses signal amplitude and zero-crossing rate to distinguish the song from background noise. The segmented song is then placed on level 1.
2. Level 1 (Syllable analysis). The SYL ANS syllable analysis KS estimates syllable length, interpause length and segment maximum amplitude. These parameters are posted back to level 1 which causes the chirp/trill (C/T) KS and FFT KS to activate. The FFT results (principal spectral component and 3 dB bandwidth) and syllable repetition frequency (SRF) are posted to levels 3 and 4 because of their significance in the recognition process.
3. Level 2 (Chirp/trill analysis). Chirp/trill detection is achieved through the

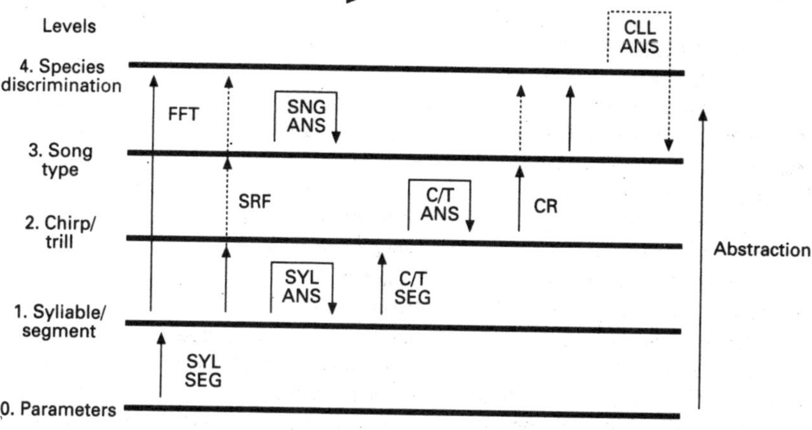

Fig. 25.9. Hierarchical diagram of the blackboard.

C/T ANS KS by examination of the syllable period length and syllable-to-pause ratio. The chirp repetition frequency (CRF) also propagates to levels 3 and 4.

4. Level 3 (song type) and level 4 (species discrimination) were not fully implemented due to a lack of time.

TES is integrated within levels 1 and 2 to provide additional discrimination parameters, for example to give more data on frequency characteristics.

Preliminary Results

The ISAR system has been tested using prerecorded songs of 17 species of British Orthoptera (Burton and Ragge, 1987) listed in Table 25.1. The cassette tape accompanies a comprehensive guide to British Orthoptera (Marshall and Haes, 1988), which should be consulted for detailed species descriptions. Table 25.2 gives data on syllable frequency features obtained from level 1 of the blackboard system, showing that useful data may be automatically extracted.

TES results for only three representative species will be presented, more results may be found in Femminella (1996). These are described below.

Mole cricket (Gryllotalpa gryllotalpa L.). Figure 25.1 shows the time and frequency domain plots for a typical calling song. Figure 25.6 shows the energy (amplitude) coded A-matrix which indicates the predominance of two symbols

Table 25.1. Orthoptera species selected for ISAR tests.

Latin name	English name	Family
Meconema thalassinum (De Geer)	Oak bush cricket	Tettigoniidae
Pholidoptera griseoaptera (De Geer)	Dark bush cricket	Tettigoniidae
Tettigonia viridissima (Linnaeus)	Great green bush cricket	Tettigoniidae
Platycleis albopunctata (Goeze)	Grey bush cricket	Tettigoniidae
Metrioptera roeselii (Hagenbach)	Roesel's bush cricket	Tettigoniidae
Metrioptera brachyptera (Linnaeus)	Bog bush cricket	Tettigoniidae
Decticus verrucivorus (Linnaeus)	Wart biter	Tettigoniidae
Acheta domesticus (Linnaeus)	House cricket	Gryllidae
Gryllus campestris (Linnaeus)	Field cricket	Gryllidae
Nemobius sylvestris (Bosc)	Wood cricket	Gryllidae
Gryllotalpa gryllotalpa (Linnaeus)	Mole cricket	Gryllotalpidae
Chorthippus parallelus (Zetterstedt)	Meadow grasshopper	Acrididae
Omocestrus viridulus (Linnaeus)	Common green grasshopper	Acrididae
Stenobothrus lineatus (Panzer)	Stripe-winged grasshopper	Acrididae
Myrmeleotettix maculatus (Thunberg)	Mottled grasshopper	Acrididae
Stenobothrus stigmaticus (Rambur)	Lesser mottled grasshopper	Acrididae
Chorthippus vagans (Eversmann)	Heath grasshopper	Acrididae

– one low amplitude (no sound) and the other at high amplitude. This is also evident in the time domain signal in Fig. 25.1a which shows an essentially binary (on-off) sound. Figure 25.1b indicates that the song is highly resonant, in fact this species digs a resonant burrow to extend communication distance.

Grey bush cricket (Platycleis albopunctata G.). Figure 25.2a is the time domain signal and Fig. 25.2b shows the frequency domain which indicates high frequency components. Examination of the A-matrix in Fig. 25.7 verifies this, where there is a predominance of short duration, low energy symbols corresponding to high frequencies.

Lesser mottled grasshopper (Stenobothrus stigmaticus R.). Figures 25.3 and 25.8 show the time and frequency domain signals and the A-matrix respectively. Again, high frequencies dominate but the A-matrix remains distinguishable from *P. albopunctata.*

The spectral estimates in Table 25.2 for the above species which are averaged over representative periods also correspond to the spectrograms, thus indicating that the level 0 and 1 knowledge sources are operating correctly.

It is also evident from Figs 25.6, 25.7 and 25.8 that the A-matrix potentially forms a basis for discrimination in its own right whereas here it is used only to support blackboard hypotheses. A-matrices are currently under investigation as preprocessors for input to artificial neural networks and preliminary results (unpublished) indicate a high degree of accuracy in species discrimination for seven species.

Table 25.2. Summary of syllable frequency features from the blackboard system.

	Carrier frequency (Hz)		Syllable repetition frequency (Hz)		3 dB bandwith (Hz)	
Species	Mean	Range	Mean	Range	Mean	Range
Mole cricket	1642.4	1640–1679	32.1	29–34	314.5	273–469
Field cricket	5114.0	5000–5195	36.1	30–44	443.0	391–508
House cricket	3957.8	3593–4279	13.8	12–17	482.7	215–664
Bog bush cricket	8066.8	4960–9414	95.0	64–180	858.5	137–2812
Dark bush cricket	9817.0	9765–9843	32.8	31–40	338.7	274–469
Grey bush cricket	8173.0	6406–9765	30.9	27–38	351.5	234–781
Meadow grasshopper	7120.0	4062–8046	56.3	21–157	562.0	157–1329
Woodland grasshopper	7646.0	7031–8398	16.8	14–19	852.5	274–1250
Great green bush cricket	9598.0	9101–9960	36.6	7–91	492.0	78–1563
Lesser mottled grasshopper	7574.3	6875–8437	59.0	24–196	786.7	156–2031

Discussion and Conclusions

The project described in this paper involved a combination of 'traditional' signal processing techniques with an expert system which shows considerable promise as a platform for automatic acoustic identification. Much work remains to be done, particularly in the implementation of levels 3 and 4 of the blackboard. An increased sampling rate of 100 kHz or higher will be implemented to enhance performance for species with high frequency content.

In addition, employing TES provides additional data for analysis but the potential power of A-matrices for species identification in their own right is only just being investigated.

Applications of ISAR are many and varied and apply directly to the categories of acoustic insect research noted in the introduction. It is important to note that the use of a blackboard system enables rapid implementation for any bandlimited acoustic signal such as amphibia, birds, mammals (especially bats) and whales.

The ultimate goal of the ISAR research is to create a species identification instrument for hand-held use. It is estimated that such a device is two to three years away.

In conclusion, the advent of high speed and powerful computers is providing new techniques for automatic acoustic species identification that were difficult or impossible to implement only a few years ago.

References

Burton, J.F. and Ragge, D.R. (1987) *Sound Guide to the Grasshoppers and Allied Insects of Great Britain and Ireland.* Harley Books, UK.

Chapman, R.F. (1982) *The Insects: Structure and Function.* 2nd edn. Hodder and Stoughton, UK.

Femminella, O.P. (1996) Acoustic Signal Analysis of Insect Sounds. Final Year Undergraduate MEng Thesis, University of Hull, UK.

Hagstrum, D.W., Vick, K.W. and Webb, J.C. (1990) Acoustic monitoring of *Rhizopertha dominica* (Coleoptera: Bostrichidae) populations in stored wheat. *Journal of Economic Entomology* 83, 625–628.

King, R.A. and Gosling, W. (1978) Time-encoded speech. *Electronics Letters* 15, 1456–1457.

Lucking, W.G. and Chesmore, E.D. (1994) Acoustical condition monitoring of a mechanical gearbox using artificial neural networks. *10th International Conference on Systems Engineering*, Coventry University, UK.

Lucking, W.G., Darnell, M. and Chesmore, E.D. (1994) Acoustical condition monitoring of a mechanical gearbox using artificial neural networks. *IEEE Conference on Neural Networks*, Florida, USA.

Marshall, J.A. and Haes, E.C.M. (1988) *Grasshoppers and Allied Insects of Great Britain and Ireland.* Harley Books, UK.

Mills, H. (1995) Automatic detection and classification of nocturnal migrant bird calls. *Journal of the Acoustical Society of America* 97, 3370–3371.

Scalabrin, C., Diner, N., Weill, A., Hillion, A. and Mouchot, M. (1996) Narrowband acoustic identification of monospecific fish shoals. *ICES Journal of Marine Science* 53, 181–188.

Shuman, D., Coffelt, J.A., Vick, K.W. and Mankin, R.W. (1993) Quantitative acoustical detection of larvae feeding inside kernels of grain. *Journal of Economic Entomology* 86, 933–938.

Simmonds, E.J., Armstrong, F. and Copland, P.J. (1996) Species identification using wideband backscatter with neural network and discriminant analysis. *ICES Journal of Marine Science* 53, 189–195.

Stephen, R.O. and Hartley, J.C. (1995) Sound production in crickets. *Journal of Experimental Biology* 198, 2139–2152.

Taylor, A., Grigg, G., Watson, G. and McCallum, H. (1996) Monitoring frog communities: an application of machine learning. *Eight Innovative Applications of Artificial Intelligence Conference*, AAAI Press.

Treherne, J.E., Berridge, M.S. and Wigglesworth, V.B. (1978) *Advances in Insect Physiology Volume 13*.

Trematerra, P. and Pavan, G. (1995) Ultrasound production in the courtship behaviour of *Ephestia cautella* (Walk.), *E. kuehniella* (Z.) and *Plodia interpunctella* (Hb.) (Lepidoptera: Pyralidae). *Journal of Stored Product Research* 31(1), 43–48.

26

A Historical Review of Identification by Computer

R.J. Pankhurst

Royal Botanic Garden Edinburgh, 20A Inverleith Row, Edinburgh EH3 5LR, UK
Fax: +44 (0)131 552 0382/E-mail: richard@rbge.org.uk

Introduction

The 1996 conference for which this chapter was prepared (see Preface) commemorates an earlier Systematics Association meeting, held in King's College, Cambridge in 1973 and summarized in the conference volume (Pankhurst, 1975a) which appeared just 21 years earlier. This work was consolidated by the appearance of two textbooks on computer methods for identification (Pankhurst, 1978a, 1991). Botanists, especially those working with higher plants, have been prominent in the development of computerized identification, but the techniques are equally applicable to zoology, so it is not clear why zoologists have been less involved. It has been suggested that it is because they are concerned with much larger numbers of taxa, but that could be argued as a reason in favour of computerization, not against it.

DELTA Format

The DEscription Language for TAxonomy was invented by Dallwitz (1980), and is the international data standard for identification data. It was one of the first standards to be adopted by the Taxonomic Databases Working Group (TDWG). DELTA is a way of coding the descriptions of taxa, their characters and states in a data matrix, using ASCII text, and is not, properly speaking, a database. There is some confusion caused by references to the 'DELTA system', meaning Mike Dallwitz's own package of programs for DELTA, since there are a number of other

 Information Technology, Plant Pathology and Biodiversity (eds P. Bridge, P. Jeffries, D.R. Morse and P.R. Scott)

DELTA systems. The examples in this paper are taken from the PANKEY DELTA system (Pankhurst, 1995). The DELTA format was adopted by PANKEY in about 1975, with the help of Mike Dallwitz who provided some of the program code.

There is no need here to describe the DELTA format in detail, which is discussed in this volume by Dallwitz *et al.* (Chapter 19), but a few illustrations will serve to show that, in the opinion of this author, it is intuitive and straightforward. The data is divided into sections by directives. The first directives give general specifications, such as:

*NUMBER OF CHARACTERS 24

Definitions of characters and states are set out with numbers for the characters, such as number 4, 'shrubbiness of stem', and with states 1 and 2, as in:

#4. Stem <shrubbiness>
1. herbaceous/
2. woody at base/

Characters can be qualitative, as in the above example, or quantitative, either integer or real. Integers can be discontinuous, to express variation such as 'petals 3, 5 or 7'.

The descriptions of actual taxa are preceded by the taxon name, e.g.

#J. pinnata/

and followed by pairs of character numbers and state numbers, such as 4,2 meaning 'stem woody at base'. Numerical characters are expressed directly, e.g. 2,10–25 might stand for 'stem 10–25 cm high'. It is very important to be able to describe variation, so that 20,3/5 might mean 'petals white or red'.

There are two important features of DELTA which require a little explanation. One is that the logical dependency of one character on another must be expressed with the directive DEPENDENT CHARACTERS. For example, if there are some species in a genus which have aerial stems, and others which just form rosettes, then characters of the stem (such as 'stem hairs') depend on whether the character 'stem presence' has the state 'present'. This is an example of a relation between a controlling character and a dependent character. The other important feature is the KEY STATES directive, which is used to show how numerical characters should be transformed into qualitative characters for certain operations, by splitting their variation into non-overlapping ranges

DELTA data can be edited directly with text editors, or with word processors in text mode, or can be manipulated directly with special purpose freeware editors such as the DELTA Editor (DEDIT, Pankhurst, 1995, see Fig. 26.1). These special editors behave like word processors for DELTA, so that users do not need to know the details of the format.

There has been considerable debate recently about the redesign of the DELTA format, inspired either by those who think that DELTA is over-complex, or by those who think that it needs to be elaborated. In the view of this author, DELTA

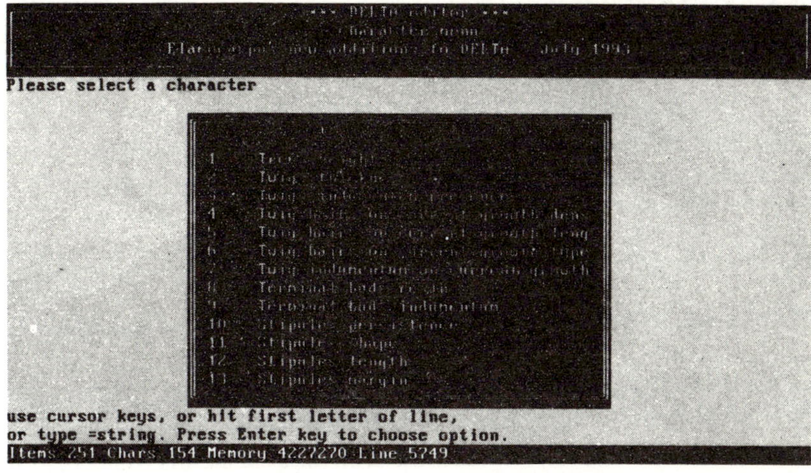

Fig. 26.1. DELTA editor with menu for selecting a character.

already neatly covers the description of most kinds of natural variation, and simpler formats would not be adequate. On the other hand, there are complex patterns of variation which are difficult or impossible to express in DELTA. Readers are encouraged to join the debate which is taking place in the DELTA list server (delta-l@nic.surfnet.nl) and the DELTA Newsletter (1988–), but not to support efforts to promote other formats in competition with DELTA.

Identification Methods

The main types of identification method are summarized in Table 26.1. As a generalization, most identification methods fall into one of two broad groups; comparison or step-by-step elimination. The comparison method is presumably that which is used when an unknown specimen is compared with a museum collection, or compared with the illustrations in a natural history guide book. Briefly, an estimate of the similarity between the unknown and a range of possible taxa is made, although this assessment may be subjective and not numerical, and the identification assigned to the taxon which agrees best. Character weighting might be involved, by giving more importance to some characters than to others. An example of this method is the matching program (Pankhurst, 1975b). The input data is a description of the unknown specimen, as complete as can be made, in the same manner as a taxon in the DELTA data matrix. The output is a list of the taxa in the DELTA data set with the highest scores. An indication of the correct subgroup can also be given, e.g. the right subgenus of a genus. The matching method is possibly the most accurate of all

Table 26.1. Different types of identification programs.

PANKEY	Key construction (batch mode) Key construction (interactive) Character analysis Online (expert) identification Punch card keys (polyclaves); now obsolete Description printing Diagnostic descriptions (heuristic or complete) Matching (comparison) Online keys (diagnostic keys in interactive mode) Format conversion, for clustering or cladistics
PANDORA	Relational database including identification data

identification methods, since it makes use of all the available information. At the same time, it also requires the most effort, and probably for this reason, it is little used. Identification methods using probability, often Bayes' theorem or maximum likelihood, as favoured by bacteriologists and in medical diagnosis, are arguably also matching methods which use a different formulation. They are discussed in this volume by Bryant (Chapter 28).

The familiar diagnostic (dichotomous) key is a good example of an elimination method. The questions ask for the states of one or sometimes several characters to be observed, and the response effectively eliminates taxa which disagree with the observation, and leads to further questions until all alternatives are eliminated and only one taxon remains. Normally only some of the total list of characters need to be considered, thus saving effort, but at the risk that the result is wrong if any of the characters have been wrongly observed.

Key Construction

Programs to construct diagnostic keys were among the first identification programs to be written (Pankhurst, 1970; Dallwitz, 1974; Morse, 1974). The programs ran in batch mode and the keys were printed out on paper. The advantages of this were:

- speed, once the data matrix has been set up.
- accuracy, since the key necessarily agrees with the data.
- repeatability; new keys can quickly be made if the data changes.

At a much later date, a program was written (Pankhurst, 1988a) to construct diagnostic keys interactively, in the style of one of the expert identification programs. This approach allows the greatest possible control by an expert author over the details of key production, whilst keeping all the advantages just listed. (See Fig. 26.2.)

```
JURINEA TEST
BEST CHAR DELE DIAG DIFF EXAM EXPA FINI HELP RENU SAVE SCOP TAXA VIEW
Next Taxa Quit Recombine Delete Order Skip

 1     (10)Basal leaves (=1)subglabrous above.
       TAXA 8 9 10 11 12 13 17

 2     (10)Basal leaves (=2)arachnoid-tomentose above.
       TAXA 1 4 5 6 7 10 14 15 16

 3     (10)Basal leaves (=3)setose above.
       TAXA 2 3

This will be lead number      1


Accept(A), abandon(Q) display again(D)? Tax    17 Cha   18   Sco 1   Leads     0
```

Fig. 26.2. Evaluation of character in interactive key construction.

Although diagnostic keys continue to be written and published in floras and monographs generally, and remain the principal means by which working biologists do their identifications, computerized key construction programs are still not very often used. One reason which has been given for this is a surprising one; the programs are designed to be precise, and do not permit users to produce logically imperfect keys. This might seem to be a highly desirable feature, but there is evidence that users resent it, preferring to gloss over difficulties such as the lack of available characters for making exact distinctions between taxa.

Online Diagnostic Keys

A few programs have been written which present diagnostic keys online. Possibly the first of these was published by Wilson and Partridge (1986), as part of a learning program. The leads of an existing key are displayed in turn as character choices are typed in. Apart from the convenient form of presentation, there seems little point in an online diagnostic key for its own sake, since all the disadvantages of diagnostic keys are preserved.

Online Identification

By far the most popular type of identification program is the interactive, online program or multi-access key. The first such program of this kind was due to Boughey *et al.* (1968), followed by Morse (1974) and Pankhurst and Aitchison (1975). The latter program has gone through many versions; the current one is called ONLIN7 Mike Dallwitz has his own expert program, called INTKEY (Dallwitz *et al.*, Chapter 19, this volume). Code from ONLINE version 3 contributed to the development of INTKEY, just as DELTA has contributed to PANKEY. There are numerous other interactive identification programs. Approximately 40 of these are known to the author, and many of these have been developed in the last 5 to 10 years. Some of them (CABIKEY, GENCOMEX, Poisonous plants, LINNAEUS II, LucID, DIABOLIS) were presented during the conference for which this chapter was prepared. Some of this development work is directed to making versions for newer operating systems such as Windows 95, or towards the Internet by using JAVA, but frankly, there is much reinvention and duplication of effort here. This is to be regretted. Increased taxonomic research is vital to meeting the demands of the biodiversity crisis, and scarce resources are being wasted. This energy would better be directed towards the many aspects of computerized identification which still need research and which have been neglected (see below).

Dallwitz *et al.* (Chapter 19, this volume) explain the details of an online program, but briefly, its most important advantages are:

1. The ability to choose from all available characters and in any order, without having to follow a fixed character set in a fixed order, as in a diagnostic key. This freedom brings with it the need for commands to choose the best character to use next, in various senses. These might be: (i) the best character, i.e. the one with the most information content (the famous BEST command in ONLIN7, Fig. 26.3); (ii) the most diagnostic character for a particular taxon, distinguishing it from all the others (DIAG command); and (iii) difference commands, showing the character difference between the unknown and any taxon, or between any two taxa (DIFF command).
2. The means to adjust the error margin (the variability limit, invented by Morse (1974)) so that an unknown can still be identified correctly even if characters are wrong. If the limit is set to some value greater than zero, then taxa which do not exactly agree with the description of the unknown are still kept in the list of potentially correct identifications and then, interestingly, this ceases to be strictly a step-by-step elimination algorithm. There is therefore not a hard and fast distinction between elimination and comparison.
3. The means to go backward by removing or editing characters already entered, just as easily as going forward by adding characters.
4. The use of images of characters and taxa, for explanation purposes (Pankhurst, 1989).

DELTA datasets with ONLIN7 or INTKEY are being applied to major taxonomic projects, such as the Fruitfly CD-ROM (Thompson, 1996) and the Dipterocarps of South-east Asia (Newman *et al.*, 1996), and the datasets for World Angiosperms (Watson and Dallwitz, 1994) and genera of Grasses (Watson and Dallwitz, 1988).

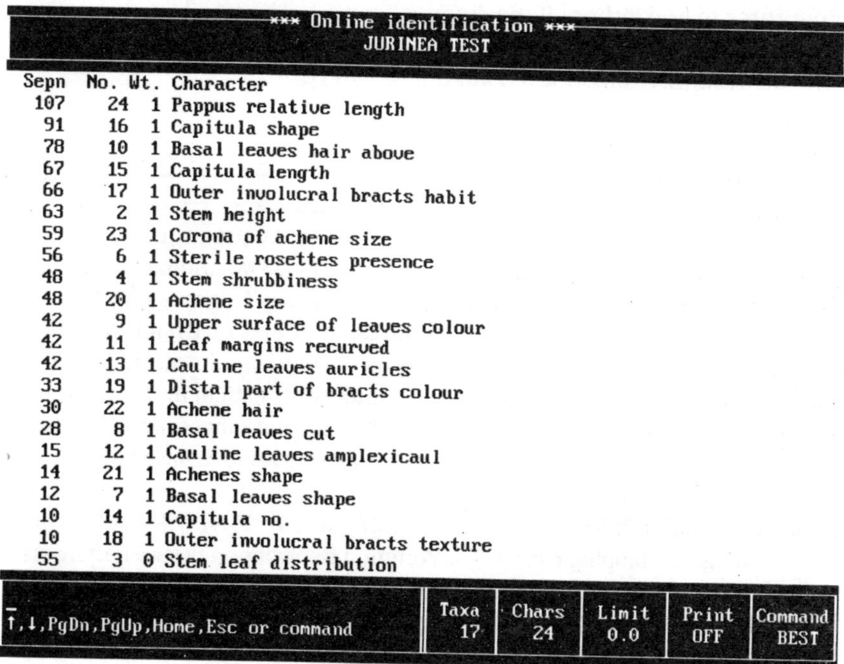

*** Online identification ***
JURINEA TEST

Sepn	No.	Wt.	Character
107	24	1	Pappus relative length
91	16	1	Capitula shape
78	10	1	Basal leaves hair above
67	15	1	Capitula length
66	17	1	Outer involucral bracts habit
63	2	1	Stem height
59	23	1	Corona of achene size
56	6	1	Sterile rosettes presence
48	4	1	Stem shrubbiness
48	20	1	Achene size
42	9	1	Upper surface of leaves colour
42	11	1	Leaf margins recurved
42	13	1	Cauline leaves auricles
33	19	1	Distal part of bracts colour
30	22	1	Achene hair
28	8	1	Basal leaves cut
15	12	1	Cauline leaves amplexicaul
14	21	1	Achenes shape
12	7	1	Basal leaves shape
10	14	1	Capitula no.
10	18	1	Outer involucral bracts texture
55	3	0	Stem leaf distribution

↑,↓,PgDn,PgUp,Home,Esc or command	Taxa	Chars	Limit	Print	Command
	17	24	0.0	OFF	BEST

Fig. 26.3. BEST command in ONLINE expert identification.

Description Printing and Diagnostic Descriptions

Although one of the major purposes of DELTA is to provide a means to encode descriptions of taxa that are published as written text, it was clear from early days that it makes a great deal of sense to use it for the opposite process. In other words, descriptions are best created initially in DELTA format and then converted to written text by a computer program (Morse, 1974; Pankhurst, 1978b; Dallwitz, 1980).

The descriptions of taxa that are printed in many taxonomic publications distinguish the diagnostic characters by italicizing them. These characters are meant to be a character or set of characters which uniquely distinguish the taxon from all others. Such character sets can readily be computed with the aid of any online identification program, using the description of the taxon as if it were the description of an unknown. This process will reach a positive identification for the known taxon, if the data matrix is adequate. The characters from the specimen description are next deleted one by one. If the taxon ceases to be uniquely identified after a character is removed, then that character must be restored. At the end of this process, the set of characters remaining is a diagnostic character set for the taxon. This is a heuristic algorithm, which does not guarantee to find either all the diagnostic sets or the smallest character set (Barnett and Pankhurst, 1974). It is possible to find all diagnostic character sets with an exhaustive algorithm (Pankhurst, 1983).

Character Analysis

As an aside to identification programs based on DELTA, programs for character analysis deserve to be mentioned. Much exploration of taxonomic data is aimed at finding correlations between characters. These may be positively significant, because they are evidence of common ancestry, or may be trivial, because the same information is being represented by more than one character. Either way, most characters are qualitative, and so standard statistical measures such as correlation coefficients do not apply. The DELTA format makes it possible to test the correlation of non-numerical characters, using an idea based on information theory (Estabrook, 1967). The CHANAL program in the PANKEY system computes the common information in characters. In Fig. 26.4, characters a and b are shown as overlapping circles. The common information between them is the area H(a,b) and the ratio S will be zero for independent characters and one for totally dependent characters. The term 'symmorphy' is suggested for this measure, in order to avoid any presumption that either synapomorphies or symplesiomorphies are involved.

Weighting and Probability in Identification Methods

We have reviewed all the extant methods of identification without saying very much about character weighting. The DELTA format allows weighting to be assigned to every character and taxon. The directives for naming and setting groups of characters and of taxa are also forms of weighting. Some, but not all, of the DELTA programs can exploit this. For example, in the batch mode construction of diagnostic keys, character weights are used to encourage the selection of weighted characters before unweighted ones, and similarly for taxa. Opinions differ among taxonomists about whether *a priori* weighting should be used. In online programs, the initial weighting of characters is sensible enough, but as soon as some characters have been given, the set of possible taxa changes, and the weighting scheme ceases to be appropriate and can even be obstructive. The use of weighting in online programs is therefore not generally recommended.

While probabilistic methods are used in microbiology and in medical diagnosis, it must be said that botanists working with higher plants have little enthusiasm for identification with probability. It is expected and desired that every specimen observed or collected can be named with 100% confidence, otherwise the specimen may be destroyed as being inadequate, and the collector

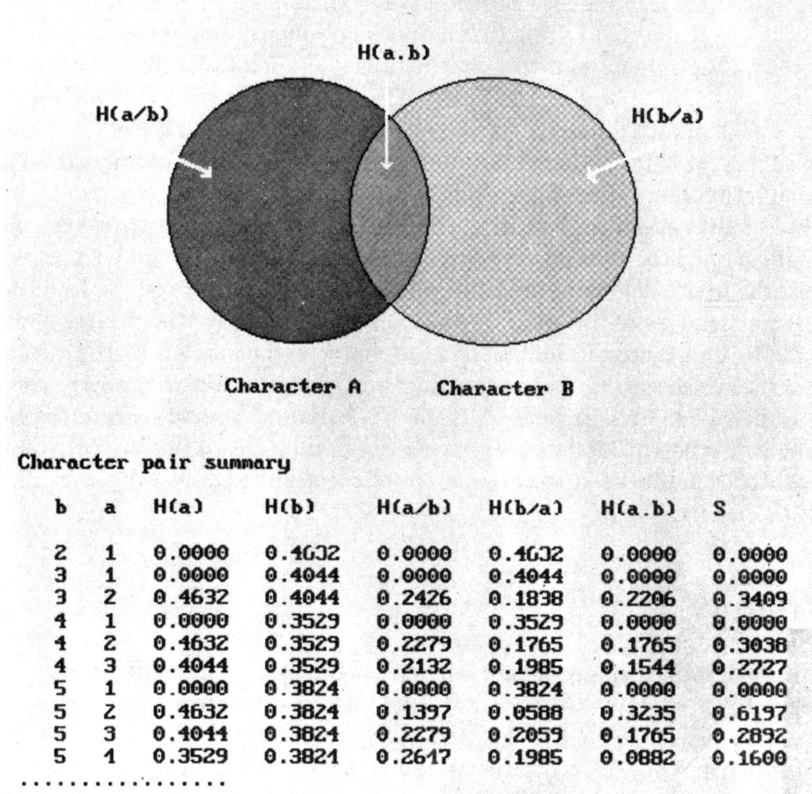

Character pair summary

b	a	H(a)	H(b)	H(a/b)	H(b/a)	H(a.b)	S
2	1	0.0000	0.4632	0.0000	0.4632	0.0000	0.0000
3	1	0.0000	0.4044	0.0000	0.4044	0.0000	0.0000
3	2	0.4632	0.4044	0.2426	0.1838	0.2206	0.3409
4	1	0.0000	0.3529	0.0000	0.3529	0.0000	0.0000
4	2	0.4632	0.3529	0.2279	0.1765	0.1765	0.3038
4	3	0.4044	0.3529	0.2132	0.1985	0.1544	0.2727
5	1	0.0000	0.3824	0.0000	0.3824	0.0000	0.0000
5	2	0.4632	0.3824	0.1397	0.0588	0.3235	0.6197
5	3	0.4044	0.3824	0.2279	0.2059	0.1765	0.2892
5	4	0.3529	0.3824	0.2647	0.1985	0.0882	0.1600

....

Fig. 26.4. Character analysis with table of correlations

instructed to take more care next time. Some specimens may be kept until an expert comes by who is able to decide, or who will publish new taxa based on them. A more positive view can be taken of the use of probability to test percentage likeness, as in matching programs, or to some degree in online programs, or by estimating degrees of belief (Atkinson and Gammerman, 1987), but most identification programs do not provide this option. There are also difficulties in assigning probabilities to the occurrence of characters or to the chances of encountering a particular taxon, which are really variables, and not constants.

Expert Systems

Expert systems, in the sense meant by computer scientists, are general purpose systems for problem solving. Why not use them for identification in biology?

Expert systems technology is rather loaded with obscure terminology, and for a time it was tempting to think that it must contain some essential and special features that were missing from the identification programs created by biologists. Edwards and Morse (1995) argue that expert systems for identification are fundamentally different. However, Chapter 8 in Pankhurst (1991) argues that expert systems contain essentially the same ideas, even if couched in different terms. Since then the author's views have not changed, and Table 26.2 presents the equivalent features of online identification and expert systems. Identification programs do not generally decide for the user about which character the user will have to use next, which is what a control system would do. Botanists and zoologists expect to be allowed to decide for themselves about that! The data matrix formulation of the character data is a far more practical and efficient way of organizing data than any system of rules, since every character state combination for every taxon would require a separate rule. In conclusion, it appears that the evolution of identification algorithms has been quite separate from that of expert systems, but has paralleled it over the same time period.

Databases

The use of hypertext is now very familiar via browsers of the World Wide Web on the Internet. Hypertext is an excellent tool for the presentation of encyclopaedic information, and ETI at Amsterdam (Schalk, Chapter 20, this volume) have shown us how to exploit this technique to the full for the presentation and distribution of taxonomic information. However, the links between pages of hypertext are often only present because somebody has manually done the work to place them there, and this is a fundamental disadvantage of hypertext. There is still nothing to better relational databases, when it comes to structuring information. This is not the place to discuss taxonomic databases in general, but only to review identification information in relational databases.

Table 26.2. Comparison of online identification and expert systems.

Online ID	Expert system
(Usually absent)	Control system
Variability limit	Graceful degradation
BEST character command	Data driven search
Diagnostic character command	Goal driven search
DIFFerence command	Explanatory power
Character replace/delete	Reversibility
DELTA data matrix	Rules, frames, etc.

Several database systems have the means to include descriptive information, e.g. ALICE (Allkin, Chapter 9, this volume) and BRAHMS (Filer, 1994), but only one database includes a complete DELTA system. The present author began work in 1986 to provide a fully specified taxonomic database, now called PANDORA (Pankhurst, 1988b, 1993), which includes a section for DELTA data (Pankhurst and Pullan, 1996). PANDORA is programmed in Advanced Revelation, which is an extremely sophisticated database management system, one of the few which has sufficient power to do justice to the complexities of taxonomic data. DELTA data files can be imported and exported, and since the main database can include data on taxa at many levels, the database can store several DELTA datasets at once, although at any one time only one DELTA data set is active. The characters and states of a DELTA data set are stored in a database table which is indexed on the names of characters and searchable in the usual way (Fig. 26.5).

All the different types of DELTA character are accommodated, as well as the dependency rules that connect them. The logical rules for data types and connections are enforced by the database. The names used for taxa in the DELTA file are stored in their own table, and linked to the main taxon definition table. Taxa which are in the main database do not have to be in the DELTA data sets, and vice versa, nor do they necessarily have to be referred to by the same name strings. This allows for DELTA data sets to be either created entirely outside the database and then imported or, conversely, they can be created from scratch inside the database. There are data entry and editing programs for setting up descriptions of taxa or specimens (Fig. 26.6), or of published descriptions.

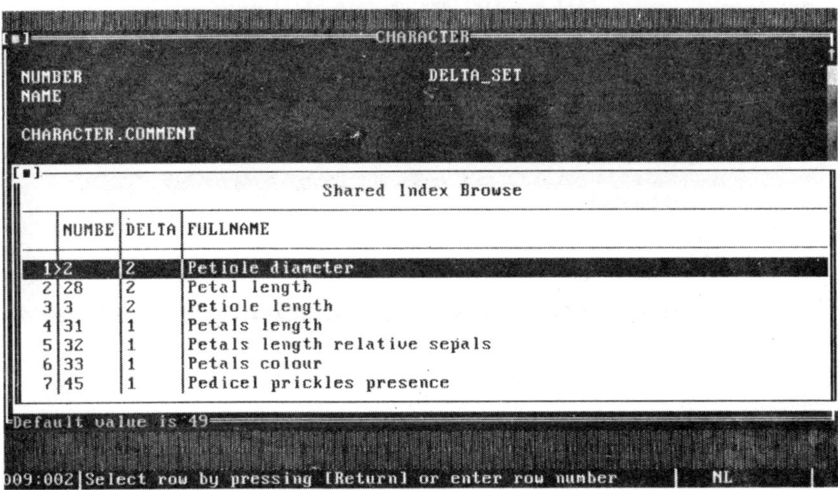

Fig. 26.5. Index to look up characters in PANDORA database.

Descriptions of such items can be combined by a special program to generate new descriptions of taxa, which are firmly based in stored information about the source of the descriptive data, whether from preserved specimens or from publications. There is even an identification program within the database, but mostly the DELTA processing is carried out by calling the PANKEY programs from within the database.

New Research in Identification

Looking back over the years to the publication of the first Systematics Association volume in 1975, it is possible to suggest that the basic principles of computerized identification methods were already in place at that time. There has been steady progress since then, up until about 1990. The main improvements during the later 15 years have been:

1. Above all, the advent of cheap PCs and colour graphics, enabling the use of colour images in online identification. No doubt this would have happened much sooner if the technology had been available. This revolution has also allowed new operating systems with a strong emphasis on graphical user interfaces, i.e. Windows etc., but apart from providing the means to make a consistent user interface together with improved user friendliness, there have been virtually no new techniques in identification emerging. One novelty in this area has been the automated construction of line drawings to correspond with specimen descriptions, attempted by Grad *et al.* (1996).

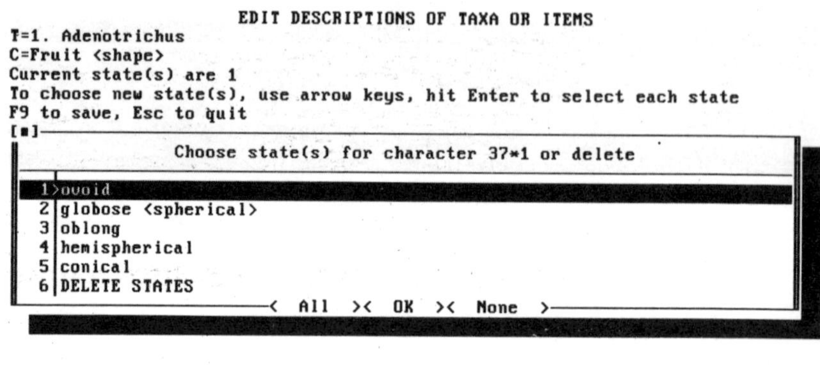

Fig. 26.6. Editing of fruit shape for *Rubus adenotrichus.*

2. Improved algorithms for diagnostic descriptions.
3. Interactive construction of diagnostic keys.
4. New ideas on how to use probability in identification (Atkinson and Gammerman, 1987).
5. DELTA data in relational databases.

The author's own conviction, dating from about 1985, is that essentially all the algorithms for identification have already been discovered, and that we already have all the basic software that is needed. Two vital things are still lacking:

1. Data. Apart from the efforts of a few pioneers (mainly in Australia), we are woefully short of computerized data for identification throughout all kingdoms of living organisms! Major national and institutional projects need to be put in place to collect identification data.
2. Teaching. The fact that computerized identification techniques are still only used by a few scattered enthusiasts, and that no major taxonomic or agronomic organization has yet endorsed them as standard procedures, shows that a vital part of the education of modern biologists is still lacking. All courses in taxonomy should include computerized identification.

Areas of research in identification that need attention are:

- data mining.
- error rates and accuracy.
- hypertext systems.
- neural networks.
- probability and weighting.

The term 'data mining' is used to refer to computerized techniques for the automatic or semi-automatic extraction of taxonomic data from text, scanned from the abundance of classical publications in libraries. The ideas in Diederich *et al.* (Chapter 5, this volume) on meta-information about characters must be relevant to this. The effect of choice of identification method and the use of weighting and of probability on accuracy and effectiveness has scarcely been touched, but see White and Sandlant (Chapter 24, this volume) and Tardivel and Morse (Chapter 23, this volume). We still have no idea whether probabilistic data (Bryant, Chapter 28, this volume) is actually valuable for identification or not – there are plenty of authors who insist that it is, but as far as the author is aware, there have never been any experiments to test it. Hypertext versions of online identification programs in JAVA on the Internet might be an advance, or might just be a reworking of an existing technique in a new operating environment. Neural networks (Boddy *et al.*, Chapter 21, this volume) seem to be a kind of matching procedure which works well, but we would like to know why, and how to design them intentionally to perform well.

Software Development for Identification Systems

Much of the software for personal computers used by an average computer user is produced by international corporations such as Microsoft. The average user therefore has reasonable expectations of ease of use, documentation, reliability and user friendliness (which even so, are not always justified!). See also Chapter 14 in this volume by Norton *et al.*, where there is a similar analysis of the current ills of identification software. The scale of resources available to produce familiar software such as word processors, spreadsheets and so forth, is gigantic. On the other hand, identification software is nearly always produced by enthusiasts, working overtime on their own, or rarely in teams of two or three people. These folk may of course be using the best modern professional methods of software development. The market for such software, in commercial terms, is microscopic. According to one major computer company, who actually assessed a taxonomic software project on behalf of the author, the commercial potential was too small, worldwide, by a factor of about 50. In other words, until taxonomic research gets the resources that its intrinsic importance merits, we are all (the author included) going to be producing software which does not match up to the expectations of our users. Bear with us!

References

Atkinson, W.D. and Gammerman, A. (1987) An application of expert systems technology to biological identification. *Taxon* 36, 705–714.

Barnett, J.A. and Pankhurst, R.J. (1974) *A New Key to the Yeasts*. New Holland, Amsterdam.

Boughey, A.S., Bridges, K.W. and Ikeda, A.G. (1968) *An Automated Biological Identification Key. Research Series No. 2*, Museum of Systematic Biology, University of California, Irvine.

Dallwitz, M.J. (1974) A flexible program for generating diagnostic keys. *Systematic Zoology* 23, 50–57.

Dallwitz, M.J. (1980) *User's Guide to the* DELTA *System: a General System for Encoding Taxonomic Descriptions*. CSIRO Division of Entomology, Canberra, Australia.

DELTA *Newsletter* (1988–) Western Australian Herbarium, PO Box 104, Corno, WA 6152, Australia.

Edwards, M. and Morse, D.R. (1995) The potential for computer-aided identification in biodiversity research. *Trends in Ecology and Evolution* 10(4), 153–158.

Estabrook, G.F. (1967) An information theory model for character analysis. *Taxon* 16, 86–96.

Filer, D. (1994) *BRAHMS – Botanical Research and Herbarium Management System. A Pocket Introduction and Demonstration Guide*. Oxford Forestry Institute, Oxford, UK. 22 pp.

Grad, P., Le Bourgeois, T. and Merlier, H. (1996) *Adventrop-DOC, Système Multimédia de Reconnaissance des Adventices d'Afrique Soudano-Sahélienne*. CIRAD-CA, Montpellier, France. 21 pp. and CD-ROM.

Morse, L.E. (1974) Computer programs for specimen identification, key construction and description printing. *Publications of the Museum of Michigan State University*, Biology, 5.

Newman, M.F., Burgess, P.F. and Whitmore, T.C. (1996) *Manuals of Dipterocarps for Foresters. Royal Botanic Garden, Edinburgh and Center for International Forestry Research*. Currently 5 manuals with software.

Pankhurst, R.J. (1970) A computer program for the generation of diagnostic keys. *Computer Journal* 12, 145–151.

Pankhurst, R.J. (ed.) (1975a) *Biological Identification with Computers*. Systematics Association Special Volume No. 7, Academic Press, London and New York, 333 pp.

Pankhurst, R.J. (1975b) Identification by matching. In: Pankhurst, R.J. (ed.) *Biological Identification with Computers*. Systematics Association Special Volume no. 7, Academic Press, London and New York, pp. 79–92.

Pankhurst, R.J. (1978a) *Biological Identification. The Principles and Practice of Identification Methods in Biology*. Edward Arnold, London. 104 pp.

Pankhurst, R.J. (1978b) The printing of taxonomic descriptions by computer. *Taxon* 27, 65–68.

Pankhurst, R.J. (1983) An improved algorithm for finding diagnostic taxonomic descriptions. *Mathematical Biosciences* 65, 209–218.

Pankhurst, R.J. (1988a) An interactive program for the construction of identification keys. *Taxon* 37, 747–755.

Pankhurst, R.J. (1988b) Database design for monographs and floras. *Taxon* 37, 733–746.

Pankhurst, R.J. (1989) A computer program with colour graphics to identify orchids. *Orchid Review* 97(1144), 53–55, 67.

Pankhurst, R.J. (1991) *Practical Taxonomic Computing*. Cambridge University Press, Cambridge, 202 pp.

Pankhurst, R.J. (1993) Taxonomic Databases: the PANDORA System. In Fortuner, R. (ed.) *Advances in Computer Methods for Systematic Biology*. Johns Hopkins University Press, Baltimore and London. pp. 229–240. See also www.rbge.org.uk

Pankhurst, R.J. (1995) *PANKEY User Manual*. Exeter Software. See also members.aol.com/exetersftw/index.html

Pankhurst, R.J. and Aitchison, R.R. (1975) An on-line identification program. In: Pankhurst, R.J. (ed.) *Biological Identification with Computers*. Systematics Association Special Volume No. 7, Academic Press, London and New York, pp 181–194.

Pankhurst, R.J. and Pullan, M.R. (1996) DELTA in PANDORA. *DELTA Newsletter* 12, 16–20.

Thompson, F.C. (1996) *Fruit Fly Expert System, Pre-beta Test Version CD-ROM*. USDA, Systematic Entomology Laboratory, Smithsonian Institution, Washington DC.

Watson, L. and Dallwitz, M.J. (1988) *Grass Genera of the World: Illustrations of Characters, Classification, Interactive Identification, Information Retrieval*. Microfiches and diskettes for IBM-PC. Research School of Biological Sciences, Australian National University, Canberra.

Watson, L. and Dallwitz, M.J. (1994) *The Families of Flowering Plants*, CSIRO Publications, Australia. 40 pp. and CD-ROM.

Wilson, J.B. and Partridge, T.R. (1986) Interactive plant identification. *Taxon* 35, 1–12.

27

GENCOMEX: a Computerized Key to Identify the Genera of *Asteraceae* of Mexico

M. Murguía[1] and J.L. Villaseñor[2]

[1]Asociación de Biólogos Amigos de la Computación, A.C. 28 de Agosto No. 32. Col. Escandón. 11870 México D.F., Mexico Fax: (5)272-1537/E-mail: mmr@hp.fciencias.unam.mx.
[2]Instituto de Biología, UNAM Departamento de Botánica, Apartado Postal 70-367. 04510 México D.F., Mexico Fax: (5)550-1760

Introduction

Mexico stands among the countries with the richest biodiversity. Its vascular flora includes about 30,000 different species and its territory is the main centre of diversification of several taxonomic groups. Among them are the members of the family *Asteraceae* (or *Compositae* as it is also known). The *Asteraceae*, cosmopolitan in distribution, consists of more than 1400 genera and about 23,000 species (Bremer, 1994). In Mexico the family is represented by 310–387 genera and 2400–3000 species, depending on the taxonomic authority consulted (cf. Rzedowski, 1991; Turner and Nesom, 1993; Villaseñor, 1993).

The family *Asteraceae* is the richest in almost any flora in the world. Except in the most tropical regions of Mexico, all areas studied floristically, either political, physiographical, or phytogeographical, include this family as the most diverse one. Because of this, and because of the characters (mostly micromorphological) traditionally used in the identification process, correct determination of its members has always been a difficult task.

There are very few tools to identify the genera of *Asteraceae* occurring in Mexico; most of them are old and obsolete. At a national level, the most important are those of Blake (1926) for the trees and shrubs, and Rzedowski (1978). Several floristic works for different regions of Mexico also provide keys to identify the genera; examples are McVaugh (1984), Rzedowski (1985) or Villaseñor (1987, 1989). Different monographic papers dealing with tribes are seminal in

 Information Technology, Plant Pathology and Biodiversity (eds P. Bridge, P. Jeffries, D.R. Morse and P.R. Scott)

helping with the identification to generic level; among them are those by Anderberg (1991), Bremer and Humphries (1993) or King and Robinson (1987).

Although there are several resources that help with the identification to generic level of the *Asteraceae* of Mexico, none of them is up to date and none covers entirely the generic flora of the country. Accordingly, we decided to produce a tool to identify the genera of *Asteraceae* of Mexico that was easy to use and as effective as possible. This computer program, written in Pascal for Windows, functions as an on-line key, and its details are described below.

The Process of Taxonomic Identification

Taxonomic identification may be defined as the process of naming or placing a specimen into a previously specified classification system. Dichotomous keys, polykeys and on-line keys are the different tools the taxonomist has to carry out this process. To make better use of them and to improve them it is necessary to know their limits and the possibilities of making wrong identifications.

We think it is important to propose a formal model of the identification process that takes into account the system of classification accepted, the specimen under identification, the user, the list of possible identities to which the specimen may belong and the list of characters and character states used in the process. If we see the identification process as a system, all the elements listed above interact. The control of this information flux is done mainly by the user (Tardivel and Morse, Chapter 23, this volume).

Edwards and Morse (1995) classify computerized keys in four main groups: multi-access keys, hypertext keys, expert systems, and neural networks. Murguía (1992) provides an alternative seven-way classification: matrical, logic sets, probabilistic, theory of preferences, trees, expert systems, and neural networks.

Dynamic Identification: a Model

There are several analyses and proposals on how to build up algorithms for taxonomic identification (see Pankhurst, 1975, 1978, 1991; Sneath, 1979; Wilson and Partridge, 1986). Gauthier and Pavé (1990) provide a discussion on how to structure a system for taxonomic identification.

In our model, given a specimen with a set of characters or character states, a subset of possible taxa is defined. On the other hand, given a taxon or subset of taxa, a subset of characters and character states is defined that can be observed in a specimen that belongs to the selected taxon or taxa. If we consider these two dimensions, we can call the first the ‘hypothesis’ about the identity of a specimen and the second the ‘characters’ relevant to its identification. It is then possible to establish a matrix of relationships between them depending on whether an affirmation (positive selection), a negation (negative selection) or a neutral selection is defined (Fig. 27.1). People doing an identification exercise can move within the 'navigation space' of this matrix.

An Example of the Model's Implementation: GENCOMEX

The model has been implemented in a computer program called GENCOMEX. The program helps to identify the genera of *Asteraceae* occurring in Mexico. The name of the program is an acronym of the **Gen**era of ***Co**mpositae* of **Mex**ico. The program's design and the algorithms are derived from the ideas of polykeys such as those by Hansen and Rahn (1969), Little (1968) and Simpson and Janos (1974). In addition, previous exercises in the computerization of taxonomic databases have been important in this development (Murguía, 1988, 1992; Villaseñor and Murguía, 1992; Murguía and Villaseñor, 1993). Originally the

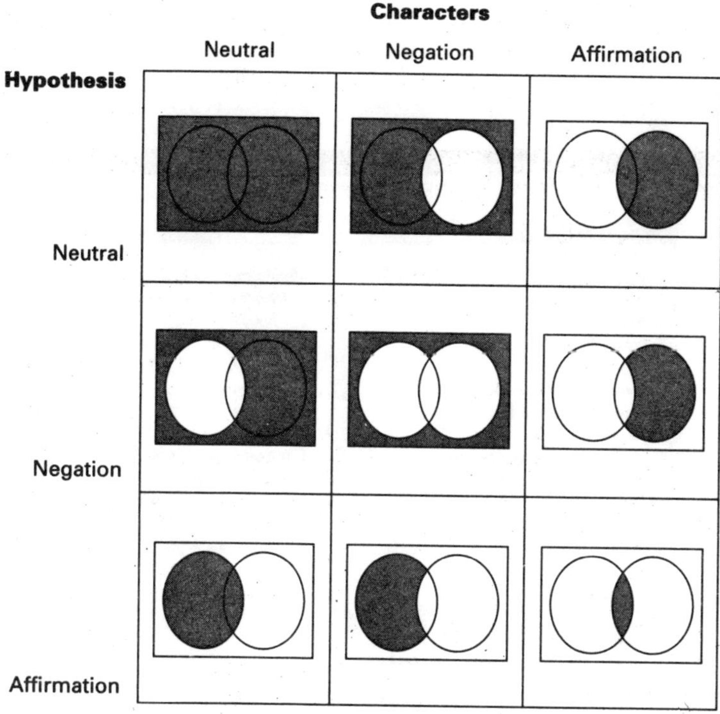

Fig. 27.1. Matrix showing the nine logic areas in the 'navigation space' of the dynamic identification model.

program had been written for a DOS environment and has proven its efficiency for the determination of the families of flowering plants of Mexico (Murguía and Villaseñor, 1993) and the genera of *Asteraceae* of Mexico (Murguia, 1992). It has now been developed further for the Windows environment (Murguía and Villaseñor, 1996).

One important feature of the program is that the user does not need to navigate among several menus. Figure 27.2 shows the main menu, giving access to the environment we call 'dynamic identification'. The menu includes two windows, one with the list of characters and character states, the other with the list of taxa ('hypotheses'). As the user selects the characters observed in the specimens with the mouse in the left window, the number of potential taxa seen in the right window reduces. At the beginning of a session, all characters are coloured blue, indicating that no character has been selected. As characters are selected, they turn red. Likewise, at the beginning of the session all the hypotheses are coloured black, indicating that all are equally possible. The user tells the computer he or she suspects the specimen belongs to a particular taxon by selecting the hypothesis with the mouse; it then turns green. The computer responds by showing in green those characters that should match those observed in the specimen if the hypothesis is correct, and in red those that should not (Fig. 27.3). The user selects and de-selects (deletes) a character just by using the left mouse button.

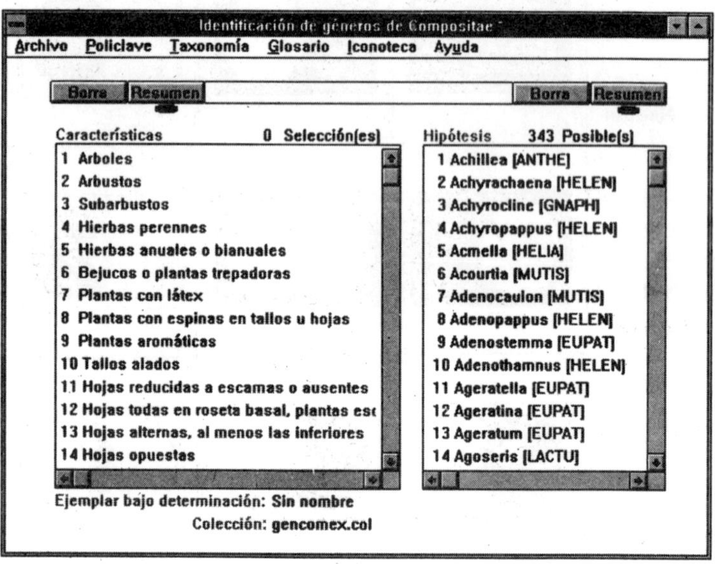

Fig. 27.2. Main menu of GENCOMEX giving access to the dynamic identification environment.

(a) Character

	Not selected	Selected
No hypothesis selected or does not fit the hypothesis	Blue	Red
Fits the hypothesis selected	Black	Green

(b) Hypothesis

	Not selected	Selected
Does not match the character selected	Blue	Red
Matches the character selected	Black	Green

Fig. 27.3. GENCOMEX: diagram showing (a) how characters change colour according to whether they are selected, and whether they fit any hypotheses selected; (b) how hypotheses change colour according to whether they are selected, and whether they match any characters selected.

Once hypotheses have been selected, a character that does not fit any of them remains red, even though it may fit another selected hypothesis. This makes it easy to detect the mistake.

Activating the appropriate Summary (Resumen) button reduces the list of characters to those in green or red, or reduces the list of hypotheses to those in green or black (Fig. 27.4). The two Erase (Borra) buttons independently de-select all characters or all hypotheses previously selected.

The program includes additional attributes, several of them considered by Dallwitz *et al.* (Chapter 19, this volume) as desirable features of on-line keys. They are a glossary (Glosario), a collections file, a list of synonymy, a family description, a file with drawings illustrating important characters (Iconoteca), a help facility, and the facility to print a summary of a session. The glossary and the drawings files (Fig. 27.5) can be accessed from the main menu or dynamically from the list of characters. The taxonomy file contains a description of the family *Asteraceae* and a list of important synonymies.

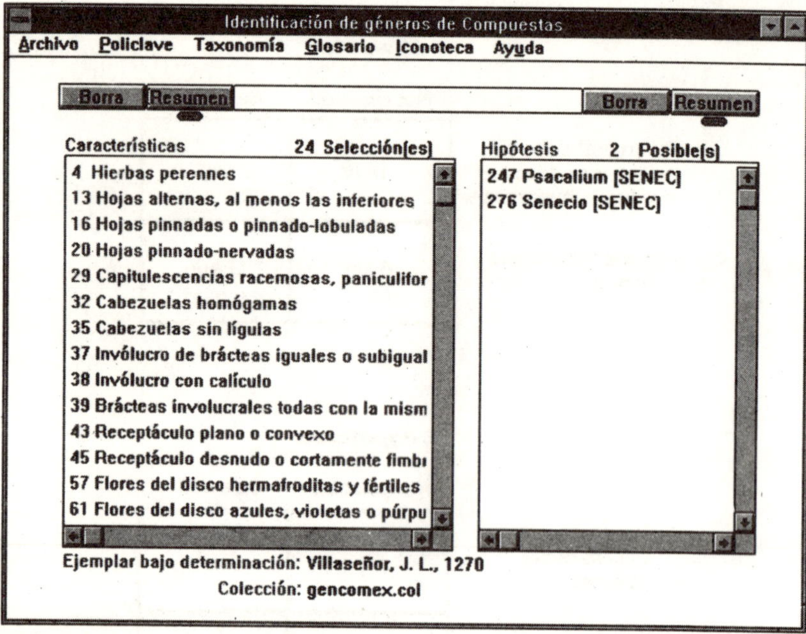

Fig. 27.4. GENCOMEX: appearance in an advanced stage of an identification session.

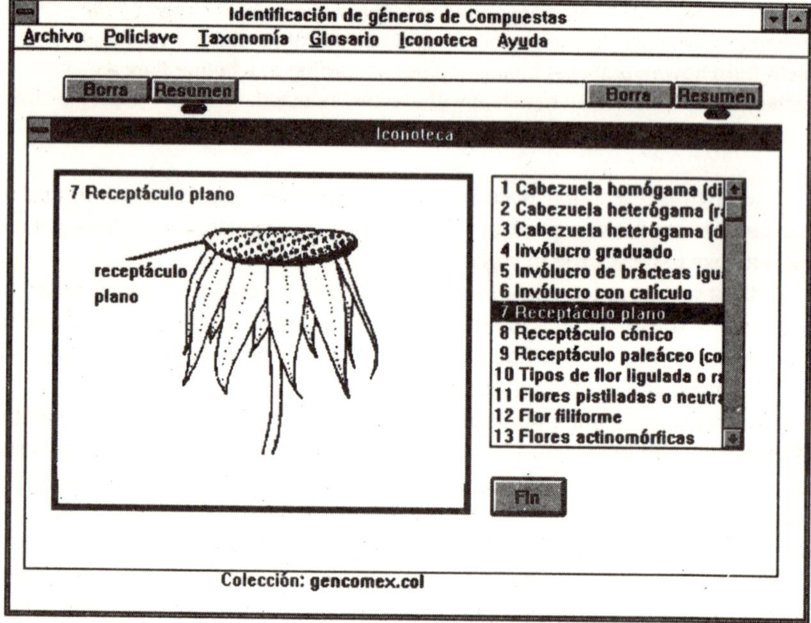

Fig. 27.5. GENCOMEX: example of a drawing of one of the character states ('Receptacle flat').

The user can store an identification session in a file. A collections file can store several identification sessions. This feature can be important in education; for example, a teacher can review later the exercises carried out by students. The user can also obtain a hard copy of the identification session; this report includes characters and hypotheses.

The GENCOMEX Program: Technical Aspects

The program uses a taxonomic data matrix based on presence/absence of 99 character states for 343 genera. For each genus, presence or absence of each character state was scored. The genera included do not represent a definitive list for Mexico, but combine practical aspects (especially the level of knowledge of the taxonomy of *Asteraceae* among the potential users) with logistic features (especially the ease with which the characters can be used to reach a successful identification).

The program was written in Turbo Pascal for Windows (version 1.5). It requires a relatively common platform for its functioning. Table 27.1 summarizes the main technical characters of the program. Installation of the program is as simple as any other Windows program, and it occupies little hard disk space (92 kB).

Figure 27.2 shows the start of an identification session. By default the Summary (Resumen) button is off (inactive) in the characters window and on (active) in the hypotheses window (genera). Figure 27.4 shows the screen after a session in which 24 characters have been selected; based on the characters chosen, there remain only two possible identities. In Fig. 27.4 both summary

Table 27.1. Technical characteristics of the GENCOMEX program.

Characteristic	Description
Programming language	Turbo Pascal for Windows [version 1.5]
Image format	Bit Map black and white [Windows]
Data matrix format	Proprietary
Characters and character states list, taxa list and glossary files format	Text
Help	Help manager system of Windows
Platform needed for operating the program	PC 386 or superior; Windows 3.11 or superior operating system; mouse; colour monitor; hard disk space of 700 kB
Distribution format	One disk $3\frac{1}{2}$" DD; paper printed manual

buttons have been activated, so that only those selected characters and the possible identities remaining are listed.

GENCOMEX has proved its usefulness at several levels. Theoretically, it has helped us acquire a better idea of the identification process; in this way we think better programs can be designed and developed. Empirically, the dynamic identification environment has proved to be an efficient tool for automating taxonomic identification.

Discussion

Computer implementation of taxonomic identification has raised interest during the last two decades. However, this process needs more theoretical study as well as modelling, in order to obtain the maximum benefits from computers. Analyses of taxonomic identification aimed at describing how information flows, the elements of information involved and how they interact are important if we want to develop more useful, friendly and better programs. Only thus will we improve the benefits we get from computers as tools to automate the process.

GENCOMEX is a useful tool to identify the genera of *Asteraceae* of Mexico, developed from theoretical principles (the dynamic identification model) and technical proposals (the dynamic identification environment), carried to a practical level. The model attempts to describe the process of taxonomic identification, clearly defining the elements involved and establishing relationships of information flux among them. The environment constitutes the partial implementation of the dynamic identification model. GENCOMEX has proved useful with a particular group of very diverse organisms whose taxonomic identification has often been regarded as difficult.

Acknowledgements

We thank Comisión Nacional para el Conocimiento y Uso de la Biodiversidad (CONABIO) for economic support to finish this project and partially to attend the Conference. We thank Dr John Goddard for revision of the preliminary version of this chapter, and Peter Scott for his detailed editorial help.

References

Anderberg, A.A. (1991) Taxonomy and phylogeny of the tribe Gnaphalieae (Asteraceae). *Opera Botanica* 104, 1–195.

Blake, S.F. (1926) Asteraceae. In: Standley, P.C. (ed.) *Trees and Shrubs of Mexico*. Part 5. Contributions from the United States National Herbarium 23, 1401–1641.

Bremer, K. (1994) *Asteraceae. Cladistics and Classification*. Timber Press. Portland, Oregon.

Bremer, K. and Humphries, C. (1993) Generic monograph of the Asteraceae–Anthemideae. *Bulletin of the Natural History Museum, Botany Series* 23, 71–177.

Edwards, M. and Morse, D.R. (1995) The potential for computer-aided identification in biodiversity research. *TREE* 10, 153–158.

Gautier, N. and Pavé, A. (1990) Object-centered representation for species systematics and identification in living systems in nature. *Computers in Applied Biosciences* 6, 383–386.

Hansen, B. and Rahn, K. (1969) Determination of angiosperm families by means of a punched-card system. *Dansk Botanisk Arkiv* 26, 1–44.

King, R.M. and Robinson, H. (1987) The genera of the Eupatorieae (Asteraceae). Monographs in Systematic Botany. Vol. 22. Missouri Botanical Garden. Saint Louis, Missouri.

Little, E.L. Jr (1968) Clave con fichas perforadas de las familias de los árboles mexicanos. *Turrialba* 18, 45–59.

McVaugh, R. (1984) Compositae. In: Anderson, W.R. (ed.) *Flora Novo-Galiciana*. Vol. 12. The University of Michigan Press. Ann Arbor, Michigan.

Murguía, M. (1988) Clave para familias (Magnoliophyta) presentes en México. Tesis, Biólogo. Facultad de Ciencias, UNAM México, D.F.

Murguía, M. (1992) Métodos en la identificación biológica automatizada. Tesis, Maestro en Ciencias. Facultad de Ciencias, UNAM México, D.F.

Murguía, M. and Villaseñor, J.L. (1993) *FAMEX v.2.0. Clave para Familias de Plantas con Flores (Magnoliophyta) de México*. Asociación de Biólogos Amigos de la Computación, A. C. México, D. F. Program (DOS) and User's manual.

Murguía, M. and Villaseñor, J.L. (1996) *GENCOMEX. Policlave para la Identificación de los Géneros de Compositae Presentes en México*. Asociación de Biólogos Amigos de la Computación, A. C. México, D. F. Program (Windows) and User's manual.

Pankhurst, R.J. (ed.) (1975) *Biological Identification with Computers*. Academic Press. New York, 333 pp.

Pankhurst, R.J. (1978) *Biological Identification*. Edward Arnold. London.

Pankhurst, R.J. (1991) *Practical Taxonomic Computing*. Cambridge University Press. Cambridge, 202 pp.

Rzedowski, J. (1978) Clave para la identificación de los géneros de la familia Compositae en México. *Acta Científica Potosina* 7, 1–145.

Rzedowski, J. (1985) Compositae. In: Rzedowski, J. and de Rzedowski, G.C. (eds) *Flora Fanerogámica del Valle de México*. Vol. II. Escuela Nacional de Ciencias Biológicas, I. P. N. e Instituto de Ecologia, A.C. México, D.F.

Rzedowski, J. (1991) Diversidad y orígenes de la flora fanerogámica de México. *Acta Botánica Mexicana* 14, 3–21.

Simpson, D.R. and Janos, D. (1974) *A Punched Card Key to the Families of Dicotyledons of the Western Hemisphere South of the United States*. Field Museum of Natural History. Chicago, Illinois.

Sneath, P.H.A. (1979) Basic program for determining overlap between groups in an identification matrix of percent positive characters. *Computers and Geosciences* 6, 267–278.

Turner, B.L. and Nesom, G.L. (1993) Biogeography, diversity, and endangered or threatened status of Mexican Asteraceae. In: Ramamoorthy, T.P., Bye, R., Lot, A. and Fa, J. (eds) *Biological Diversity of Mexico: Origins and Distribution*. Oxford University Press. Oxford, pp. 559–575.

Villaseñor, J.L. (1987) Clave genérica para las Compuestas de la cuenca del río Balsas. *Boletín de la Sociedad Botánica de México* 47, 65–86.

Villaseñor, J.L. (1989) *Manual para la Identificación de las Compositae de la Península de Yucatán y Tabasco*. Technical Report No. 4. Rancho Santa Ana Botanic Garden. Claremont, California.

Villaseñor, J.L. (1993) La familia Asteraceae en México. *Revista de la Sociedad Mexicana de Historia Natural* Vol. Esp. 44, 117–124.

Villaseñor, J.L. and Murguía, M. (1992) La computadora en la identificación botánica: una grata experiencia. *Ciencia y Desarrollo* 18, 130–137.

Wilson, J.B. and Partridge, T.R. (1986) Interactive plant identification. *Taxon* 35, 1–12.

Probabilistic Identification Systems for Bacteria

28

T.N. Bryant

Medical Statistics and Computing, University of Southampton, Southampton General Hospital, Southampton SO16 6YD, UK
Fax: +1703 794460/E-mail: tnb@soton.ac.uk

Introduction

Probabilistic identification of bacteria is based on traditional phenotypic methods for classifying and identifying bacteria. Although these are the methods that are used by many microbiologists, they are just some of a variety of methods currently available to microbiologists. Other phenotypic methods include analysis of cell wall composition, cellular fatty acids, isoprenoid quinones, whole-cell protein analysis, polyamines, pyrolysis mass spectrometry, Fourier transformation infra red spectroscopy and UV resonance Raman spectroscopy. Alongside these techniques are the genotypic methods of analysing DNA and RNA molecules: DNA base ratio (moles percent C+G), DNA–DNA hybridization, rRNA homology studies, DNA-based typing methods. The integration of all these types of information provides a consensus approach to bacterial taxonomy which is termed polyphasic taxonomy, and encompasses polyphasic identification. It must be remembered therefore, that probabilistic identification makes use of a subset of the taxonomic information available to bacteriologists.

Probabilistic identification methods are based on the likelihood that the observed pattern of results of tests carried out on an unknown bacterium can be attributed to the results of a known taxon within an identification matrix. Typically, the known taxon is a recognized bacterial species, although this does not have to be the case, and the identification matrix contains species from a small number of related genera. Few attempts have been made to produce a general purpose matrix covering all bacteria.

Theory

Beers and Lockhart (1962) are credited as being the first authors to suggest that computers could assist in the identification of bacteria. Since their suggestion, a number of probabilistic methods have been proposed (Dybowski and Franklin, 1968; Lapage *et al.*, 1970; Friedman *et al.*, 1973; Gyllenberg and Niemelä, 1975; Van Oystaeyen, 1995) however, the most commonly used is that suggested by Lapage *et al.* (1973) who developed the first routine service for computer assisted bacterial identification.

Most computer assisted identification systems are based on Willcox's implementation of Bayes theorem for use with bacteria (Willcox *et al.*, 1973, 1980):

$$P\left(t_i \mid R\right)=\frac{P\left(R \mid t_i\right)}{\sum P\left(R \mid t_i\right)} \tag{28.1}$$

where $P(t_i \mid R)$ is the probability that an unknown isolate, giving a pattern of test results R, is a member of taxon (group of bacteria) t_i and $P(R \mid t_i)$ is the probability that the unknown has a pattern R given that it is a member of taxon t_i.

Bayes theorem (Ledley and Lusted, 1959) incorporates prior probabilities with bacterial identification, these prior probabilities would be the expected frequency of incidence of strains included in the identification matrix. Most authors give strains an equal chance of being identified and therefore the prior probabilities for all strains are set to 1.0 and omitted from the equation, although Berger (1990) has suggested that for commercial identification systems, prior probabilities, or prevalence, should be considered because this avoids over reporting of rare species.

The above equation can be re-expressed as:

$$L_i^{*}=\frac{L_i}{\sum L_i} \tag{28.2}$$

where the probabilities are now referred to as identification, or Willcox, scores. The identification scores for each taxon are normalized values and L_i^* for all taxa sum to one. Identification of an unknown isolate is achieved when L_i^* for one taxon exceeds a specified threshold value.

A worked example is shown below with a probability matrix consisting of three taxa for which we have the probabilities for four tests (Table 28.1). An unknown has been isolated whose results for the first three tests are positive, negative and positive respectively.

The likelihoods that the taxa a, b and c will give the pattern of results observed for the unknown is calculated by multiplying the probability of obtaining a positive result for test 1 by the probability of obtaining a negative result for test 2 by the probability of obtaining a positive result for test 3 for each taxon in turn (Table 28.2). Note that original identification matrix only gives the

Table 28.1. Identification matrix with results of unknown.

		Tests			
		1	2	3	4
Taxa	a	0.01	0.30	0.99	0.90
	b	0.90	0.01	0.99	0.01
	c	0.99	0.10	0.85	0.99
	unknown	+	−	+	missing

Table 28.2. Calculation of likelihood of unknown.

		Tests						
		1		2		3		Likelihood
Taxa	a	0.01	×	(1–0.30)	×	0.99	=	0.00693
	b	0.90	×	(1–0.01)	×	0.99	=	0.88209
	c	0.99	×	(1–0.10)	×	0.85	=	0.75735
Sum							=	1.64637

probabilities for positive results and in order to use the probability for a negative result we must subtract the matrix entries for test 2 from 1. The identification scores are expressed as normalized likelihoods in Table 28.3.

In this small example, the unknown is not identified because a single taxon does not reach the identification threshold value. Taxa b and c are both still candidates for the identity of the unknown. An identification threshold value of 0.999 was defined by Lapage *et al.* (1973) for use with aerobic non-spore-forming Gram negative rods, but with other groups of bacteria values as low as 0.95 have been used (Bryant, 1993). In practical terms, a value of 0.999 means that the taxon which the unknown identifies with will have at least two test differences from all other taxa in the matrix being used.

Table 28.3. Willcox probabilities (normalized likelihoods).

				Identification score
Taxa	a	0.00693 / 1.64637	=	0.004209
	b	0.88209 / 1.64637	=	0.535779
	c	0.75735 / 1.64637	=	0.460012
Sum			=	1.000000

Outcomes of Identification

Whatever type of identification system is used, there are four possible outcomes.

1. The unknown is identified with the correct taxon.
2. The unknown is misidentified, i.e. attributed to wrong taxon.
3. The unknown is not identified at all, and correctly so because the taxon to which it belongs is not in the matrix.
4. The unknown is not identified, but should have been identified with a taxon that is in the matrix.

It is important that the system deals with these possibilities, although the last one is difficult to resolve. One problem with the Willcox score is that if an unknown is not represented in the matrix, but one strain within the matrix is closer to it than all others, the unknown may be identified as this strain. This is where additional criteria should be used to assist the identification process. These include, listing the differences in test results between the unknown and the strain it has been identified as, as well as the use of other numeric criteria such as taxonomic distance measures (Gyllenberg and Niemelä, 1975; Sneath, 1979a) or maximum likelihoods (Dybowski and Franklin, 1968).

The maximum, or best, likelihood (Dybowski and Franklin, 1968) is the maximum probability for a taxon calculated using those tests carried out on the unknown. The calculation uses the maximum of the probabilities of a negative and positive result of a test (Table 28.4) and copes with the problem of taxa having several test results of 0.50 in a matrix.

Some authors calculate the likelihood/maximum likelihood ratio, termed the modal likelihood fraction (Dybowski and Franklin, 1968) (see Table 28.5), or its inverse (Portyrata and Krichevsky, 1992), and use it to decide whether to accept the identification offered by a Willcox score that has exceeded the identification threshold. In the IDENT module of the MICRO-IS program (Portyrata and Krichevsky, 1992), for example, the identification score is not given if the best likelihood/likelihood is greater than 100.

Table 28.4. Maximum possible likelihoods.

		1		2		3		Best likelihood
Taxa	a	(1–0.01)	×	(1–0.30)	×	0.99	=	0.68607
	b	0.90	×	(1–0.01)	×	0.99	=	0.88209
	c	0.99	×	(1–0.10)	×	0.85	=	0.75735

Table 28.5. Modal likelihood fraction.

				Modal likelihood
Taxa	a	0.00693 / 0.68607	=	0.010101
	b	0.88209 / 0.88209	=	1.000000
	c	0.75735 / 0.75735	=	1.000000

Bayes theorem assumes that all tests in an identification matrix are independent; the result of one test does not affect the result of another test. This is not always the case, for example, if an isolate does not grow at 42°C, it cannot be motile at 42°C and, conversely, if the isolate is motile at 42°C, it must grow at this temperature. Some authors propose methods for dealing with this problem, whereas others point out that if only one or two pairs of linked tests occur in a matrix, the linking can be ignored for practical purposes.

Selection of Additional Tests

When an unknown is not identified, a set of additional tests can be selected from those tests in the identification matrix that have not been used. The results of these tests, combined with those already carried out, should enable the identity of the unknown to be established. The method often used is based on the work of Rypka *et al.* (1967) who developed the ideas of Gyllenberg (1963). A test separates a pair of taxa if one test has a matrix entry of 0.85 or greater and the other 0.15 or less (i.e. if one test is usually positive and the other is usually negative). Tests are selected one at a time, at each stage the test added to the list of selected tests is the one that separates the most pairs of taxa not separated by the tests already selected. When all taxa are separated by the minimum number of tests specified, which for bacterial identification is typically set at two, or when no further pairs of taxa can be distinguished, the process is stopped.

An additional refinement can be added to this approach (Bryant, 1991). If two tests separate the same number of taxa, then for each test the sum of the absolute probability difference between pairs of taxa is calculated. The test with the highest value is selected as the next test. The example in Table 28.6 illustrates this: if test selection were based only on the number of taxa separated, Test A would be chosen before Test B simply because it was encountered first by the selection algorithm. However, Test B separates taxa which are always positive or negative, compared to Test A where taxa are usually positive or negative. Selection based on the number of taxa separated together with the sum of absolute differences can therefore lead to selection tests with better discriminatory power. In the example, Test B would be selected before Test A.

Table 28.6. Calculation of test separation ability.

		Tests		
		A	B	C
Taxa probabilities	1	0.85	0.99	0.99
	2	0.15	0.99	0.99
	3	0.15	0.01	0.70
	4	0.85	0.01	0.01
Taxa results	1	+	+	+
	2	–	+	+
	3	–	–	V
	4	+	–	–
Number of taxa separated		4	4	2
Sum of absolute difference		2.80	3.92	2.65

The sum of absolute difference is given by:

$$\Sigma p = |p_1 - p_2| + |p_1 - p_3| + |p_1 - p_4| + |p_2 - p_3| + |p_2 - p_4| + |p_3 - p_4|$$

$$\begin{aligned}
A &= |0.85-0.15| + |0.85-0.15| + |0.85-0.85| + |0.15-0.15| + |0.15-0.85| + |0.15-0.85| \\
&= 0.70 + 0.70 + 0.0 + 0.0 + 0.70 + 0.70 \\
&= 2.80 \\
B &= |0.99-0.99| + |0.99-0.01| + |0.99-0.01| + |0.99-0.01| + |0.99-0.01| + |0.01-0.01| \\
&= 0.0 + 0.98 + 0.98 + 0.98 + 0.98 + 0.0 \\
&= 3.92 \\
C &= |0.99-0.99| + |0.0| + |0.99-0.01| + |0.0| + |0.99-0.01| + |0.70-0.01| \\
&= 0.0 + + 0.98 + + 0.98 + 0.69 \\
&= 2.65
\end{aligned}$$

Identification Software

Several probabilistic identification programs have appeared in the literature. A list of these is presented in Table 28.7 and Sackin (1987) give details of programs that were published earlier. In most instances the programs are non-specific and can be used with various identification matrices. Examples from one program, Bacterial Identifier are shown in Figs 28.1, 28.2 and 28.3.

One interesting system is that produced by the BioInformatics Centre, Pune, India who offer probabilistic identification via the Internet. Using a form and a CGI script a user can enter results via their Web browser and an identification score for the unknown is returned. With the development of programming languages, such as JAVA, on the Web one can envisage more sophisticated identification applications appearing.

Table 28.7. Programs for bacterial identification.

Name of program	Organisms that can be identified	Author	Comments
Bacterial Identifier	various matrices	T.N. Bryant	http://medstats.soton.ac.uk/0c:/microbio/microsof.html
BBACTID	*Bacillus* species	(Bryant *et al.*, 1985)	
BACTID	*Enterobacteriaceae*	(Jilly, 1988)	
CIBAC	lactic acid bacteria	(Döring *et al.*, 1988)	
Gideon	various bacteria	Berger	Commercial, uses prior probabilities, e-mail: mberger@ccsg.tau.ac.il
IDENTIFY	various matrices	(Jahnke, 1995)	Based on the MicroIS system
MICRO-IS	various matrices	(Portyrata and Krichevsky, 1992)	Data Management as well as identification
MATIDEN	various matrices	(Sneath, 1979a)	
The Identifier	various matrices	(Gibson *et al.*, 1992)	
Identmpm	various matrices	(Maradona, 1994)	
no-name	mycobacteria	(Tortoli *et al.*, 1992)	
Unpublished programs	various programs	see (Canhos *et al.*, 1993)	
Identification via the World Wide Web	various matrices	Niak BioInformatics Centre, Pune	http://bioinfo.ernet.in/www/prob.html

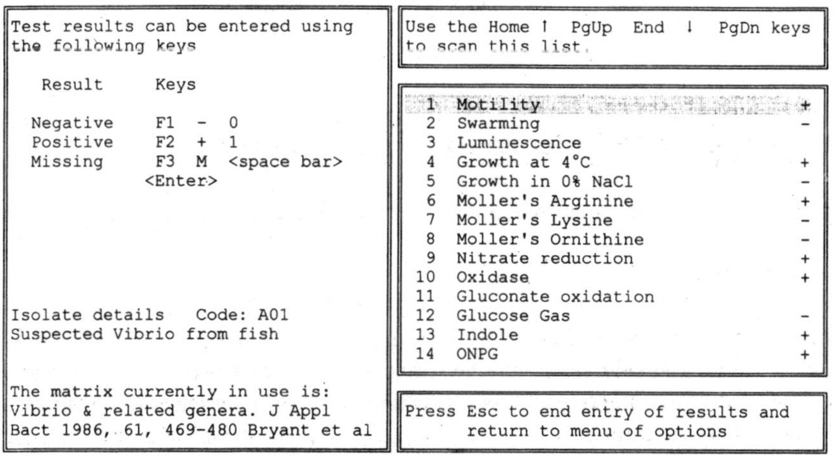

Fig. 28.1. The entry screen for tests results in the program Bacterial Identifier.

```
Identification threshold (0.980) not reached. The most likely taxa are:

                                                  Identification
             Taxon                                   Score

     1   Vibrio anguillarum                          0.8412
     2   Phenon 10   (West et al 1986)               0.1527
     3   Vibrio diazotrophicus                       0.0040
     4   Photobacterium species                      0.0014
     5   Sucrose -ve Aeromonads (West et          0.0003


              Press the SPACE BAR to continue
```

Fig. 28.2. The unknown has not been identified, no single taxon has exceeded the identification threshold. The five most likely taxa are listed.

```
Expected results for Tests distinguishing
Vibrio anguillarum from the other 4 taxa by at least 2 tests
                                   Likely taxa - abbreviated names below
                                   1 2 3 4 5

  1   Cellobiose cs                + + + - -
  2   Gelatinase                   + + - - +
  3   D-glucuronate                - + + - -
  4   L-arabinose acid             + - + - v
  5   Gluconate oxidation          v - - - +
  6   Sucrose acid                 + + + - -

                      Press the SPACE BAR to continue

 1 Vibrio anguillar  2 Phenon 10  (West  3 Vibrio diazotrop  4 Photobacterium s
 5 Sucrose -ve Aero
```

Fig. 28.3. Selection of additional tests to distinguish *Vibrio anguillarum,* the most likely taxon, from the other four taxa.

Methods Used to Generate Identification Matrices

Three approaches are used by authors to generate identification matrices.

Cluster analysis

This involves a numerical taxonomic analysis of fresh isolates and reference strains. Phena are selected from a dendrogram and their properties are summarized to create a starting matrix containing all characters used in the study. The identification matrix is developed by including the most useful characters and rejecting those that do not distinguish between phena. Some clusters produced by the cluster analysis are omitted from the starting matrix because they contain few members, one, two or three isolates, or comprise a group that cannot be identified.

Grouping known strains

Known strains are characterized using a battery of tests and the percentage of each strain exhibiting a character is calculated. The identification of each strain is assumed to be accurate and no taxonomic analysis is carried out. There are potential problems with this approach. Isolates that have traditionally been treated as strains of one species might be a grouping of two or more species or, if characterization is carried out on strains that have been repeatedly subcultured these strains may show less metabolic activity compared to fresh isolates. This could result in a matrix that is biased against fresh isolates.

Data collected from the literature

In this less reliable approach, data is collected from a variety of publications and merged to create a single matrix. The authors must resolve any conflicting test results and assign probabilities for positive, negative and variable results. If the characterization methods are not adequately described in the literature, some characters may be misinterpreted and the results matrix may contain erroneous probabilities.

Whatever method is used to create the probability matrix, it is important that the characterization of unknowns is performed using the same techniques. For example it would be inappropriate to create a matrix using miniaturized tests and use it for the identification of an unknown characterized using conventional tests.

Preparing the starting matrix

Whichever of the above methods are used, an initial or starting matrix of probabilities for a large number of taxa and tests is created. This will be a table of the percentage frequency of characters for each phenon defined from a numerical classification. Typically taxa that are always positive (100%) for a character are coded as 0.99 and those that are always negative (0%) as 0.01. This was suggested by Lapage *et al.* (1973) as the way to allow for test error and atypical strains. This 1% error rate has been adopted by most producers of identification matrices. However Boeufgras *et al.* (1988) found that for the API system, the probability that a positive test result was wrong was 0.1% and for an incorrect negative result 1%. Using this modification, the accuracy of the API system was improved although other systems have not adopted this.

Programs for providing summary statistics for the phena defined by a numerical taxonomic analysis can be used to monitor the quality of the starting matrix and identify atypical isolates. If isolate data is available the OUTLIER program (Sneath and Langham, 1989) can be used to identify aberrant isolates within a cluster. The program uses objective criteria by testing the fit of the data to a χ^2 distribution. The GPROPS program (Bryant, 1987) can create a starting matrix from raw data by aggregating isolate information. The program provides information about the distance of isolates from the centre of the phenon to which they have been assigned. At this stage, it is also advisable to check that any two phenons in the matrix do not overlap with each other in taxonomic space. If overlap is found, then either consideration should be given to merging the two phenons, or additional characteristics should be sought for discriminating between them. Sneath (1978, 1979b) has provided two programs for doing this.

Identifying useful tests

The starting matrix is used to identify tests with good discriminatory ability. Tests with no, or poor, diagnostic value are rejected and are therefore removed from the starting matrix leading to the creation of an identification matrix. Although some tests may, individually, have reasonable discriminatory ability, they are rejected when they are considered in conjunction with other tests that discriminate better. As pointed out by Priest and Williams (1993), there is some controversy concerning the minimum number of tests that are required to separate taxa effectively. A good diagnostic test is one that has an equal distribution of taxa that are positive and negative; however, tests that are only positive, or negative for one or two taxa also have a diagnostic value and may be desirable. For example a positive coagulase test is typical of *Staphylococcus aureus* since very few related strains possess this attribute. Several separation indices that identify good diagnostic tests have been proposed, for example see (Gyllenberg, 1963; Niemelä *et al.*, 1968) and are included in the CHARSEP program (Sneath, 1979c). Once a matrix has been constructed the DIACHAR program (Sneath, 1980a) can be used to rank the tests by their discriminatory ability for each taxon. Use of these two programs provides a semi-automated approach to the creation of an identification matrix and these have been used in a number of studies, for example (Williams *et al.*, 1983; Priest and Alexander, 1988). The program BEST (Bryant, 1991) uses similar methods to identify useful tests and prepares an identification matrix. It incorporates the modification for test selection described earlier (see Selecting additional tests). Another method for optimal selection of tests has been proposed by Claus (1993).

Evaluating the identification matrix

Identification matrices can be evaluated by checking that each taxon in the matrix can be identified. This is done by creating an 'unknown' set of results using matrix entries of above 0.50 as positive and below 0.50 as negative. Two programs have been published that do this: MOSTTYP (Sneath, 1980b) and IDSC

(Bryant, 1991). If any test probability for a taxon in a matrix of 0.50 are encountered, the IDSC calculates the identification score three times, once with all entries of 0.50 for the taxon under test treated as missing values, then as positive and finally as negative. Another check can be applied to the matrix using the OVERMAT program (Sneath, 1980c) which calculates the overlap between taxa in an identification matrix. Finally, the matrix should be evaluated using an identification program to test the isolates that were used in the matrix construction and also against fresh isolates.

Probabilistic Identification Matrices

Table 28.8 gives details of identification matrices that have been published since 1980. Matrices created and used by manufacturers of commercial identification kits are not included as very few have published their data or the algorithms that they use for identification; details of some of these were presented by Stager and Davis (1992). A bar chart of the number of matrices published, starting from 1981, is shown in Fig. 28.4. This shows that publication activity was high in the period from 1982 to 1992 but fewer matrices have appeared since then. This may reflect more recent interest in the use of sequence analysis for identification.

The Future

Probabilistic identification has provided bacteriologists with an accurate system for identifying bacteria using phenotypic methods. The technique has been successful. This is shown by the number of commercial identification kits that make use of it and the use of these kits, together with in house systems, for routine bacterial identification. However, the approach is dependent on good quality matrices and this means that only certain groups of bacteria can be identified. Those bacteria that cannot be identified, either because of insufficient phenotypic characters, or simply the lack of a matrix, must be classified and identified using other techniques. Although with these probabilistic approaches might be appropriate. Plikaytis *et al.* (1992) showed that an identification matrix, built using restriction fragment length polymorphism data, could be used for the identification of slowly growing mycobacteria.

Busse *et al.* said:

> The most promising strategy for the classification of bacteria leading to a reliable identification would include as a first step the analysis of the 16s rRNA (300–1500 nucleotides). Comparison with published data (EMBL, GenBank) should clearly indicate close relatives of the strain under investigation.
>
> (Busse *et al.*, 1996)

Table 28.8. Identification matrices published since 1980.

Species/groups covered	Taxa	Tests	Data pub.	Matrix eval.	Clus. anal.	Method	% id	ID coef.	Media	Authors
Gram-negative, aerobic, non-fermenters	66	83	y	y	n	2	91.5	0.999	Conventional	Holmes *et al.* (1986a)
Gram-negative, aerobic rod-shaped fermenters	110	66	y	y	n	2	89.2	0.999	Conventional	Holmes *et al.* (1986b)
Enterobacteriaceae	41	16	y	y	n	2	91.8	0.95	Conventional Multipoint inoculation	Clayton *et al.* (1986)
Enterobacteriaceae	90	50	y	y	n	2			Conventional	Holmes and Costas (1992)
Pseudomonas species	26	70	y	y	n	2			Conventional	Costas *et al.* (1992)
Vibrios	31	50	y	y	n	2	79.4	0.99	Conventional API 20E	Dawson and Sneath (1985)
Vibrio and related species	38	81	y	y	y	1	74.0	0.999	Conventional	Bryant *et al.* (1986)
Vibrios	48	44	y						Conventional	Alsina and Blanch (1994)
Aeromonas species	14	30	y	y	y	1	71.5	0.90	microtitre	Kämpfer and Altwegg (1992)
Aeromonas hybridization groups	15	32	y	y	n	3			Conventional	Oakley *et al.* (1996)
Campylobacteria	23	42	n	y	n	2			Conventional	Holmes *et al.* (1992)
Campylobacter, Helicobacters and others	37	67	y	y	y	1		0.999	Conventional	On *et al.* (1994)
Lactic acid bacteria	37	49	n	n	n	3	79.0	0.90	API 50CH	Cox and Thomsen (1990)
Lactic acid bacteria	59	27	y	n	n	3	98.0	0.70	Conventional	Döring *et al.* (1988)
Lactic acid bacteria	11	53	y	n	n	2		0.999	API 50CH	Maissin *et al.* (1987)

Streptomyces species	23	41	y	y	y	1	81.3	0.85	Conventional	Williams *et al.* (1983)
Streptomyces species (minor clusters)	26 28	50 39	y	y	y	1	77.0	0.85	Conventional	Langham *et al.* (1989)
Streptomyces species	52	50	y	y	y	1	78.1	0.90	Miniaturized	Kämpfer and Kroppenstedt (1991)
Streptoverticillum species	24	41	y	y	y	1		0.95	Conventional	Williams *et al.* (1985)
Actionplanes species	15	53	y		y				Conventional	Long (1994)
Phytopathogenic corynebacteria (subclusters)	3 5	19 27	y	y	y	1			Conventional	Firrao and Locci (1989)
Slowly growing mycobacteria	24	33	y	n	y	1			Conventional	Wayne *et al.* (1991)
Non-tuberculous mycobacteria	27	23	y	n	n	3	94.5		Conventional	Tortoli *et al.* (1992)
Bacillus species	44	30	y	y	y	1	70.0	0.95	Conventional	Priest and Alexander (1988)
Bacillus sphaericus	14	29	y	y	y	1	100	0.999	Conventional	Alexander and Priest (1990)
Aerobic Gram-positive cocci catalase positive	39	60	y	y	y	1	93.0	0.999	API 50CH API 20E	Feltham and Sneath (1982)
Aerobic Gram-positive cocci catalase negative	33	60	y	y	y	1	85.0	0.999	API 50CH API 20E	Feltham and Sneath (1982)
Micrococcus	29	35	n	y	y	1				Alderson *et al.* (1991)
Staphylococcus species	12	15	y	y	n	2	94.6		API Staph	Geary *et al.* (1989)
Alaskan marine bacteria	86	61	n	n	y	1	93.0	0.999	Conventional	Davis *et al.* (1983)
Medical bacteria to genus level	60	20	y	y	n	3		0.99	various	Feltham *et al.* (1984)

Data pub. = identification data given in publication; Matrix eval. = identification matrix evaluated for overlap or uniqueness of taxa; Clus. anal. = data analysed using cluster analysis prior to creation of starting matrix; Method 1 = Numerical taxonomic study of reference strains and fresh isolates; Method 2 = data from identified isolates; Method 3 = data collated from the literature; %id = percentage of strains identified when matrix evaluated; ID coef. = threshold (Willcox score) used for identification.

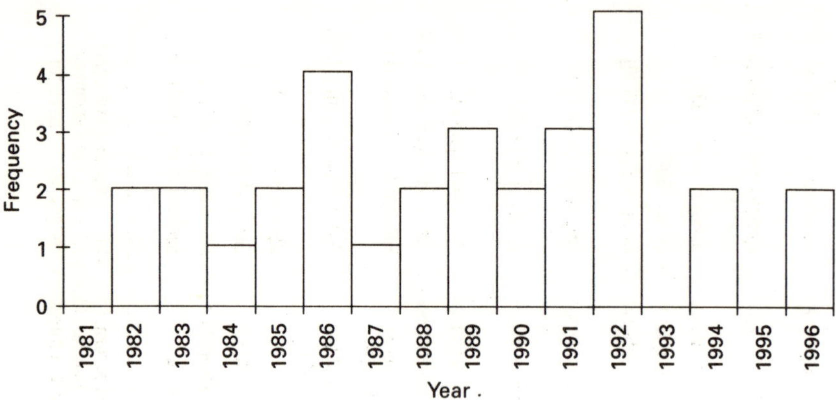

Fig. 28.4. Probabilistic identification matrices by year of publication.

In a well established laboratory, 16s rRNA sequencing currently takes about two days. So it is possible that in future, probabilistic identification could give way to computer assisted identification using sequence data.

Information technology provides us with the ability to process large amounts of data. The ultimate goal, for those involved in bacterial identification, is a polyphasic system where phenotypic or genotypic data can be used to identify an unknown. Projects are underway to build systems that can store bacterial information in a single database. What bacteriologists will require from the database is the ability to identify their unknown from it. Probabilistic methods will probably have a role to play in this.

References

Alderson, G., Amadi, E.N., Pulverer, G. and Zai, S. (1991) Recent advances in the classification and identification of the genus *Micrococcus*. In: Jeljaszewicz/Ciborowski (ed.) *The Staphylococci, Zentralblatt für Bakteriologie Supplement 21*, Stuttgart: Gustav Fisher Verlag, pp. 103–109.

Alexander, B. and Priest, F.G. (1990) Numerical classification and identification of *Bacillus sphaericus* including some strains pathogenic for mosquito larvae. *Journal of General Microbiology* 136, 367–376.

Alsina, M. and Blanch, A.R. (1994) A set of keys for biochemical identification of environmental *Vibrio* species. *Journal of Applied Bacteriology* 76, 79–85.

Beers, R.J. and Lockhart, W.R. (1962) Experimental methods in computer taxonomy. *Journal of General Microbiology* 28, 291–315.

Berger, S.A. (1990) Lack of precision in commercial identification systems: correction using Bayesian analysis. *Journal of Applied Bacteriology* 68, 285–288.

Boeufgras, J.M., Blazer, J.L., Allard, F. and Diaz, I. (1988) A new computer program for routine interpretation of API identification systems. In: Schindler, J. and Chyle, M. (eds) pp. 125–137. Czechoslovak Academy of Sciences, Prague.

Bryant, T.N. (1987) Programs for evaluating and characterising bacterial taxonomic data. *Computer Applications in the Biosciences* 3, 45–48.

Bryant, T.N. (1991) Software for the development and evaluation of probabilistic identification matrices. *Computer Applications in the Biosciences* 7, 189–193.

Bryant, T.N. (1993) A compilation of probabilistic bacterial identification matrices. *Binary – Computing in Microbiology* 5, 207–210.

Bryant, T.N., Capey, A.G. and Berkeley, R.C.W. (1985) Microcomputer assisted identification of bacillus species. *Computer Applications in the Biosciences* 1, 23–27.

Bryant, T.N., Lee, J.V., West, P.A. and Colwell, R.R. (1986) A probability matrix for the identification of species of *Vibrio* and related genera. *Journal of Applied Bacteriology* 61, 469–480.

Busse, H.J., Denner, E.B.M. and Lubitz, W. (1996) Classification and identification of bacteria: current approaches to an old problem. Overview of methods used in bacterial systematics. *Journal of Biotechnology* 47, 3–38.

Canhos, V.P., Manfio, G.P. and Blaine, L.D. (1993) Software tools and databases for bacterial systematics and their dissemination via global networks. *Antonie van Leeuwenhoek, International Journal of General and Molecular Microbiology* 64, 205–229.

Claus, H. (1993) Optimal selection of biochemical tests to identify microbial species. *Zentralblatt für Bakteriologie–International Journal of Medical Microbiology, Virology Parasitology and Infectious Diseases* 278, 522–528.

Clayton, P., Feltham, R.K.A., Mitchell, C.J. and Sneath, P.H.A. (1986) Constructing a database for low cost identification Gram negative rods in clinical laboratories. *Journal of Clinical Pathology* 39, 798–802.

Costas, M., Holmes, B., On, S.L.W. and Stead, D.E. (1992) Identification of medically important *Pseudomonas* species using computerized methods. In: Board, R.G., Jones, D. and Skinner, F.A. (eds) *Identification Methods in Applied and Environmental Microbiology*. Blackwell Scientific Publications, Oxford, pp. 1–27.

Cox, R.P. and Thomsen, J.K. (1990) Computer-aided identification of lactic acid bacteria using the API 50 CHL system. *Letters in Applied Microbiology* 10, 257–259.

Davis, A.W., Atlas, R.M. and Krichevsky, M.I. (1983) Development of probability matrices for identification of marine bacteria. *International Journal of Systematic Bacteriology* 33, 803–810.

Dawson, C.A. and Sneath, P.H.A. (1985) A probability matrix for the identification of vibrios. *Journal of Applied Bacteriology* 58, 407–423.

Döring, B., Ehrhardt, S., Lücke, F. and Schillinger, U. (1988) Computer-assisted identification of lactic acid bacteria from meats. *Systematic and Applied Microbiology* 11, 67–74.

Dybowski, W. and Franklin, D.A. (1968) Conditional probability and the identification of bacteria: a pilot study. *Journal of General Microbiology* 54, 215–229.

Feltham, R.K.A. and Sneath, P.H.A. (1982) Construction of matrices for computer-assisted identification of aerobic Gram-positive cocci. *Journal of General Microbiology* 128, 713–720.

Feltham, R.K.A., Wood, P.A. and Sneath, P.H.A. (1984) A general-purpose system for characterizing medically important bacteria to genus level. *Journal of Applied Bacteriology* 57, 279–290.

Firrao, G. and Locci, R. (1989) Identification by probabilistic methods of plant pathogenic corynebacteria. *Annals of Microbiology* 39, 81–92.

Friedman, R.B., Bruce, D., MacLowry, J. and Brenner, V. (1973) Computer-assisted identification of bacteria. *American Journal of Clinical Pathology* 60, 395–403.

Geary, C., Stevens, M., Sneath, P.H.A. and Mitchell, C.J. (1989) Construction of a database to identify *Staphylococcus* species. *Journal of Clinical Pathology* 42, 289–294.

Gibson, L.F., Clarke, C.J. and Khoury, J.T. (1992) Identification of bacteria by a semi-interactive personal computer program: Its use in teaching and in the hospital/diagnostic laboratory. *Binary – Computing in Microbiology* 4, 25–30.

Gyllenberg, H.G. (1963) A general method for deriving determination schemes for random collections of microbial isolates. *Annales Academiae Scientiarum Fennicae A,IV* 69, 1–23.

Gyllenberg, H.G. and Niemelä, T.K. (1975) New approaches in computer-assisted identification of micro-organisms. In: Pankhurst, R.J. (ed.) *Biological Identification with Computers.* Academic Press, London, pp. 121–136.

Holmes, B. and Costas, M. (1992) Identification of Enterobacteriaceae by computerized methods. In: Board, R.G., Jones, D. and Skinner, F.A. (eds) *Identification Methods in Applied and Environmental Microbiology.* Blackwell Scientific Publications, Oxford, pp. 127–149.

Holmes, B., Dawson, C.A. and Pinning, C.A. (1986a) A revised probability matrix for the identification of Gram-negative, aerobic, rod-shaped, fermentative bacteria. *Journal of General Microbiology* 132, 3113–3135.

Holmes, B., Pinning, C.A. and Dawson, C.A. (1986b) A probability matrix for the identification of Gram-negative, aerobic, non-fermentative bacteria that grow on nutrient agar. *Journal of General Microbiology* 132, 1827–1842.

Holmes, B., On, S.L.W., Ganner, M. and Costas, M. (1992) Some new applications of probabilistic identification. In: Schindler, J. (ed.) *Proceedings of the Conference on Taxonomy and Automated Identification of Bacteria*, Prague, pp. 6–9.

Jahnke, K.D. (1995) IDENTIFY – A menu-driven computer program compatible with MICRO-IS 5.0 for the probabilistic identification of bacteria. *Journal of Microbiological Methods* 21, 133–142.

Jilly, B.J. (1988) Microcomputer application of Bayesian probability testing for the identification of bacteria. *International Journal of Biomedical Computing* 22, 107–119.

Kämpfer, P. and Altwegg, M. (1992) Numerical classification and identification of *Aeromonas* genospecies. *Journal of Applied Bacteriology* 72, 341–351.

Kämpfer, P. and Kroppenstedt, R.M. (1991) Probabilistic identification of *Streptomyces* using miniaturized physiological tests. *Journal of General Microbiology* 137, 1893–1902.

Langham, C.D., Williams, S.T., Sneath, P.H.A. and Mortimer, A.M. (1989) New probability matrices for identification of *Streptomyces. Journal of General Microbiology* 135, 121–133.

Lapage, S.P., Bascomb, S., Willcox, W.R. and Curtis, M.A. (1970) Computer identification of bacteria. In: Baillie, A. and Gilbert, R.J. (eds) *Automation, Mechanization and Data Handling in Microbiology*, Academic Press, London, pp. 1–22.

Lapage, S.P., Bascomb, S., Willcox, W.R. and Curtis, M.A. (1973) Identification of bacteria by computer: General aspects and prospectives. *Journal of General Microbiology* 77, 273–290.

Ledley, R.S. and Lusted, L.B. (1959) Reasoning foundations of medical diagnosis. *Science* 130, 9–21.

Long, P.F. (1994) Identification of some industrially important *Actioplanes* species. *Journal Industrial Microbiology* 13, 300–310.

Maissin, R., Bernard, A., Duquenne, V., Baeten, S., Gerard, G. and Decallonne, J. (1987) Characteristics of a computer-aided procedure for the identification of lactobacilli. *Belgian Journal of Food Chemistry and Biotechnology* 42, 176–183.

Maradona, M.P. (1994) Integrated software for probabistic identification of microorganisms. *Computer Applications in the Biosciences* 10, 71–73.

Niemelä, S.I., Hopkins, J.W. and Quadling, C. (1968) Selecting an economic binary test battery for a set of microbial cultures. *Canadian Journal of Microbiology* 14, 271–279.

Oakley, H.J., Ellis, J.E. and Gibson, L.F. (1996) A biochemical protocol for the differentiation of current genomospecies of *Aeromonas*. *Zentralblatt für Bakteriologie–International Journal of Medical Microbiology Virology Parasitology and Infectious Diseases* 284, 32–46.

On, S.L.W., Holmes, B. and Sackin, M.J. (1996) A probability matrix for the identification of Campylobacters, Helicobacters and allied taxa. *Journal of Applied Bacteriology* 81, 425–432.

Plikaytis, B.D., Plikaytis, B.B. and Shinnick, T.M. (1992) Computer-assisted pattern recognition model for the identification of slowly growing mycobacteria including *Mycobacterium tuberculosis*. *Journal of General Microbiology* 138, 2265–2273.

Portyrata, D.A. and Krichevsky, M.I. (1992) MICRO-IS, A microbiological database management and analysis system. *Binary – Computing in Microbiology* 4, 31–36.

Priest, F.G. and Alexander, B. (1988) A frequency matrix for the probabilistic identification of some bacilli. *Journal of General Microbiology* 134, 3011–3018.

Priest, F.G. and Williams, S.T. (1993) Computer-assisted identication. In: Goodfellow, M. and O'Donnell, A.G. (eds) *Handbook of New Bacterial Systematics.* Academic Press, London, pp. 361–381.

Rypka, W., Clapper, W.E., Bowen, I.G. and Babb, R. (1967) A model for the identification of bacteria. *Journal of General Microbiology* 46, 407–424.

Sackin, M.J. (1987) *Computer Programs for Classification and Identification in Methods in Microbiology*, 19th edn. Academic Press, London.

Sneath, P.H.A. (1978) Basic program for a significance test for two clusters in Euclidean space as measured by their overlap. *Computers and Geosciences* 5, 143–155.

Sneath, P.H.A. (1979a) Basic program for identification of an unknown with presence–absence data against an identification matrix of percent positive characters. *Computers and Geosciences* 5, 195–213.

Sneath, P.H.A. (1979b) Basic program for a significance test for clusters in UPGMA dendrograms obtained from squared euclidean distances. *Computers and Geosciences* 5, 127–137.

Sneath, P.H.A. (1979c) Basic program for character separation indices from an identification matrix of percent positive characters. *Computers and Geosciences* 5, 349–357.

Sneath, P.H.A. (1980a) Basic program for the most diagnostic properties of groups from an identification matrix of percent positive characters. *Computers and Geosciences* 6, 21–26.

Sneath, P.H.A. (1980b) Basic program for determining the best identification scores possible from the most typical examples when compared with an identiification matrix of percent positive characters. *Computers and Geosciences* 6, 27–34.

Sneath, P.H.A. (1980c) Basic program for determining overlap between groups in an identification matrix of percent positive characters. *Computers and Geosciences* 6, 267–278.

Sneath, P.H.A. and Langham, C.D. (1989) Outlier: A Basic program for detecting outlying members of multivariate clusters based on presence–absence data. *Computers and Geosciences* 15, 939–964.

Stager, C.E. and Davis, J.R. (1992) Automated systems for identification of microorganisms. *Clinical Microbiology Reviews* 5, 302–327.

Tortoli, E., Boddi, V. and Penati, V. (1992) Development and evaluation of a program and probability matrix for the computer-aided identification of non-tuberculous mycobacteria. *Binary – Computing in Microbiology* 4, 200–203.

Van Oystaeyen, B. (1995) Bacterial identification – an improvement on the classical approach. *Annales de Biologie Clinique* 53, 429–434.

Wayne, L.G., Good, R.C., Krichevsky, M.I., Blacklock, Z., David, H.L., Dawson, D., Gross, W., Hawkins, J., Vincent Levy-Frebault, V., McManus, C., Portaels, F., Rusch-Gerdes, S., Schroder, K.H., Silcox, V.A., Tsukamura, M., Van Den Breen, L. and Yarkrus, M.A. (1991) Fourth report of the cooperative, open-ended study of slowly growing mycobacteria by the international working group on mycobacterial taxonomy. *International Journal of Systematic Bacteriology* 41, 463–472.

Willcox, W.R., Lapage, S.P., Bascomb, S. and Curtis, M.A. (1973) Identification of bacteria by computer: Theory and programming. *Journal of General Microbiology* 77, 317–330.

Willcox, W.R., Lapage, S.P. and Holmes, B. (1980) A review of numerical methods in bacterial identification. *Antonie van Leeuwenhoek* 46, 233–296.

Williams, S.T., Goodfellow, M., Wellington, E.M.H., Vickers, J.C., Alderson, G., Sneath, P.H.A., Sackin, M.J. and Mortimer, A.M. (1983) A probability matrix for identification of some streptomycetes. *Journal of General Microbiology* 129, 1815–1830.

Williams, S.T., Locci, R., Vickers, J.C., Schofield, G.M., Sneath, P.H.A. and Mortimer, A.M. (1985) Probabilistic identification of *Streptoverticillium* species. *Journal of General Microbiology* 131, 1681–1689.

Identification of Yeasts through Computer-based Systems

29

R.W. Payne

Statistics Department, IACR-Rothamsted, Harpenden, Herts AL5 2JQ, UK
Fax: +44(0)1582 467116/E-mail: roger.payne@bbsrc.ac.uk

Introduction

Computers were already being used in the identification of yeasts at the time of the Systematics Association Symposium on *Biological Identification with Computers* (see Preface to this volume). However, the aim then was more to support (and enhance) the traditional pencil-and-paper methods than to provide direct help with identification. For example, Payne (1975) described the computer program GENKEY for constructing identification keys, with examples from a key to the yeasts. The identification of yeasts certainly presents a worthwhile and challenging application for computer-based identification systems. There are many species and varieties to consider (680 in the 1996 version of the *Yeast Identification Program* of Barnett *et al.*), and over 100 tests. These involve such attributes as the ability of the yeast to ferment certain types of sugar or to grow in particular media, as well as some morphological characters such as the formation of spores and filaments. The basic data for the identification can be regarded as a species-by-test table indicating the possible responses that each yeast may give to each test. Table 29.1 shows the data for *Schizosaccharomyces pombe* obtained from the VIEW command of the *Yeast Identification Program.*

The response may be fixed: all strains of the yeast always give the same result (positive or negative) to the test. However, there are many variable responses (code V), where different strains produce different results. Some responses are delayed (code D) or weak (code W) and thus must also be treated as variable, and others are unknown (code ?).

 Information Technology, Plant Pathology and Biodiversity (eds P. Bridge, P. Jeffries, D.R. Morse and P.R. Scott)

Table 29.1. Characteristics of *Schizosaccharomyces pombe.*

+	F1	D-Glucose fermentation
V	F2	D-Galactose fermentation
V	F3	Maltose fermentation
+	F5	Sucrose fermentation
–	F8	Lactose fermentation
+	C1	D-Glucose growth
V	C2	D-Galactose growth
–	C3	L-Sorbose growth
–	C4	D-Glucosamine growth
–	C5	D-Ribose growth
–	C6	D-Xylose growth
–	C7	L-Arabinose growth
–	C8	D-Arabinose growth
–	C9	L-Rhamnose growth
+	C10	Sucrose growth
+	C11	Maltose growth
V	C12	alpha,alpha-Trehalose growth
V	C13	Me alpha-D-glucoside growth
–	C14	Cellobiose growth
–	C15	Salicin growth
–	C16	Arbutin growth
V	C17	Melibiose growth
–	C18	Lactose growth
+	C19	Raffinose growth
–	C20	Melezitose growth
V	C21	Inulin growth
V	C22	Starch growth
–	C23	Glycerol growth
–	C24	Erythritol growth
–	C25	Ribitol growth
–	C26	Xylitol growth
–	C27	L-Arabinitol growth
–	C28	D-Glucitol growth
–	C29	D-Mannitol growth
–	C30	Galactitol growth
–	C31	*myo*-Inositol growth
V	C32	D-Glucono-1,5-lactone growth
V	C33	2-Keto-D-gluconate growth
–	C34	5-Keto-D-gluconate growth
V	C35	D-Gluconate growth
–	C36	D-Glucuronate growth
–	C37	D-Galacturonate growth
–	C38	DL-Lactate growth
–	C39	Succinate growth
–	C40	Citrate growth
–	C41	Methanol growth
–	C42	Ethanol growth
–	C43	Propane 1,2 diol growth
–	C44	Butane 2,3 diol growth
–	C45	Quinate growth
–	C46	Saccharate growth
–	C47	Galactonate growth
–	N1	Nitrate growth
–	N2	Nitrite growth
V	N3	Ethylamine growth
V	N4	L-Lysine growth
V	N5	Cadaverine growth
–	N6	Creatine growth
–	N7	Creatinine growth
–	N8	Glucosamine (nitrogen) growth
–	N9	Imidazole growth
–	V1	Growth w/o vitamins
–	V2	Growth w/o *myo*-Inositol
V	V3	Growth w/o Pantothenate
V	V4	Growth w/o Biotin
V	V5	Growth w/o Thiamin
V	V7	Growth w/o Pyridoxine
–	V9	Growth w/o Niacin
+	V10	Growth w/o PABA
+	T1	Growth at 25°C
+	T2	Growth at 30°C
+	T3	Growth at 35°C
+	T4	Growth at 37°C
V	T5	Growth at 40°C
V	T6	Growth at 42°C
–	T7	Growth at 45°C
V	O1	0.01% Cycloheximide growth
–	O2	0.1% Cycloheximide growth
+	O3	1% Acetic acid growth
+	O4	50% D-Glucose growth
V	O5	60% D-Glucose growth
–	O6	16% NaCl + 10% D-Glucose
–	M1	Starch formation
–	M2	Acetic acid production
+	M3	Urea hydrolysis
–	M4	Diazonium Blue B reaction
–	E1	Pink colonies
–	E2	Budding cells
–	E3	Lemon-shaped cells
–	E4	Buds on stalks
+	E5	Splitting cells
–	E6	Filamentous
–	E7	Pseudohyphae
–	E8	Septate hyphae
–	E9	Arthroconidia
–	E10	Ballistoconidia
+	A1	Ascosporogenous
+	A2	Ascospores round, oval, conical or reniform
–	A3	Ascospores cap-, hat-, Saturn- or walnut-shaped
–	A4	Ascospores needle-shaped or whip-like

Identification Keys

Conventionally, yeasts have been identified using identification keys. Table 29.2 shows an example, for the yeasts found in brewing, taken from Barnet *et al.* (1990). The first test in the key requires the user to check whether the unknown yeast can utilize citrate for growth (test C40 in Table 29.1). If a negative response is obtained, the next test is Sucrose fermentation (F5) at the line numbered 2; for a positive response, it is Nitrate growth (N1) at line 39. The key thus provides a sequence of tests which culminate in the identification of the unknown yeast. For example 1 Citrate growth (C40) −, 2 Sucrose fermentation (F5) +, 22 Nitrate growth (N1) −, 23 Cadaverine growth (N5) −, 24 2-Keto-D-gluconate growth (C33) −, 25 L-Lysine growth (N4) −, 26 Growth w/o Niacin (V9) − would identify *Schizosaccharomyces pombe*.

Physiological tests for the yeasts can, however, take up to a month to observe. Thus, if a specimen was known to be one of the brewing yeasts, all the tests in the key would usually be performed simultaneously and the key would be used merely to look up the identification. An alternative aid to identification by non-sequential testing is to arrange the test results lexicographically, as a diagnostic table (see for example Payne and Preece, 1980, Section 2).

Key Construction

When the tests in a key are applied sequentially, its efficiency is usually measured by the expected cost of identification:

$$\kappa = \sum_{i=1}^{d} \left(\gamma_i \times \rho_i \right) \tag{29.1}$$

where d is the number of endpoints in the key, γ_i is the cost of the tests on the branch to endpoint i and ρ_i is the probability of reaching endpoint i.

For any particular set of data, to find the optimal key essentially requires the construction and comparison of all possible keys to the species in the set, that could be formed with the available tests. This is impracticable even for moderate numbers of tests and species. Thus, heuristic algorithms are used which construct the key sequentially, selecting first the test that 'best' divides the species into sets (where set k for test i contains all the species that can give result k to test i), then selecting the best test to use with each set, continuing until the sets each contain only one species – or until no further separation is possible. These algorithms are very efficient at finding a near-optimal key: for example, to construct a key to all the yeasts in Barnett *et al.* (1990) using the program Genkey (Payne, 1993) takes only 9.95 seconds on a Microvax 4400.

The 'best' test can be defined using a selection criterion function (Gower and Payne, 1975). Two examples, from Payne (1981), are

Table 29.2. Key to the yeasts found in brewing.

	Negative	Positive
1 Citrate growth	2	39
2 Sucrose fermentation	3	22
3 2-Keto-D-gluconate growth	4	15
4 0.01% Cycloheximide growth	5	12
5 L-Lysine growth	6	7
6 Nitrate growth	*Saccharomyces cerevisiae*	*Saitoella complicata*
7 Ethanol growth	8	9
8 60% D-Glucose growth	*Hanseniaspora vineae*	*Zygosaccharomyces rouxii*
9 1% Acetic acid growth	10	11
10 60% D-Glucose growth	*Issatchenkia orientalis* *Pichia membranaefaciens*	*Zygosaccharomyces rouxii*
11 Growth w/o vitamins	*Zygosaccharomyces bailii*	*Issatchenkia orientalis*
12 D-Galactose growth	13	14
13 Ethanol growth	*Hanseniaspora valbyensis*	*Dekkera custersiana*
14 L-Lysine growth	*Candida humilis*	*Saccharomyces unisporus*
15 0.01% Cycloheximide growth	16	20
16 Cadaverine growth	*Torulaspora delbrueckii*	17
17 1% Acetic acid growth	18	*Zygosaccharomyces bailii*
18 60% D-Glucose growth	19	*Zygosaccharomyces rouxii*
19 D-Galactose growth	*Candida solani*	*Candida sake*
20 Nitrate growth	21	*Candida versatilis*
21 Ethanol growth	*Hanseniaspora uvarum*	*Candida tropicalis*
22 Nitrate growth	23	36
23 Cadaverine growth	24	29
24 2-Keto-D-gluconate growth	25	28
25 L-Lysine growth	26	27
26 Growth w/o Niacin	*Schizosaccharomyces pombe*	*Saccharomyces cerevisiae* *Saccharomyces exiguus*
27 Growth at 37°C	*Candida stellata*	*Schizosaccharomyces pombe*
28 Ethanol growth	*Schizosaccharomyces pombe*	*Torulaspora delbrueckii*
29 0.01% Cycloheximide growth	30	33
30 1% Acetic acid growth	31	32
31 60% D-Glucose growth	*Candida sake*	*Zygosaccharomyces rouxii*
32 Ethanol growth	*Schizosaccharomyces pombe*	*Zygosaccharomyces bailii*
33 Growth w/o Niacin	34	35
34 Ethanol growth	*Schizosaccharomyces pombe*	*Kluyveromyces marxianus*
35 Starch growth	*Dekkera anomala* *Dekkera bruxellensis*	*Candida tropicalis*
36 0.01% Cycloheximide growth	37	38
37 Erythritol growth	*Candida vartiovaarai*	*Pichia subpelliculosa*
38 60% D-Glucose growth	*Dekkera anomala* *Dekkera bruxellensis*	*Candida versatilis*
39 Nitrate growth	40	59
40 Raffinose growth	41	53
41 Growth at 37° C	42	49
42 Erythritol growth	43	48
43 L-Rhamnose growth	44	*Candida tenuis*
44 D-Galactose growth	45	46

45 2-Keto-D-gluconate growth .	.*Filobasidium capsuligenum*	.... *Candida beechii*
46 2-Keto-D-gluconate growth .	.*Filobasidium capsuligenum*	 47
47 0.01% Cycloheximide growth	.... *Candida sake*	.... *Candida oleophila*
48 L-Rhamnose growth	... *Debaryomyces marama*	.... *Candida tenuis*
49 D-Galactose growth	 50	 51
50 Growth w/o vitamins	.... *Pichia fermentans*	...... *Issatchenkia orientalis*
51 Erythritol growth	 52	.... *Pichia farinosa*
52 Starch growth	...... *Candida parapsilosis*	.... *Candida tropicalis*
53 2-Keto-D-gluconate growth .	 54	 55
54 D-Galactose growth	.... *Pichia onychis*	.. *Kluyveromyces marxianus*
55 Erythritol growth	 56	.... *Debaryomyces hansenii*
		..... *Debaryomyces marama*
56 Xylitol growth	 57	 58
57 Growth at 37° C	.... *Candida intermedia*	.... *Pichia ohmeri*
58 Growth at 37° C	.... *Candida intermedia*	.... *Debaryomyces hansenii*
	... *Debaryomyces hansenii*	.... *Pichia guilliermondii*
59 Sucrose fermentation	 60	 63
60 D-Galactose growth	61	.... *Candida versatilis*
61 Raffinose growth	 62	.... *Saitoella complicata*
62 Starch growth	.... *Candida norvegica*	.... *Candida ernobii*
63 Starch growth	 64	 65
64 D-Galactose growth	...... *Candida vartiovaarai*	.... *Candida versatilis*
65 Growth w/o vitamins	 66	.... *Pichia anomala*
66 Erythritol growth	.... *Pichia fabianii*	.... *Pichia subpelliculosa*

$$\left(CM_e\right)_i = c_i + \left(\bar{c}/\log\bar{m}\right)\left\{\sum_{k=1}^{m_i}\left(p_{ik}+r_i/m_i\right)\log\left(p_{ik}+r_i/m_i\right)+r_i\log m_i\right\} \quad (29.2)$$

$$\left(CM_v'\right)_i = c_i + \bar{c}\left\{\sum_{k=1}^{m_i}\left(p_{ik}+r_i/m_i\right)\left(n_{ik}^2+n_{ik}-2\right)/2n_{ik}\right\} \quad (29.3)$$

where c_i is the cost of test i, $\bar{c}$ is the average cost of the available tests, m_i is the number of possible responses to test i, $\bar{m}$ is the average number of possible responses to the available tests, p_{ik} is the proportion of the species at the current point of the key that always give result k to test i, and r_i is the proportion that are variable. Criterion $(CM_e)_i$ is an estimate of the expected cost of completing the identification from the current point of the key, assuming that test i is used and that, below this point, the key is completed optimally. Criterion $(CM'_v)_i$, which was used for the yeasts, is a less optimistic estimate. This assumes that the key is completed by simple binary tests (i.e. tests for each of which one particular species always gives a positive response and other species give negative responses).

Irredundant Test Sets

If, as with the yeasts, sequential testing is not feasible, the appropriate measure of efficiency is the total cost of the tests in the key or table. The set of tests with minimum cost or of minimum size are both irredundant, that is they are such that if any of the component tests is deleted, there will be some pair of species that can no longer be distinguished.

An algorithm, reviewed by Payne and Preece (1980, Section 6) is available for finding all irredundant sets of tests. A implementation in Fortran is given by Payne (1991). The method is illustrated below using the hypothetical results shown in Table 29.3.

First, lists are made of tests that can distinguish each pair of species as shown in Table 29.4: for example, species 1 and species 2 in Table 29.3 differ in their responses to tests B and E, so the first entry in Table 29.4 is B,E. The lists are then expressed as a sum (e.g. B + E for species 1 and 2) and multiplied together according to the rules of Boolean algebra. These rules state that $A \times A = A$, etc., and that if a term contains all the elements of some other term in the sum, it should be deleted, for example $AB + ABC = AB$. Thus, multiplying the lists for the pairs of species (1,2) and (1,3) gives

$$\begin{aligned}(B + E) \times (B + C + D) &= BB + BC + BD + BE + CE + DE \\ &= B + CE + DE\end{aligned} \tag{29.4}$$

Table 29.3. Artificial data showing results of tests A–E with species 1–4.

	Test				
Species	A	B	C	D	E
1	+	+	+	–	+
2	+	–	+	–	–
3	+	–	–	+	+
4	–	+	+	+	–

Table 29.4. List of tests that can distinguish each pair of species.

	Species			
Species	1	2	3	4
1		B,E	B,C,D	A,D,E
2			C,D,E	A,B,D
3				A,B,C,E
4				

Each term in this list is a set of tests than can be used to distinguish between species 1 and species 2, and between species 1 and species 3, so the sets are: (i) test B, (ii) tests C and E; and (iii) tests D and E. (This indicates the rationale behind the rules of the algebra. BB becomes B because there is no need to record a test more than once in the same set. The fact that B is a set in its own right implies that BC, BD and BE contain redundant tests C, D and E, respectively; if these redundant tests are deleted, three more instances of the set B would be obtained. This set is already in the list, so BC, BD and BE can be deleted.)

Continuing the process, irredundant sets to distinguish all the species 1–4 are given by

$$
\begin{aligned}
&(B + CE + DE) \times (A + D + E) \times (C + D + E) \times (A + B + D) \times (A + B + C + E) \\
&= (AB + BD + BE + CE + DE) \times (C + D + E) \times (A + B + D) \times (A + B + C + E) \\
&= (ABC + BD + BE + CE + DE) \times (A + B + D) \times (A + B + C + E) \\
&= (ABC + BD + BE + ACE + DE) \times (A + B + C + E) \\
&= ABC + BD + BE + ACE + DE \qquad (29.5)
\end{aligned}
$$

As the example shows, there can be more than one irredundant set, and the sets can be of different sizes. The number of sets available for the keys in Barnett *et al.* (1990) varied from six, for their key to all the yeasts using only physiological tests, to over 4500 for their key to the yeasts that utilize methanol.

The algorithm can also be used to construct sets of tests to distinguish each species from the other species, and thus confirm a suggested identification. Table 29.5, from Barnett *et al.* (1990, p. 878) shows some sets for *Schizosaccharomyces pombe*. Each column of +s or −s indicates the results of a suitable set of tests. Thus for *Schizosaccharomyces pombe* there are five sets of test results: {F1+, C23−, O3+ and M3+}, {F1+, C29−, O3+ and M3+}, {C6−, C23−, O3+ and M3+}, {C11+, C23−, V2− and O3+} and {F1+, C10+, C23−, O4+ and M3+}. If any of the five sets of tests is used and the specified results are obtained, we can identify the yeast as *Schizosaccharomyces pombe*. The different sets are offered to cater for individual preferences for tests.

Table 29.5. Sets of tests to distinguish *Schizosaccharomyces pombe* from other yeasts.

F1 D-Glucose fermentation	+	+			+
C6 D-Xylose growth			−		
C10 Sucrose growth					+
C11 Maltose growth				+	
C23 Glycerol growth	−		−	−	−
C29 D-Mannitol growth		−			
V2 Growth w/o *myo*-Inositol				−	
O3 1% Acetic acid growth	+	+	+	+	
O4 50% D-Glucose growth					+
M3 Urea hydrolysis	+	+	+		+

These sets can be formed using a similar method to that described earlier. For species n, the sets of tests are obtained by multiplying the lists of tests to distinguish pairs of species $(1, n) \dots (n-1, n) \dots (n, N)$ where N is the total number of species. Thus, sets of tests to distinguish species 3 from the other species in Table 29.3 would be given by

$$\begin{aligned}&(B + C + D) \times (C + D + E) \times (A + B + C + E)\\&\quad = (BE + C + D) \times (A + B + C + E)\\&\quad = BE + C + AD + BD + DE \qquad (29.6)\end{aligned}$$

The methodology can be extended to allow extra tests to be included to guard against errors in testing: nth-order irredundant test sets (Payne and Preece, 1980, p. 290) contain at least n tests to distinguish between each pair of species. These can be constructed by multiplying together lists of the possible combinations of n tests from each of the entries in the table. For example, second-order irredundant test sets to distinguish species 3 from the other species would be given by

$$\begin{aligned}&(BC + BD + CD) \times (CD + CE + DE) \times (AB + AC + AE + BC + BE + CE)\\&\quad = (BCE + BDE + CD) \times (AB + AC + AE + BC + BE + CE)\\&\quad = (ACD + BCD + BCE + BDE + CDE) \qquad (29.7)\end{aligned}$$

For some data sets, particularly where there are many more tests than species, this method can be very time consuming – there may even be too many irredundant sets to store within the computer. Alternatively, sequential algorithms are available for constructing a single irredundant set (Payne and Preece, 1980, Section 6.6), or a small selection of sets (Barnett *et al.*, 1983), but these do not guarantee to find a set of minimum size or cost. The methods operate by selecting suitable tests, one at a time, until each pair of species can be distinguished. Usually, the process starts with an initial set containing all the essential tests. (A test is essential if it is the only test which can distinguish between the members of a particular pair of species.) The test selected at each stage is the one that distinguishes between the members of as many pairs of species as possible, of those pairs not distinguished by tests already in the set. An improvement, suggested by Barnett *et al.* (1983), is to weight each pair of species by the reciprocal of the number of tests that could distinguish between them. Thus, in the example above, the pair of species (3, 4) would have weight $\frac{1}{4}$, as they can be separated by tests A, B, C or E. The test selected is then that distinguishing the maximum weighted number of pairs of species. This procedure causes tests that distinguish 'difficult' pairs of species (those with nearly identical characteristics) to be selected earlier during the construction of the set. It can also be useful to multiply the weighted number of pairs of species by a preference rating for the test (which may, for example be the reciprocal of its cost or some measure of convenience).

Computer-based Identification of Yeasts

The methods of keys and irredundant test set construction form the basis of many computer-based identification systems, and indeed of many expert systems (Payne, 1988). In most systems, we are essentially following one branch of an identification key – the appropriate branch for the specimen in hand. Some systems merely work their way through a stored key, and would seem to offer few advantages over the use of a printed key. Others provide the ability to choose the tests as we progress (equivalent to constructing the branch of the key as it is required). This provides much more flexibility, and the same selection criterion functions can be used to advise on the usefulness of the available tests and prevent time being wasted on those that provide little additional information. For further details see Pankhurst (1991, Section 5.2) who gives an example using the program of Pankhurst and Aitchison (1975) and reviews some other similar systems. Most of these are intended for botanical identification, and assume that the tests will be observed one at a time.

For the yeasts however, as already discussed, several tests are generally applied at once. Consequently, the *Yeast Identification Program* assumes that the user will have observed a batch of tests and will be seeking to ascertain the yeasts that most closely match the unknown specimen, then (if necessary) to apply a further batch of tests to complete the identification. The test selection methods are thus those relevant to irredundant tests sets rather than to keys. The program can be run in either a command-based, conversational mode or by using pop-up menus. The available commands are listed in Table 29.6. For clarity here we shall use the conversational mode.

Figure 29.1 shows part of a session, with input characters in bold type. First the results that have been entered already (using DATA or LIST) are listed using VIEW, then the IDENTIFY command is used to generate a list of the yeasts that can give these results. Only one species can give results identical to those recorded and so (assuming that the yeast is not aberrant and that all the tests for which results have been entered were applied and interpreted correctly) the unknown yeast would be identified as *Schizosaccharomyces pombe*. The figure printed under the heading 'Pr.' gives an indication of the reliability of the identification. This is the probability that the unknown yeast belongs to the listed species, given the observed results and assuming that all species are a priori equally likely: that is, it is the posterior probability (calculated using Bayes theorem) that the unknown belongs to the listed species, given the results and observations that have been entered. The component probabilities required for the calculation are stored in a configuration file, and can be changed according to the user's own beliefs. Thus, for example, the file defines 0.99 to be the probability that a yeast listed as being positive for a particular test will actually give a positive result when tested, and 0.01 to be the probability that it will be negative. The default values, distributed with the program, are chosen in order to

Table 29.6. Available commands for the *Yeast Identification Program.*

BREAK	to use a DOS command.
COMMENT	to label input.
DATA	to enter test results in any order.
FORM	to take results from a data form already stored on the computer.
HELP	to give information about the program and commands.
IDENTIFY	to identify species that match the entered results.
LIST	to enter results; tests are listed in turn and a result can be entered (or deleted) or the test can be skipped.
MISTAKE	to allow for mistakes in the results.
NEXT	to clear previous entries before entering a new set of results.
OPEN	to open a file of data about extra species, or a transcript file, or a form containing results to be read by subsequent FORM commands, or a report file.
QUIT	to finish running the program.
REPORT	to produce a report containing details of the test results that have been entered, followed by the list of species that can give them.
SELECT	to select further tests to complete the identification of all species able to give the results already entered; some choice of these tests is possible.
TRANSCRIPT	to record input and output.
VIEW	to display the characteristics of a particular yeast, or the test results that you have entered, or the differences between the unknown yeast and a particular species, or the differences between 2 species.
WINDOW	to switch between the use of commands and pop-up menus.

be cautious. So for example 0.99 is defined as the probability that a result listed as unknown will be observed as either positive or negative (representing perhaps a measure of 'plausibility' instead of probability). The printed probabilities are intended only as a guide to the reliability of the identification, and it is recommended that, however close to 1.0 the probability may be, the identification should be also checked with the appropriate photograph and description in Barnett *et al.* (1990). Descriptions of the species that differ by only a few characteristics from those listed should also be examined. These are listed on the screen in order of the number of differences, until either you tell the program that no more are required or until all the species have been listed.

In Fig. 29.1, only one species could give all the results specified. If several species are listed, you can ask the program to suggest a set of tests to complete the identification. For example, if we delete the results for test O3 there are now four possible species, as shown in Fig. 29.2. The SELECT command can then be used to select an irredundant set of tests to distinguish between these species, and thus complete the identification. The program lists the set of tests on the screen and then gives you the opportunity to request alternatives to be selected for any test or tests that are disliked.

Fig. 29.1. Output from the *Yeast Identification Program.*

```
Next command? > VIEW<ENTER>
Do you want to view:
the characteristics of a particular species (answer C)
the test results that have been entered (answer R)
differences between two particular species (answer S)
or between a species and the unknown species (answer U) > R<ENTER>
+     F1 D-Glucose fermentation
–     F2 D-Galactose fermentation
+     F3 Maltose fermentation
+     F5 Sucrose fermentation
–     F8 Lactose fermentation
+     C1 D-Glucose growth
–     C2 D-Galactose growth
–     C3 L-Sorbose growth
–     C23 Glycerol growth
+     03 1% Acetic acid growth
+     M3 Urea hydrolysis
Entered results all printed.

Next command? > IDENTIFY<ENTER>
   1 species can give the results specified
Species                              Pr.
Schizosaccharomyces pombe            0.936
   4 species differ by 1 result. Do you want them to be printed?
   (Y/N) > Y<ENTER>
                                        Pr.          Discrepant test(s)
Dekkera bruxellensis                    0.016        M3
Schizosaccharomyces japonicus           0.016        03
Schizosaccharomyces octosporus          0.016        03
Xanthophyllomyces dendrorhous**         0.012        03
   13 species differ by 2 results. Do you want them to be printed?
(Y/N) > N<ENTER>
```

The program forms the set of tests sequentially, as described earlier. The preference ratings of the tests can, if required, be modified by the user within the configuration file. The program reports if some pairs of species are indistinguishable by means of the available tests, or if you have deleted all the tests that distinguish other pairs (in which case you will be asked whether you will make them available for reselection).

Unless you specify otherwise, the set of tests is selected solely to distinguish between those species that can give exactly the results entered. The test results entered are assumed to be correct and the yeast is assumed not to be aberrant; these same assumptions are also applied to the results of the selected tests.

Fig. 29.2. Continuation of the output in Fig. 29.1.

Next command? > **DATA<ENTER>**
Next test (or E to end DATA)? > **O3<ENTER>**
Result to test C23 ? (+/-/?) > **?<ENTER>**
Next test (or E to end DATA)? > **E<ENTER>**
Next command? > **IDENTIFY<ENTER>**
4 species can give the results specified

Species	Pr.
Schizosaccharomyces japonicus	0.290
Schizosaccharomyces octosporus	0.290
Schizosaccharomyces pombe	0.167
Xanthophyllomyces dendrorhous**	0.220

12 species differ by 1 result. Do you want them to be printed?
(Y/N) > **N<ENTER>**

Next command? > **SELECT<ENTER>**

Do you wish to consider only physiological tests (answer P),
physiological tests and tests of reproduction and appearance (answer R)
or all tests, including those involving ascospores, (answer A)? > **A<ENTER>**

3 test(s) selected:
C10 Sucrose growth
C12 alpha,alpha-Trehalose growth
O4 50% D-Glucose growth
Do you wish to replace any tests? (Y/N) > **Y<ENTER>**

Please specify test number(s) > **O4<ENTER>**

Do you wish to replace any other tests? (Y/N) > **N<ENTER>**
3 test(s) selected:
C10 Sucrose growth
C12 alpha,alpha-Trehalose growth
O3 1% Acetic acid growth
Do you wish to replace any tests? (Y/N) > **N<ENTER>**

Final set stored (for LIST), selection ends.

Should an isolate correspond to a species for which only one or two, rather than many, strains were examined to provide its description, there is a greater likelihood that the isolate may differ from the description in some characteristics. The command MISTAKE allows you: (i) to specify the number of test results that may be 'wrong' in the results already entered; and (ii) to request that extra tests should be included, by SELECT, to guard against future mistakes. If errors occur, it may be impossible to distinguish some pairs of species (because their results differ for fewer tests than the allowed number of errors). Sometimes it may be practicable to resolve this difficulty by repeating some of the original tests (i.e. those whose results were entered by DATA or LIST) and the program will ask if you are prepared to do this. However, this will be worth doing only if the 'errors' arise from mistakes in testing or interpreting test results; if the strain is aberrant, the same ('wrong') answer will again be obtained. This is similar to the methodology used in the construction of the check keys of Payne and Preece (1977). Alternatively, having identified the unknown yeast, you could use a confirmatory sets of tests, as illustrated in Table 29.5.

Conclusions

The brief example above illustrates the flexibility that the computer provides for yeast identification. There is complete freedom in the tests that are used at the first stage while, at later stages, the program can supply a customised set of tests to complete the identification in an efficient way. Furthermore, it is easy to generate additional information, such as the full list of characteristics of a yeast (Table 29.1) or the differences between two species. For the first stage the user could, for example, choose some or all of the sets of tests in the keys of Barnett *et al.* (1990). Alternatively, Payne (1998) discusses methods of test selection for two-stage identification and these will be used to advise on suitable initial sets of tests in future releases of the program.

Acknowledgement

IACR receives grant-aided support from the Biotechnology and Biological Sciences Research Council of the United Kingdom.

References

Barnett, J.A., Payne, R.W. and Yarrow, D. (1983) *Yeasts: Characteristics and Identification.* Cambridge University Press, Cambridge, 821 pp.

Barnett, J.A., Payne, R.W. and Yarrow, D. (1990) *Yeasts: Characteristics and Identification.* 2nd edn. Cambridge University Press, Cambridge, 1012 pp.

Gower, J.C. and Payne, R.W. (1975) A comparison of different criteria for selecting binary tests in diagnostic keys. *Biometrika* 62, 665–671.

Pankhurst, R.J. (1991) *Practical Taxonomic Computing.* Cambridge University Press, Cambridge, 202 pp.

Pankhurst, R.J. and Aitchison, R.R. (1975) An on-line identification program. In: Pankhurst, R.J. (ed.) *Biological Identification with Computers.* Systematics Association Special Vol. No. 7. Academic Press, London, pp. 181–194.

Payne, R.W. (1975) Chapter 4: Genkey: a program for constructing diagnostic keys. In: Pankhurst, R.J. (ed.) *Biological Identification with Computers.* Systematics Association Special Vol. No. 7, Academic Press, London, pp. 65–72.

Payne, R.W. (1981) Selection criteria for the construction of efficient diagnostic keys. *Journal of Statistical Planning and Inference* 5, 27–36.

Payne, R.W. (1988) Identification keys, diagnostic tables and expert systems. In: Edwards, D. and Raun, N.E. (eds) *COMPSTAT 88 Proceedings in Computational Statistics.* Physica-Verlag, Heidelberg, pp. 199–204.

Payne, R.W. (1991) Algorithm AS263 Construction of irredundant test sets. *Applied Statistics* 40, 213–229.

Payne, R.W. (1993) *Genkey: A Program for Constructing and Printing Identification Keys and Diagnostic Tables. Mark 4.* Harpenden: Rothamsted Experimental Station.

Payne, R.W. (1998) Construction of efficient identification schemes using batches of discrete-valued tests. *Journal of Classification* (in press).

Payne, R.W. and Preece, D.A. (1977) Incorporating checks against observer error into identification keys. *New Phytologist* 79, 201–207.

Payne, R.W. and Preece, D.A. (1980) Identification keys and diagnostic tables: a review (with discussion). *Journal of the Royal Statistical Society, Series A* 143, 253–292.

30

Electronic Teaching Aids for Students and Practitioners

G.L. Schumann

Department of Microbiology, University of Massachusetts, Fernald Hall, Amherst, MA 01003, USA
E-mail: schumann@pltpath.umass.edu

Introduction

This paper examines some of the pedagogical advantages and disadvantages of newly emerging teaching technologies for plant pathology students. The discussion is guided by the teaching experiences of the author in several introductory level courses and, although it includes some broader considerations, it focuses specifically on how plant pathology students may or may not benefit from these technologies. Some of the limitations of teaching technologies are technological, but the more important problems are probably associated with possible misinterpretations of how technology may help students learn.

Advantages of New Teaching Technologies for Plant Pathology Students

Probably the greatest benefit from the use of teaching technologies for plant pathology students is access to quantities of high quality images. If one has already been educated in plant pathology, it is sometimes difficult to remember that much of our discipline-specific vocabulary is essentially meaningless to students who are initiating their studies, in large part because the vocabulary does not yet trigger mental images. Part of that is because most plant pathogens are microscopic and are not familiar to people outside our discipline. Another part is a lack of familiarity with 'typical' symptoms of plant diseases. Textbooks rarely have colour photographs at all because of cost limitations and generally contain sparse collections of even black and white photographs. Plant pathology texts, because of the relatively small sales potential, are particularly lacking in good images, especially books on more advanced and specialized subjects.

High quality images are important for students who need to become familiar with the basic structures and characteristics of the organisms which cause plant diseases. Even though prepared slides and infected plant materials are important, it is difficult to guarantee that each student will make a full range of observations during laboratory classes. Students may also be absent from certain laboratories due to illness or other reasons.

Archives of images can also help students visualize the biological interaction between hosts and parasites in ways that are generally unavailable in a classroom setting. Important examples include time course studies of disease epidemics or infection processes. Electron micrographs often reveal pathogens and host–parasite interactions in detail that is invisible with a light microscope. After study of such photographs, many students find that they can better appreciate what they are observing in squash mounts and even prepared slides observed with a light microscope. Many plant pathology courses are offered during seasons that preclude collection of infected plants or materials that are at each necessary stage for study. Many students would also like to become familiar with disease problems from distant regions and foreign climates.

Teaching technologies offer the potential for self-paced, interactive lessons that can be individualized for student interests. Students can access the information at nearly any time and as often as is needed for mastery. Interactive technologies can be formatted to help ensure that students proceed when their attention is engaged and that they have accomplished certain learning goals before moving on to new material. This can be difficult to accomplish in a traditional classroom lecture format where an instructor may go too quickly or too slowly for individual students.

As an example, every introductory student should become familiar with common disease symptoms. Rather than tediously projecting slides and naming symptoms in a lecture setting, students can be given a list of symptoms and a page of bar codes that code for images of the symptoms on a videodisc. Each student is then expected to match the symptoms with the images. In a lecture setting, it has been estimated that approximately one-third of the audience is not engaged at any particular moment. Some would suggest that this is a conservative estimate. In accomplishing the symptom matching exercise, each individual must make a conscious decision about what word describes the photographed symptom. The response from students using this exercise has been very positive. Many students have asked for an answer key to check their results even though the activity was not a course requirement.

The new technologies can also provide illustrated glossaries which can be valuable study tools. To appreciate how important accompanying photographs or diagrams can be, instructors might consult some of the definitions in the glossary that they recommend to their students and try to imagine how meaningful the words would be to someone without the appropriate images to bring to mind.

Technologies with Great Potential for Plant Pathology

There are many teaching technologies with obvious applications for plant pathology. Most of the technologies that are in current use or are in development focus on making images available to students. Applications are occurring so rapidly that specific recommendations are certain to overlook valuable materials. The following discussion considers some general types of technologies and some of their relative advantages and disadvantages.

Videodisc technology

Laser videodiscs use analogue recording to create collections of still and moving images. A current advantage of a videodisc is the potential size of the collection of images. A single side of a videodisc can contain about 30 minutes of moving video, 65,000 still images, or some combination of these. For instructors who are teaching general courses and wish to present images of a wide variety of pathogens and diseased host plants to capture the interest of students with diverse interests, a videodisc can be a valuable resource. Each image has a 5-digit numerical address which can be used to rapidly access the image. If the original images are high quality, their quality on the videodisc will also be excellent.

A major advantage of videodiscs over slides is that each image can be accessed repeatedly and in any order one wishes. For example, an image might be used in tutorial lessons on topics related to the pathogen group, the type of disease, or commodity-specific training and again for identification practice or review programs. This repeated use of an image in many applications is a major advantage over the auto-tutorial slide sets popular in the 1960s and early 1970s.

Collecting quality images is a time-consuming and sometimes difficult task, but an even more daunting task must be accomplished if the collection is to be of optimal valuable. That task is to create a database of information about each image so that instructors may find the images they want to use. This requires that each image is accurately described with respect to as many possible subjects as possible. For images related to plant pathology, subjects might include: host plant, pathogen, disease name, geographical region, pathogen structures, host plant part, photographer, and source of image. A software package is then necessary to rapidly and simply search the database using a range of topics, from the most broad to the most narrow, to help instructors select images of interest.

Once a set of images has been selected for instructional purposes, there are several applications of videodisc technology for teaching (Tainter *et al.*, 1995). Images are accessed using their numerical addresses. One very simple accessing mechanism is through the use of bar code technology. Bar codes can be created on a personal computer using software that converts the image number into a bar code that is printed with a laser printer. Some software includes word and/or number labels on the barcodes. Depending on the kind of word processing that is used, the bar codes can be placed on text using the computer program and printed in a completed form or the bar codes can be printed separately, cut out with scissors, and placed where needed in the text. Multiple copies of materials with bar codes can be photocopied for students. The creation of teaching materials with bar codes to access images is limited only by the imagination of an instructor. Instructors can essentially produce their own easily modifiable text materials with extensive use of colour images which could never be made available in an affordable textbook. Self-study exercises for identification practice and vocabulary improvement can also be easily made.

More formal teaching modules can be prepared using interactive authoring software that combines text with images. A special videoboard must be placed in the computer to allow the program to communicate between the computer and videodisc to bring together the images and text overlays. Finding compatible hardware to go with the programs is not always easy.

One of the biggest disadvantages of analogue images of videodiscs compared to a digitized image is that videodisc images cannot be sized and placed in different places on the computer screen. The videodisc image fills a computer screen and the text overlays of the interactive software open 'windows' to show all or only portions of the image. Another major disadvantage of videodiscs is that individual videodisc players and videodiscs can be used by only one or two students at a time. Because the images are analogue rather than digitized, they cannot be shared by multiple users with file servers. On the other hand, if one or two study stations can be made available for extended hours in a department or general computer room, most plant pathology instructors should be able to provide access to their students at a reasonable cost.

CD-ROM technology

CD-ROM technology solves many of the limitations of videodisc technology but does not yet offer the ability to access rapidly very large collections of images. On the other hand, smaller specialized collections of digitized images can be used for lessons on particular topics. CD-ROM collections of digitized images are becoming very common and will serve as important sources of images for teaching.

Disc space can be conserved by making images smaller and using the lowest resolution necessary for what needs to be observed. Multiple images can be placed on a single screen. Images can be accessed using hypertext links from highlighted words. An example with wide applications in plant pathology teaching is the disease cycles which, many instructors would probably agree, are often not used to full advantage by students. Explanatory text can be placed in a scrolling text box on the same screen as the diagrammatic disease cycle. Using point and click technology with a mouse, highlighted words can create a temporary box around the part of the disease cycle to which the text is referring. Highlighted terms within the disease cycle can also be linked to images of the structures and/or symptoms diagrammed in the life cycle. Such technology can encourage students to connect the life cycle diagram with detailed text and with images of what symptoms and signs really look like. Keys to diseases or pathogens which are suitable for professionals can be made in simplified forms for instructional purposes. Images at each choice help students learn to visualize the words describing the diagnostic or taxonomic choice.

We have used these applications in the new CD-ROM 'Turfgrass Diseases: Diagnosis and Management' which used the 'Compendium of Turfgrass Diseases', a text and photograph publication, as a base of information (Schumann and MacDonald, 1997). We considered the limitations of space and presentation in the paper publication and how we could use CD-ROM technology to improve access to the information for a wide range of users from scientists to students to turfgrass managers. One important advantage in CD-ROM technology is the ability to layer access to information so that users can find the materials that are most useful to them. Our project includes keys to pathogens which might be useful to professional diagnosticians, but not to most turfgrass managers. Users would access this technical information by clicking on buttons. Many turfgrass managers might never access the pathogen identification information at all, rather than being forced to page through it in their search for more applied information. We have used highlighted words with hypertext links to images and arrows where we felt that technical terms might make understanding difficult. Each disease section contains a list of links to diseases with similar symptoms. Disease management guides, which were previously limited by the length restrictions of the written Compendium, could be easily expanded and even modified for different management environments such as golf courses vs. lawns and grounds.

Most importantly, we have been able to add guides to the appropriate sections. Too often, practitioners are left to flip through pages of photographs in hopes of finding symptoms to match their problem. In the Turfgrass Disease CD-ROM, we have added three ways to determine which disease sections to consult when a problem arises: an illustrated key to diseases in which the user answers a series of questions and is brought to a likely disease or a short list of diseases for comparison, a series of host-specific monthly calendars in which diseases are listed at the times when they are most likely to occur with direct links from the disease names to the disease sections, and an 'image browser', for those who still wish to compare symptoms, with links from the image pages to the appropriate disease sections. In many sections of this CD-ROM, keys are used, but most are presented step-wise with one choice at a time and with explanations of unfamiliar terms and photographs and/or illustrations to guide the user. Text-only keys are generally unusable except by experienced and well-educated users, and even these people greatly benefit from illustrations and photographs to accompany the choices.

CD-ROM teaching materials have some technological limitations including student access to modern computers with high quality graphics, appropriate speed and memory, and CD-ROM drives. However, CD-ROM servers make it possible for instructors to make lessons available to large numbers of students through computer classrooms when appropriate hardware is available.

Web pages

Some instructors use Web pages as a way to communicate more easily with their students concerning the day to day operation of their courses: providing course schedules, access to misplaced handouts, or old exams for preparatory study. Other instructors find Web pages a convenient place to post lessons and images that they have created for their students. This can give students access to images that they may have seen in class and wish to review for additional study. A major advantage to Web pages is that instructors, within reasonable limitations, can update and expand information over time. Once on-line, students also have access to incredible databases of information for research topics which have never been so easily available. It remains to be seen which of these will be most valuable in teaching and how instructors can help guide students in the use of this vast new world of information.

Web pages are an excellent means to reach non-traditional students and will probably become more commonly used for distance learning. They can also be used to post information designed for the general public as a means of promoting plant pathology and providing disease information for gardeners and homeowners.

One problem with Web pages is that they are generally accessible by anyone with the Web site address, which can introduce the problem of copyright. Many instructors use slides in their teaching from a variety of sources. Even with disclaimers, the legal technicalities of posting slides and diagrams from various textbooks that are not the instructor's own can be complex.

Another potential problem of Web pages for teaching is that of creating unequal access for students. Some students have computers of their own, while others must use public (university) facilities which may or may not be easily available when a student requires their use. If instructors post materials that need to be printed, students with their own printers again have an advantage. Some of these inequalities will diminish with time, but it is important that instructors consider them to avoid the perception, and the reality, of unfair treatment of students.

Web pages are already becoming an important outreach instrument for practitioners. They can be sources of technical information, illustrations of disease problems, and pesticide recommendations. They also serve to provide timely reports of current problems and management recommendations. Although these are only now in development, their importance as a real-time link to agricultural advisers will be commonplace as practitioners obtain computers and gain access to this information.

Hyper-lecturing

Various software packages are available for lecturing using digitized and/or video images. More and more universities are equipping classrooms with such teaching technologies. Some instructors are digitizing their slide collections as well as using CD-ROM image resources. The organization of such images is critical for their efficient use, and the initiation of such lecturing technology can be very time-consuming.

If videodisc projection equipment is available, bar codes can also be used for lecture presentations similar to slide presentations with the added advantages of a much larger set of images to choose from and the ability to easily modify the order of the presentation of the images according to questions and discussion from the audience.

Simulations for decision-making

Several programs have been developed that can be used in plant pathology teaching programs. One excellent program on diagnosis will be discussed in this volume (see Stewart, Chapter 33). Several involve simulations so that students can practice making management decisions for various diseases. Such simulations allow students to try management without risk. They can participate in the equivalent of many growing seasons in a single afternoon in the computer laboratory. Several simulation games have been available for some time. TURFBLIGHT, developed by T.K. Danneberger at Ohio State University, requires students to manage turf with nitrogen applications, irrigation, and fungicides to prevent Pythium blight and anthracnose. In APPLESCAB, parameters can be modified to present ascospore reports, weather forecasts, and fungicide applications with the potential for fungicide resistance to develop (Blaise *et al.*, 1987). LATEBLIGHT is a similar simulation and is available at the plant pathology teaching Web site which includes an electronic plant pathology teaching journal maintained by P.A. Arneson at Cornell University (http://ppathw3.-cals.cornell.edu/arneson/journal.html). Resistan, available at the same Web site, focuses on fungicide resistance management. A very simple program, Logistic, was developed by R.V. James, now at the University of Wisconsin. This program allows students to easily modify initial inoculum and epidemic rate and see the resulting polycyclic epidemic graphs. DISTRAIN, a program for assessing diseases of cereal crops and DISEASEPRO, a program for assessing peanut foliar diseases, were created as training programs for field assistants who need to assess disease levels accurately, but they can also be interesting teaching tools (Tomerlin and Howell, 1988; Nutter and Schultz, 1995). These programs demonstrate how difficult accurate disease assessment is, which leads to discussions on why it is important for integrated management programmes.

All of these programs require a tremendous amount of information and preparation time. They offer, however, some feeling of real world field experience even in the dead of winter. My own students, many of whom have a strong aversion to pesticide use, feel a sense of panic as they watch the exponential curve of apple scab when they mis-time their fungicide applications and begin to spray repeatedly in a futile attempt to stop the epidemic. Many remark that they have a sense of what real growers must experience.

Obstacles, Misinterpretations and Limitations

Pedagogical issues

For students, plant pathology on CD-ROM or a Web page may initially cause excitement and appeal similar to a new video game (Holzberg, 1994). Even poorly designed teaching materials may have value to students simply because they gain access to many colour images. But 'point and click' technology that allows students to quickly find terms and simple facts may deceive them into thinking that they have plant pathology 'at their fingertips'. It may portray an image that the mastery of the body of knowledge that is plant pathology does not require the effort or focus we have all had to make. There is no short-cut to learning. It takes time to digest a new vocabulary and new concepts before one can begin to organize one's thoughts and begin critical thinking.

The design of teaching materials is critical to their value. It is my opinion that clarity and simplicity are important in presentation. Complicated mixes of fonts and colours can jar the eyes and tire them quickly, or fail to indicate the relative importance of information because everything is presented in large bold letters or clashing colour schemes. Because most of us have not had the opportunity to present our teaching materials in colour before, we need to be educated in suitable presentation methods that are most effective and do not challenge users with various forms of colour blindness.

In general, long segments of text are uncomfortable to read on a computer screen and are best presented in clearly printed textbooks. Short, concise summaries and identifying information are most valuable in computerized presentations. The simplified versions of text used to accompany images on the computer screen are no substitute for the more detailed text materials in which concepts and theories are presented. On the other hand, the ability to update and improve computerized materials gives them a great advantage for presentation of a rapidly changing science where textbooks can only be expected to be printed in new editions perhaps once a decade.

One of the most powerful aspects of computerized teaching technologies is that information takes on a three-dimensional form because of the ability to link segments to each other *ad infinitum*. The danger here is that students can easily become lost in these webs and fail to grasp the basic framework that we live with every day in our professional lives. Navigational tools that guide students back to main menus and submenus are necessary to prevent this confusion.

Some readers may remember when televised education was predicted to make it possible for a single lecturer to present information to hundreds or even more students. Classrooms were equipped with wall-mounted televisions along the side walls. These were abandoned fairly quickly, collected dust, and were eventually removed. Some of us took our introductory biology courses with auto-tutorial laboratories, listening to tapes and watching slides or even doing simple hands-on activities. Large numbers of students were given the opportunity of ready access to laboratories during extended hours with only a few teaching assistants necessary to check students in and out of the laboratory and be available for questions. I wonder how many students in those laboratories shared my feelings of isolation as I listened to the faceless voice on the tape and struggled with my microscope to see what I was supposed to see in the prepared slides.

Many administrators will be tempted to view the new teaching technologies as offering the opportunity to serve larger numbers of students with smaller numbers of lecturers. Only those with expertise in the discipline which they are judging will be able to determine what teaching technology has substance and value and what is a flashy show. Even though many good teachers will be able to improve their teaching with computerized technology, unimaginative or unconcerned teachers will be no better. Technology is not teaching, and it is important for lecturers to make that fact clear to those who might believe otherwise.

Distance learning and Web page courses may be useful for continuing education for select adults under limited circumstances, but they are no substitute for education in its broadest sense. The magic of excellent teaching comes from the interactions between human beings. Most students come to higher education with a broad range of interests, but if you ask people how they became interested in their chosen discipline, many will trace that interest back to an individual teacher.

Teachers have many roles for students. They may be the person who first creates the intense interest in that area because of their own personal excitement about the subject. They may be the person who first shows an individual interest in a student and helps a student believe in their own abilities. Students need to feel that the teacher cares that they are learning. Students gain a lot from interacting with other students in laboratory classes, sharing the excitement and beauty of what they are observing. And students gain from discussions in classrooms in which information is connected to what they are learning in other courses.

Students need to be able to meet face-to-face with instructors individually, and they need to be taught by instructors who are interacting with other faculty members to create a learning community. At the risk of expressing too idealistic an image of modern university life, these human factors are what lead students to one discipline over another and explain why courses in a particular discipline are over-enrolled at one institution and dwindling at another.

Technical support issues

Even though technology is not the equivalent or a substitute for real teaching, many courses can be enriched by technology-based learning programs. A major obstacle is the time required to create high quality materials. The software available for creation of materials is rapidly changing. Expertise in its use is a speciality in itself. Few institutions have made the commitment to provide the professionals in teaching technology who can help lecturers determine the best way to present their information and then quickly and efficiently do the routine and tedious programming necessary to create interesting and useful teaching modules. Lecturers should spend their limited time deciding on what and how to present the information and collecting the appropriate images. When lecturers are required to become proficient in a quickly out-of-date software package, they end up using simple programs and simplistic presentations because they cannot master and program more complex presentations in a reasonable amount of time. Administrators must recognize and fulfil this need for a new kind of professional at colleges and universities who can work with lecturers and who will provide the expertise in authoring rather than programming.

Conclusions

A final question continues to perplex anyone who has teaching experience and has considered the value of teaching technologies. How well does a good image – even an excellent one – relate to the real three-dimensional pathogen or infected plant? Perhaps other instructors have had the experience of presenting lectures with 35 mm slides to orient students and then seen them struggle with identification of the same organisms and structures in prepared microscope slides. The problem is increased when students begin to examine living cultures and infected plant materials and attempt to make slides of their own for microscopic examination. This is not to suggest that images presented in lectures, on CD-ROM, videodisc, or Web pages are not valuable. But just as a compendium of disease photographs is no substitute for a professional diagnostic laboratory, teaching technologies are no substitute for the traditional hands-on laboratories where students explore for themselves. Such human interactions are time-consuming, expensive to offer, and exhausting for dedicated teachers, but they are the path to an educated plant pathologist. We must now begin to consider how to integrate teaching technologies into the full range of intellectual skills necessary for the education of future plant pathologists.

References

Blaise, P., Arneson, P.A. and Gessler, C. (1987) APPLESCAB, a teaching aid on microcomputers. *Plant Disease* 71, 574–578.

Holzberg, C.S. (1994) The new multimedia: what researchers say about multimedia. *Electronic Learning* 13 (8), 55–62.

Nutter, F.W., Jr and Schultz, P.M. (1995) Improving the accuracy and precision of disease assessments: selection of methods and use of computer-aided training programs. *Canadian Journal of Plant Pathology* 17, 174–184.

Schumann, G.L. and MacDonald, J.D. (1997) *Turfgrass Diseases: Diagnosis and Management CD-ROM*. American Phytopathological Society Press, St Paul, Minnesota.

Tainter, F.H., Schumann, G.L. and Evans, T.A. (1995) Development of the plant pathology laser videodisc. *Natural Resources and Life Sciences Education* 24, 48–52.

Tomerlin, J.R. and Howell, A. (1988) DISTRAIN: A computer program for training people to estimate disease severity on cereal diseases. *Plant Disease* 72, 455–459.

31

Making Books Interactive: an Electronic Experiment

P. Jones

Agricultural and Biological Department, University of Florida, PO Box 110570, Rogers Hall, Gainesville, FL 32611-0570, USA
Fax: +1 352-392-4092/E-mail: ez@agen.ufl.edu

Introduction

In the modern world, it is not unusual for people to have a paradoxical attitude toward information. On the one hand, daily life's complex interactions seem to require ever increasing access to specific, accurate information. On the other hand, there seems to be an expanding overabundance of information. As John Naisbett said in his 1982 book, *Megatrends*, 'we are currently drowning in information, but we are starved for knowledge'. The dilemma is how to efficiently sift through all of the unwanted information to access precisely what is needed, when it's needed.

Pest and Disease Management Information in Electronic Form

Naisbett's observation certainly rings true in the area of management of pests and diseases. One response to the current challenge is to use information technology in an innovative way to combine diverse information and bring it to the attention of users, in a form in which it is readily accessed, understood and applied. This is a process of creating knowledge out of information, and thereby precisely addressing Naisbett's dilemma.

The University of Florida was among the first to recognize the power of CD-ROM and Internet technology to solve agricultural problems through knowledge management. The Florida Agricultural Information Retrieval System

(FAIRS) is now firmly established as a source for decision support, serving the needs of farmers, growers and extension staff. For example, the FAIRS (1997) World Wide Web site lists a wide range of resources that are accessible interactively:

- Alternative Crops
- Citrus Pest Management Guide
- Commercial Citrus Handbook
- Commercial Farming Regulations
- Commercial Nursery Handbook
- Commercial Vegetable Production
- Commercial Turfgrass Production
- Florida Tomato Scouting Guide
- Rice Production Handbook
- Small Farms Handbook
- Sugarcane Handbook
- Vegetable Gardening Handbook

It is of course a feature of World Wide Web technology that the information is immediately accessible through a standard Web browser such as Netscape or Explorer from anywhere in the world, through a standard telephone socket. FAIRS also features a special list of pest management topics:

- Aquatic Weed Management Guide
- Beneficial Insects
- Citrus Pest Management Guide
- Florida Weeds
- Insect Management Guide
- Nematode Management Guide
- Pesticide Poisoning
- Weed Management Guide
- The Plant Disease Management Guide

A mouse click on any of these takes the user immediately to the relevant section of information on the Web site, where further clickable connections are highlighted. For example, clicking on Citrus Pest Management Guide takes the user to a further list of topics:

- Pesticide Resistance and Research Management
- Spray Application Technology
- Management of Soil Applied Agricultural Chemicals
- Micronutrients
- Citrus Rust Mite
- Spider Mites
- Scale Insects
- Eastern Subterranean Termite

- Flower and Orchid Thrips
- Root Weevils
- Citrus Leafminer
- Other Insect Pests
- Greasy Spot
- Melanose
- Citrus Scab
- Alternaria Brown Spot
- Postbloom Fruit Drop
- Phytophthora Foot Rot and Root Rot
- Brown Rot of Fruit
- Preharvest Decay
- Nematodes
- Growth Regulators
- Weeds
- Citrus Sprouts
- Pesticides Registered for Citrus
- Interpreting PPE Statements on Pesticide Labels
- Useful Telephone Numbers

Clicking on the section on Citrus Scab leads to a data sheet on the disease caused by *Elsinoe fawcetti* with the following headings:

- Approaches to Processed and Fresh Fruit Management
- The Use of Pesticides in Citrus IPM
- Worker Protection Standard (WPS)
- Interpreting PPE Statements on Pesticide Labels
- Citrus Scab

The last of these leads in turn to a brief account of the biology and epidemiology of the disease, with a Table of recommended treatments, tailored to the needs of the local citrus industry.

The contents of any of these files can be downloaded, and can also be printed with full formatting through the use of a Portable Document Format (PDF) file. Furthermore, the entire contents of the Web site are also available on CD-ROM, and are supported by an e-mail enquiry service. The CD-ROMs contain printable extension publications, colour photographs and illustrations, detailed line art, tables, graphs and equations. Everything is accessible through a full-text search feature, and is updated frequently to stay abreast of current research and new developments. The Web site also provides a search tool, and offers even more flexibility than the CD-ROM alternative in terms of updating. Furthermore, the user is offered advice and training on establishing a personal or corporate Web page to make even fuller use of the information that can be obtained from the FAIRS resource.

It is immediately apparent from this example that an interactive electronic system of this sort can not only mimic but also far exceed the capability of a book with similar content.

There are many other examples of the provision of book-like information resources through the application of IT. The content is typically not a determining factor for the medium or even the structure. The principles can therefore be illustrated from disciplines other than plant pathology, and the next part of this chapter addresses a related topic in which clarity and immediacy are imperative – occupational safety and health.

An Example from a Related Field: the Hazard Communications Problem

In 1983 (the year following publication of *Megatrends*) the US Occupational Safety and Health Administration (OSHA) promulgated the Hazard Communication Standard (HCS) (29 CFR 1910.1200) and applied it to manufacturing. Simply stated, the HCS required that employers:

- develop a written Hazard Communications Standard (HCS) program.
- maintain current lists of products containing hazardous chemicals on the worksite.
- properly label containers of hazardous products.
- maintain an archival file of Material Safety Data Sheets (MSDSs) for each of the hazardous products used on the worksite.
- provide training for workers in the proper procedures for storing, handling, and disposing of hazardous products.

When the HCS was promulgated, the breadth of the information management task was simply not anticipated and, as the Standard was implemented over the next decade, most manufacturers were overwhelmed to some degree by the scope of their HCS information management tasks. This was especially true with respect to the management of Material Safety Data Sheets (MSDSs) (Lindsay, 1994). In spite of their efforts to comply with the HCS, most employers (and employees) did not feel a commensurate increase in access to the specific, accurate information actually needed (Hansen, 1994).

In the construction industry, the result is that failures to comply with different aspects of the HCS top the list of OSHA compliance infractions. For 1990 and 1991, *The 100 Most Frequently Cited OSHA Construction Standards in 1991: A Guide for the Abatement of the Top 25 Associated Physical Hazards* (OSHA, 1993) identifies the lack of a written HCS program, inadequate employee training, lack of MSDSs, lack of OHSA posters, and inaccessibility of MSDSs as the top five citations, respectively. It is noteworthy that the publication provides supplemental information on how to recognize and eliminate physical hazards, but offers no guidance on how to properly institute the HCS Standard. It is a difficult problem.

Although regular unannounced site visits by OSHA inspectors can insure that minimal compliance is a compelling issue on construction sites, inspections cannot address the simple organizational and logistical problems of HCS program maintenance on construction sites. The problem is how to streamline and simplify procedures from the superintendent's perspective. One must determine: how to maintain several simple databases, how to provide targeted, meaningful training, how to produce an appropriate hazard warning label with minimal hassle, and how to quickly find critical information in an emergency.

A Partial Solution

By conceptually recreating Hazard Communications Standard (HCS) programmes as integrated electronic databases of information for use on field office PCs, many of the inconveniences and inefficiencies of the current system could be greatly reduced. In addition to compact storage, an electronic database makes locating and retrieving specific information easier. Information from an electronic HCS database could be accessed either on the computer screen or printed on demand. Multiple menus based on common names, active ingredients and scientific names would avoid questions about where to find information on a particular material. In addition to the menus, several standard search techniques (such as word wheels, hyperlinks, and full-text indexing) would be applied to the database to facilitate finding specific HCS information.

In addition to providing simple information access, an HCS database could readily generate labels and posters describing specific chemical hazards which could be printed on colour stock and/or on colour printers. Such labels and posters would use standard graphical symbols (Petreycik, 1996) that indicate the type and degree of hazards associated with specific chemicals. Likewise, multi-media training programs derived from a HCS database could use the same graphical symbols on screen in conjunction with English (or second language) text and audio. This type of access to multi-media HCS programs would directly address some of the safety training problems discussed in a *Wall Street Journal* front page report (Labor Letter – December 21, 1993). Specifically, the article concluded that safety programming needs to incorporate more '*visually appealing materials*' and opportunities for '*participatory training*' to increase its effectiveness. All of the information delivery capabilities described above are consistent with the presence of personal computers in construction site field offices.

In practice, most specific HCS programme information currently used in the construction industry comes from Material Safety Database Sheets (MSDSs). Unfortunately, MSDSs are provided on paper (or as facsimile images) and are not generally available in digital form. Even worse, MSDSs are neither consistent nor uniform in their organization or appearance. The problem is further complicated by the sheer volume of MSDSs. The net result is that MSDSs are not a suitable source for building a HCS database.

At the opposite end of the spectrum, the International Chemical Safety Cards (IPCS, 1991) produced by the International Programme on Chemical Safety (IPCS) are highly structured even to the extent that individual phrases used to characterize chemical properties and hazards are explicitly defined in English as well as a core group of other languages. The ICSC system is fundamentally different from the MSDS system mandated by the HCS standard. ICSCs cover individual hazardous chemicals while MSDSs cover individual products that contain hazardous chemicals. MSDSs are individually produced by manufacturers with minimal centralized coordination while the ICSC system is highly coordinated by the IPCS.

This paper reports on the development of a prototypical electronic HCS database that addresses the issues described above. For the last two years, this project's major goal was to develop procedures for porting information on hazardous chemicals from the IPCS International Chemical Safety Card (ICSC) system into a graphics-enhanced electronic HCS database to serve as a supplement to typical construction industry HCS programs. The database was designed specifically for delivering HCS information and training on typical construction site PC computers. The project's specific objectives were to develop and implement procedures to generate a properly structured electronic database of multi-lingual ICSC information that includes appropriately incorporated icons (or symbols), graphically indicating various types of hazards associated with chemicals. The graphical icons indicate the potential degree of severity associated with each specified hazard. In contrast to currently used, conventional methods of delivering information about hazardous chemicals in MSDSs, the HCS database prototype addresses issues of language and literacy through integration of visual, audio and textual materials.

A Prototype: *HazCom Interactive*

Text files

Text information from all currently available International Chemical Safety Cards (ICSC) contain information on: Types of Hazards (Fire and Explosion); Types of Exposure (Inhalation, Skin, Eyes and Ingestion); Spillage Disposal; Storage; Packaging & Labeling; and Chemical Facts (Physical State, Chemical Dangers ...). In the ICSC database, the information is stored as a series of descriptive phrases that are explicitly linked to each information category. For example for the chemical hydrazine under the category 'Physical State: Appearance' the following descriptive phrases are listed; 'colourless', 'fuming', 'hygroscopic liquid', 'with pungent odour'. Text data (descriptive phrases) are available in many languages. The ICSC text data were imported into an object oriented database management system to support flexible access to the information and forms the foundation for *HazCom Interactive.*

Pictographs

The fields of information in the ICSCs were evaluated to prioritize data according to their need for graphical enhancement. For example pictographs for hazards such as fire and explosion and prevention measures such as 'NO open flames' and 'ventilation' were deemed essential. Conceptually, a pictograph (graphical symbol) database structure was developed to convey information that linked, as explicitly as possible, to IPCS glossary terms (phrases) as they were applied in the ICSCs to characterize chemicals. The pictographs adhere to international conventions intended to be understandable across cultures. To date, approximately 100 pictographs for the database have been developed in priority order.

Audio files

In addition to the databases of text and graphical information, a database of audio files was created that linked explicitly to high priority text files. For example digital audio files for phrases such as 'Highly explosive' and 'First rinse with plenty of water' were recorded and stored.

User interface

This project depends to a large extent on the quality of the user interface that displays the hazardous chemicals described. In particular the interface needs to present a visually appealing and participatory program that engages workers quickly and effectively delivers pertinent information (Szudy and Arroyo, 1994). The interface was designed to quickly and easily communicate ICSC information in a way that addresses the language and literacy requirements of its users (specifically through a combination of graphics, multi-language audio and multi-language text). It is a simple point and shoot interface with a variety of supplemental navigational features. Figure 31.1 shows the system's opening screen that defines major areas of information using text and pictographs. The figure cannot show the system's capability to activate audio files when the cursor is dragged across the various buttons. For example if the cursor was dragged across the uppermost pictograph, an audio file would be activated saying 'Chemical Search'.

HazCom Interactive

Chemical Search (shown in Fig. 31.1) allows a user to search for a specific chemical by common name, scientific name or any of five chemical numbers

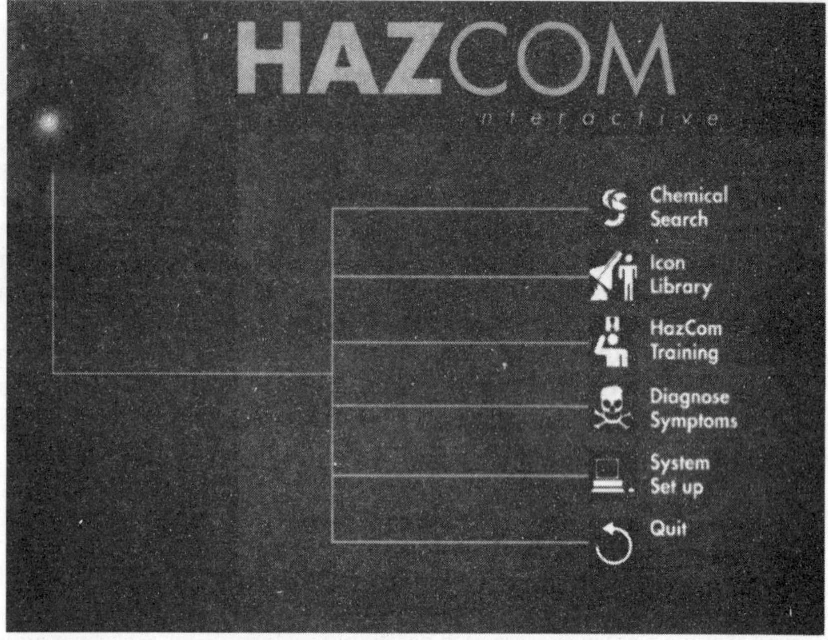

Fig. 31.1. *HazCom Interactive* opening screen.

(CAS, RTECS, ICSC, UN, or EC). Once a chemical is selected, such as hydrazine, a general information screen allows the user to make specific information requests. Figure 31.2 shows the opening screen for hydrazine. In addition to various chemical numbers, the opening menu offers access to seven categories of information: Explosion Hazards, Emergency Medical, Fire Hazards, Exposure Hazards, Spillage Disposal, Storage, Packaging & Labeling, Chemical Facts & Additional Information. Note that the seven category pictographs correspond to figures in the dark region on the screen's left which can be used as shortcuts to specific information.

To obtain more information on the effects of ingesting hydrazine, the user selects Exposure Hazards; the program moves to a screen that displays the four exposure routes of Inhalation, Skin, Eyes, and Ingestion; Ingestion is chosen and a screen displays Acute Symptoms, Prevention and First Aid; and finally, Acute Symptoms is selected to display the descriptive phrases from the ICSC database that correspond to ingestion hazards associated with hydrazine. Figure 31.3 shows the Ingestion screen with Acute Symptoms selected. The screen shows that hydrazine can cause nausea, unconsciousness, vomiting, weakness, and/or convulsions.

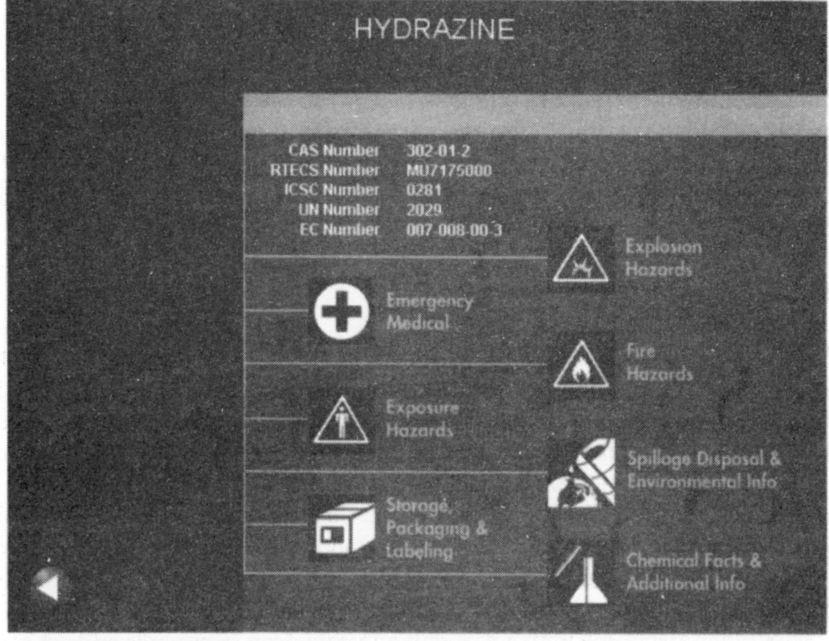

Fig. 31.2. The opening screen for accessing information on hydrazine.

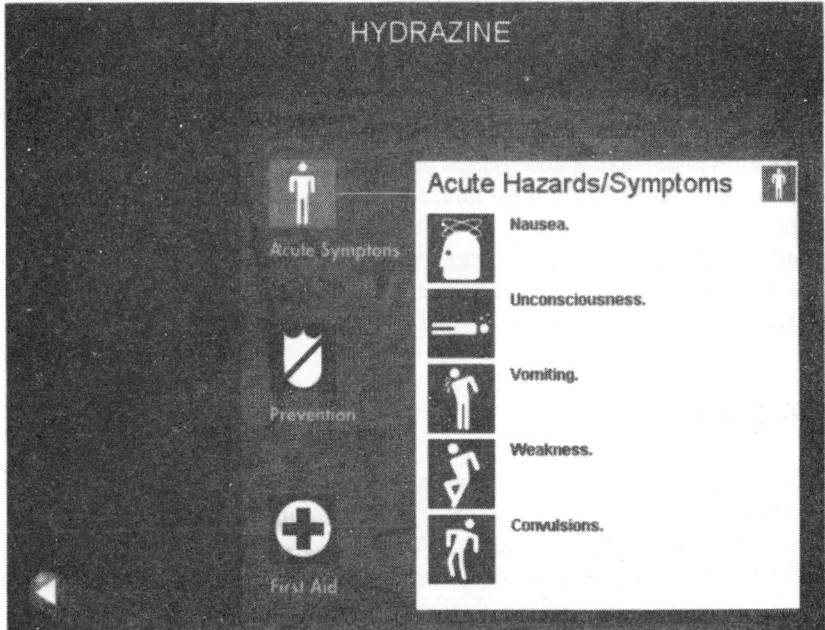

Fig. 31.3. A screen showing the acute symptoms of hydrazine inhalation.

Multiple languages

Because of the underlying design of the ICSC system, translation of the information between languages is greatly facilitated. Specifically, the phrases that correspond to the various categories are available in more than 18 languages including: Dutch, French, German, Japanese, Spanish, and others. When fully implemented *HazCom Interactive* will display text, deliver audio and print in all of these languages.

Supporting programs

HazCom Interactive depends on users understanding the meaning of the graphical symbols. This can be partially accomplished through good, systematic design of the pictographs, however, some concepts cannot be simply represented symbolically. To support the pictographs, games are being developed to reinforce the meanings of symbols. The first game is intended to raise general awareness of the pictographs. Twelve pictographs are displayed, corresponding words appear at the top of the screen. As the words appear the user clicks on the appropriate pictograph. A timer challenges the user to move as quickly as possible. The second game is a chemical specific matching game. The user is shown four pictographs associated with a particular chemical for 15 seconds and asked to memorize them. After 15 seconds the screen is filled with 28 pictographs with the original four mixed in. The user is asked to identify them as quickly as possible. Once the four pictographs are all found, a multiple choice quiz reinforces the meaning of the symbols.

Other features

In addition to displaying information on-screen, *HazCom Interactive* includes features to print information on demand. Specifically, graphical versions of International Chemical Safety Cards for specific chemicals can be printed. The graphical information card for hydrazine is shown in Fig. 31.4. In addition, container labels and posters can be printed. All of these print products rely heavily on the pictographs to convey information visually.

A Better Solution

Even though *HazCom Interactive* represents a conceptual improvement over current HCS program implementations, it still has a major flaw. *HazCom Interactive* is based on chemicals not products. Furthermore, the underlying database is not exhaustive and even if the database contained all hazardous chemicals that would not address the problem of mixtures. OSHA requires that manufacturers

Fig. 31.4. A print on demand information card for hydrazine that heavily uses pictographs to convey hazards and preventative measures.

be accountable for their products and be responsible for producing MSDSs on a product by product basis. Under the current regulations, even products that contain a single hazardous chemical could not be covered by *HazCom Interactive*.

Although *HazCom Interactive* was developed specifically to deliver the information in a particular database, the International Chemical Safety Card system, it also represents a concept for how information from other databases might be delivered. One very interesting possibility is to use *HazCom Interactive* as a template for developing information about classes of products. For example, flat latex wall paint has a fairly standard formulation that is consistent across all manufacturers. It contains two active chemicals: ethylene glycol and zinc oxide. This means that if a taxonomy of paint products were developed that allowed large numbers of paint products to be covered adequately in relatively few categories, tremendous efficiencies could result. This model is currently under development.

Interactive Electronic Books

HazCom Interactive is not based on a 'book', it is based on a highly modularized series of printed cards – the International Chemicals Safety Cards produced by the International Programme on Chemical Safety. In fact, as a print product it looks very much like a 'database' highly suitable for a traditional computer application. Of course, the project described in this paper is not a traditional database application, rather it is concerned with modifying an existing database to improve actual access.

The system's navigational schemes are intended to allow users to quickly (and in a nonlinear fashion) find specific information. The use of graphics and audio files is targeted to expanding access to semi-literate users. The print-on-demand capabilities are intended to expand the impact into the physical workplace. The system also creates a broad foundation on which training programmes can be developed in an environment where a critical mass of information can be found.

It has all of the advantages of other electronic database systems. Nothing is ever misfiled. The same information can be accessed from many different directions (menus). Version control can be ensured. It takes up very little space and is cheap to replicate and distribute. The primary disadvantage is that users need computer hardware to access the information. As computers proliferate that is unlikely to be a serious constraint. The project is expected to continue for three more years during which time its usefulness and practicality are expected to steadily increase until it is delivered to and in use on construction worksites.

Meanwhile, as we have seen from the FAIRS example above, similar principles are being applied in the discipline of pest management. While it may be some time before hardware is developed that physically replaces that remarkably durable technology of communication, the printed book, the challenge of applying the capabilities of modern software to the solution of practical problems in applied disciplines such as crop protection is certain to go on evoking innovative solutions of the kind that are a feature of the contents of this chapter, and indeed of many other parts of this volume.

References

FAIRS (1997) *Florida Agricultural Information Retrieval System*. University of Florida, Gainesville, USA. World Wide Web site at http://hammock.ifas.ufl.edu/.

Hansen, M.D. (1994) Environmental safety and health compliance in the 1990s: using technology to manage the paper chase. *Professional Safety* 39(2), 29–32.

International Programme for Chemical Safety (1991) *International Chemical Safety Cards*. Commission of the European Communities, Telecommunications, Information Industries and Innovation. Luxembourg.

Lindsay, R. (1994) Getting a handle on MSDS records. *Industrial Hygiene News* 17(4).

Naisbett, J. (1982) *Megatrends*. Warner, New York.

Occupational Safety and Health Administration (1993) *The 100 Most Frequently Cited OSHA Construction Standards in 1991: A Guide for the Abatement of the Top 25 Associated Physical Hazards*. US Department of Labor, Washington, DC.

Petreycik, R.M. (1996) EU cracks down on chemical labeling. *Safety + Health* March, 68–71.

Szudy, E. and Arroyo, M.G. (1994) *The Right to Understand: Linking Literacy to Health and Safety Training*. Labor Occupational Health Program, University of California at Berkley.

32

Crop Protection, Information Technology and Ecosystem Health

Z.R. Shen

Department of Plant Protection, College of Plant Science and Technology, The China Agricultural University, 2 Yuanmingyuan Xilu, Beijing 100094, China
Fax: +86 10 6258 2332/E-mail: zrshen@public.bta.net.cn

The author's laboratory is conducting a project to develop a software system BJ-CABBAGIS, for vegetable production information management and decision support. It is designed to be GIS-based and consists of four subsystems: social and natural information management, vegetable production information management, vegetable protection information management, and pest diagnosis and IPM decision support. The data used comes from monitoring vegetable production and pest control practice in the Beijing region over many years.

PQ-INFORMIS is a multi-media system from which information on plant quarantine insects (58 species) can be retrieved and managed. The information includes text for many items such as English and Latin names, host plants, geographical distribution (countries), quarantine treatments, and also images of identifying morphological characteristics. It offers a Prolog program to help quarantine staff to distinguish similar species. The software is proving popular at about 70 importing port quarantine stations throughout China.

These two cases illustrate the successful application of information technology to plant protection. An important component of the methodology is the participatory approach which unites university researchers, government administrators, crop protection practitioners at plant quarantine stations and farms, and information experts.

Introduction

Agenda 21 endorsed by the 1992 Earth Summit in Rio de Janeiro attempted to define a map to sustainable development. Following up the spirit of Agenda 21

 Information Technology, Plant Pathology and Biodiversity (eds P. Bridge, P. Jeffries, D.R. Morse and P.R. Scott)

is not easy. Agricultural sustainability is a popular word in China. Is it a dream or a myth for the Chinese people? There is pride that China can feed 22% of the global population with 7% of the arable lands in the world. However, between now and the year 2030 their population is estimated to rise from 1.2 billion to 1.6 billion, while the area of arable land will shrink from 132 billion ha to 125 billion ha. Consequently there is a need to increase per-hectare yield of crops and to decrease food loss.

Chapter 14 of Agenda 21 observes that the amount of food lost to pests has been estimated at 25% of harvests. Chemical control of food pests has been dominant, but over-use of pesticides is costly and has adverse effects on human health and on the environment. The best option is integrated pest management, which combines biological controls, host-plant resistance and appropriate farming practices, to minimize pesticide use (Keeting, 1993).

The Strategy of Precaution Priority and Integrated Management has been implemented in China. Recently, over 800 participants assembled at consecutive conferences in Beijing and Kunming to celebrate the twentieth anniversary of the declaration of this Strategy. At these meetings I collected pertinent materials from my colleagues about information technology and its application to crop protection.

It is difficult for people to understand the strategy and to firmly believe in the usability of information technology in crop protection. To demonstrate how the strategy can be used in the context of the information revolution, case studies and general experiences from China are presented here.

Ecosystem Health Assessment Requires Information

Before presenting the case studies, it is necessary to give a general assessment of agro-ecosystem health in China. Since entering the 1990s China's agriculture has been hit frequently by pests (Guo, 1996). In 1990, wheat stripe rust (*Puccinia striiformis*) occurred throughout the country resulting in the loss of 2600 tons of wheat. In 1991, 2500 tons of rice were lost to planthoppers (*Nilaparvata lugens* and *Sogatella furcifera*). In 1992, a severe outbreak of cotton bollworms (*Helicoverpa armigera*) caused the loss of 60 tons of cotton, equal to 30% of the usual national yield, and continued for four years until 1995. In 1993, rice blast (*Pyricularia oryzae*) was prevalent in southern China and 15,000 tons of rice were lost; and aphids (*Macrosiphum avenae, Schizaphis graminum, Rhopalosiphum prunifoliae* and *Acyrthosiphum dirhodum*) infested 10 million ha of wheat. In 1995, verticillium wilt (*Verticillium alboatrum*) caused a loss of 60 tons of cotton. Other pests and diseases such as wheat midge (*Cecidomyia mosellama*), wheat powdery mildew (*Erysiphe graminis* f.sp. *tritici*), corn leaf blight (*Helminthosporium turcicum*) and leaf spot (*H. maydis*), corn borer (*Ostrinia furnacalis*) and soybean cyst nematodes (*Heterodera glycines*) also caused significant damage to crops.

Such situations spring from complex scenarios. However, we can conclude that it is due to problems of ecosystem health (Nielsen, 1994). High productivity in agriculture is achieved by enhancing the multiple cropping index, changing cultivation systems, and large scale growing of high yielding varieties with a simple gene pool. Such high production is heavily subsidized with energy, nutrients, pesticides, and other economic and technological inputs. These are putting ecosystems under heavy pressure. The continued high stress is beginning to express itself in the fragility and vulnerability of ecosystems, or ecosystem pathology; and a consequence is susceptibility to pests. The concept of phytomedicine which focuses on ecosystem health has become acceptable in China (Guan, 1996).

Evidence shows that the Strategy of Precaution Priority and Integrated Management cannot achieve complete sustainability without ample information on ecosystem health. This is concerned with crops, insect pests, plant diseases, weeds, other harmful organisms, their natural enemies, so-called 'resource biota', components of the physical environment, and structures connecting and relating all of them in food webs and biogeographical cycles. In addition, history reveals that to achieve sustainability by this strategy the scale of the ecosystem monitored, diagnosed, and protected should be substantial. The strategy should be enlarged from the level of physiological ecology, individuals, populations, communities and fields to the level of farms and even whole regions. Information technology will play an increasingly important role in this scaling-up and will be used in more diverse and advanced ways.

The laboratory of IPM Information and Software Technology (IPMIST) at the China Agricultural University (CAU) investigates the application of computer technology to sustainable crop protection. Now it is undertaking a number of research projects, undergraduate and graduate education programmes, and public information services. Our two case studies are selected from this laboratory, at the website <http://ipmchina.cn.net/ipmist>.

Case Study: BJ-CABBAGIS Project

The government and people are concerned about vegetable production, the so-called 'vegetable basket engineering' programme. This programme has been undertaken in many urban-centred regions in China. Beijing, the capital metropolis, has a total area of 16,800 km^2 with 62% mountainous and hilly territory, 1370 km^2 of urban land and 82 km^2 of city. The total population is 10.45 million, with an urban population of 2.43 million. During the last 5 years, the area of vegetable production in Beijing has been enlarged to 40,000 ha and a strategic shift of vegetable production bases has been achieved from the outskirts of Beijing to its outer suburbs and surrounding counties.

IPMIST is conducting a project to develop a software system, BJ-CABBAGIS, for vegetable production information management and decision support. The project is funded and supported by Beijing Bureau of Agriculture (BBA). The data used comes from monitoring vegetable production and pest control practice in the Beijing region over many years.

BJ-CABBAGIS is developed with CityStar, a Windows tool to develop desktop GIS-based systems, and consists of four subsystems.

1. Social and natural information management. This subsystem provides maps and databases of Beijing and its six suburban divisions and eight surrounding counties. The data stored for the divisions and counties includes topographical, pedological and meteorological data. More detailed data have been specified for some 'xiang', subdivisions or complexes of villages within a division or county. The typical area of each xiang is about 35 km^2. From these counties or xiang, data can be retrieved as maps, text, tables, images, and audio and video clips.
2. Vegetable production information management. This includes multi-media information about: types and varieties of vegetable crops planted in Beijing; policies and regulations of vegetable production; 20 villages specialized in, and four demonstration zones for, vegetable production; seasonal production; and market dynamics. In Beijing, about 180 types of vegetables (30 major types) belonging to 29 families are grown. The system has incorporated data relating to vegetable production, including distribution of varieties, growing areas and yields of each varicty, sowing and harvesting times, irrigation and fertilization.
3. Vegetable protection information management. Multi-media information is stored and managed on major pests and diseases, such as imported cabbage worms (*Pieris rapae*), tomato fruitworms (*Helicoverpa armigera*), cabbage aphids (*Myzus persicae* and *Brevicoryne brassicae*), greenhouse whitefly (*Trialeurodes vaporariorum*), soft rot, downy mildew, the virus disease of Chinese cabbage, downy mildew and powdery mildew of cucumber, and tomato mosaic virus. Crops in greenhouses and fields covered with plastic film are susceptible to diseases such as grey mould, sclerotinia rot and late blight. Information on common and new pesticides can also be retrieved from this system.
4. In the fourth subsystem there are two modules: Pest diagnosis and IPM decision support. The module of Pest diagnosis, called PESTDIAG, is ready for use. It is written in Visual BASIC and has an interface to the framework programme of BJ-CABBAGIS. Its main screen is divided into four boxes. Box 1 is used to question the user about morphological characteristics of an insect or disease, or the symptoms caused by an insect or disease. The question is chosen from a knowledge base developed by plant protection experts. Box 2 shows a colour image of the pest or disease to help the user in answering the question in Box 1. Box 3 is used when an identification is reached to give the Chinese, English and Latin names of the disease or insect. Box 4 advises on control measures.

For IPM decision support, the technological systems should be based on communications between the ecosystem, man and machine. Several philosophical points are raised by phytomedicine, a new medicine which treats health and disease problems of ecosystems.

1. Evolution can be regarded as 'intelligence' of the natural ecosystem, so mankind should learn from natural ecosystems when making and managing artificial ecosystems.
2. Plant diseases and pests are mostly man-made, resulting when mankind lacks understanding of natural processes.
3. We should develop information systems and use artificial intelligence techniques, not only by mimicking the intelligence of experienced men, but also simulating the natural intelligence of healthy ecosystems.
4. Diagnosing ecosystem health involves communicating with the ecosystem and requires further technological developments, just like human and veterinary medicines.

In order to apply information technology to IPM, the technology for information acquisition from natural ecosystems needs urgent improvement and innovation. Progress will be made through long-term research.

Case Study: PQ-INFORMIS Project

According to the Strategy of Precaution and Priority and Integrated Pest Management, plant quarantine is a precautionary measure. In the field of phytomedicine, it is considered equivalent to preventative medicine. Since 1978, both the amount of imported plant materials and the number of ports through which they have been imported has increased rapidly. Now the number of institutions and organizations involved in animal and plant quarantine has reached 363, over eight times more than in 1978. The number of quarantine staff has reached 4500, which is a ten-fold increase in the same period. The new staff require training in quarantine regulations and technology. In 1995–1996, under the auspices of, and in collaboration with, the Beijing Bureau of Animal and Plant Quarantine (BBAPQ) IPMIST conducted a project to develop a multimedia system, PQ-INFORMIS. Now the software has been commercialized and is used in quarantine stations at about 70 importing ports throughout China. It is also popular for field training and school teaching.

PQ-INFORMIS is developed with WDBase, a Windows-based development kit for making text and image database systems, and includes other tools such as Scantools, Photostacker, Animator, Turbo C and Prolog. PQ-INFORMIS provides information on 57 species and one genus of insects, which were declared by the Chinese government to be a high risk in plant quarantine terms. Each species has its own identity card, with the following information: the Chinese, English and Latin names, host plants, morphological description, geographical distribution (countries, or states in the USA), method of transmission, quarantine treatments, and images of the identifying morphological characteristics. The 58 cards are stored in a database and can be browsed using a scroll bar.

Within the menu bar there are five sets of menu options.

- Options concerned with hardware settings, scanner and printer options.
- Options for developing and maintaining the multi-media database.
- Options for searching and performing statistical calculations on the database.
- Options concerned with image processing.
- An option to help with species identification.

The fifth menu option provides a Prolog program to help quarantine staff to distinguish between similar species. Information about 24 of the species was acquired from specialist taxonomists and used in the development of a small expert system. The expert system can be used to identify 47 species from five groups: bean weevils (11 species), hide skin beetles (7 species), bark beetles (5 species), fruit flies (21 species) and mango weevils (3 species). The knowledge base consists of rules which form a dichotomous identification key. An identification is reached by an interactive question and answer session which results in a rule being satisfied. The users of this expert system need a good training in insect taxonomy. In the next version, this option will be similar to PESTDIAG of the BJ-CABBAGIS project, which is much easier to use.

A Practical Approach to Information Technology for IPM

These two examples illustrate China's successful application of information technology to plant protection. A vital component of this methodology is the participatory approach which unites university researchers, government administrators, plant protection practitioners at plant quarantine stations and farms, and information experts.

Our experience proves that plant protection practitioners have to learn, and can learn, modern information technology techniques. Since 1994 I have run a graduate course called 'A pragmatic introduction to software engineering in plant protection'. The main points in the course are to ensure the students are:

- directed by the concepts of software engineering.
- trained in object-oriented programming.
- aware of reusable data and software modules.
- familiar with several software development tools.
- part of a study group of at least four people.

IPMIST usually has five graduates and two undergraduates. Other software developed by the group includes ECOPEST and WGTOOLS. ECOPEST is a system for quantitative analysis of basic ecological problems in pest management, such as the spatial pattern of an insect population, growth curves, the demography of an age-structured population, modelling of population dynamics either alone or under predation or competition, and community analysis. WGTOOLS is a tool to build a weather generator, which can generate simulated daily weather data for a location. It has been used to drive a dynamic model of insect population and an epidemic model of plant disease, which can assist in the risk assessment and management of crop pests (Shen, 1992; Shen and Luo, 1993; Luo *et al.*, 1993). Examples from other Chinese institutions can be found in Zeng and Yang (1986) and Pu (1990). I was recently invited to write a report for the Chinese government, which summarized Chinese experiences with applying information technology to IPM and compared them with developments in the west, and included suggestions on policy for developing information technology in IPM (Shen, 1996).

International Cooperation is Required

The world needs a unified approach to sustaining ecosystem health and sharing achievements in information technology. Sustainable crop protection requires international cooperation in information technology. In the case of PQ-INFORMIS, we are looking for more data in the form of text, images and videos from overseas for pests which do not occur in China. The same problems can be found in all countries.

As a result of growth in the world's economy and increasing international trade, the Agreement on the Application of Sanitary and Phytosanitary Measures (SPS) was signed at the Uruguay Negotiation. To implement the SPS, an internationally exchangeable and easily accessible information system for plant quarantine is required. This should be a comprehensive and sophisticated project conducted in an atmosphere of harmonious cooperation.

Acknowledgements

The author is grateful for the support provided by the Chinese National Natural Science Foundation Grants Nos. 39470126 and 39610211.505, and grants from the Beijing Bureau of Agriculture, and the Beijing Bureau of Animal and Plant Quarantine. Thanks also for the invitation and financial assistance from Dr Peter R. Scott and the local organizers, which allowed me to present this paper at the conference. I would also like to thank Dr Marion Edwards and the editors for their helpful comments on my manuscript and correction of my English.

References

Guan, Z.H. (ed.) (1996) *Introduction to Phytomedicine*. The Beijing Agricultural University Press, Beijing (in Chinese with English abstracts of each chapter).

Guo, Y.Y. (1996) Review and perspective of implementation of the strategy of IPM in China. *Proceedings of the 3rd National Conference on Crop Pest Management*, 12–15 November, Beijing (in Chinese).

Keeting, M. (1993) *The Earth Summit's Agenda for Change*. The Centre for our Common Future, Geneva.

Luo, Y., Shen, Z.R. and Zeng, S.M. (1993) Risk analysis of disease epidemics on wheat by simulation studies. *Agricultural Systems* 43, 67–89.

Nielsen, N.O. (ed.) (1994) *Agroecosystem Health*. University of Guelph Press, Guelph, Ontario.

Pu, A.L. (ed.) (1990) *Mathematical Models and its Applications in the Management of Crop Pest Insects*. Guangdong Science and Technology Publishing House, Guangzhou (in Chinese).

Shen, Z.R. (1992) The development and use of models and expert systems to the control of crop pests in China. *Mathematics and Computers in Simulation* 33, 403–414.

Shen, Z.R. (1996) Impacts of modern information technology on sustainable plant protection. *Proceedings of the 3rd National Conference on Crop Pest Management*. Chinese Science and Technology Publishing, Beijing, pp. 73–77 (in Chinese).

Shen, Z.R. and Luo, Y. (1993) Insect and phytopathogen population models randomized with a stochostic weather simulator, *Polish Ecological Studies* 19(1–2), 59–72.

Zeng, S.M. and Yang, Y. (1986) *Epidemics of Plant Diseases*. Agricultural Press, Beijing (in Chinese).

33

Computer Games and Other Tricks to Train Field Pathologists

T.M. Stewart

Department of Plant Science, Massey University, Palmerston North, New Zealand
Fax: +64 6350 5606/E-mail: t.stewart@massey.ac.nz

Introduction

As preceding chapters have shown, the use of computer technology has brought a wealth of exciting tools to assist in training plant protection practitioners. These include illustrated textbooks, easily accessible resource material, expert systems, keys, databases and simulations. They can be delivered to students in a variety of ways; via networks, on CD-ROMs, disks or over the Internet. Indeed, even whole courses can now be delivered, using these electronic media. Previous authors have looked at the technologies involved, and how they might be used. In this chapter, the author would like to focus specifically on computer games, as learning tools for the budding plant pathologist. The first part of the paper will discuss the use of games in learning and conclude with a short summary of plant pathology-related examples. The second part will discuss embedding such programs into a course, with reference to the author's own experience in developing and utilizing such a game, and in particular, the pedagogical reasons for doing so.

Games and their Role in Education

From the earliest stages of growth, playing games is a significant part of the learning process. Even simple games, such as tic-tac-toe, can help a very young child develop logic, and explore relationships. Older children might use 'Hangman' to assist with spelling. In general, as we move through the school system, educational games become less common in the formal learning environment, but retain a role in informal learning and entertainment. Despite this, there are still many examples where games are used to illustrate various principles in a formal curriculum at the post-primary level. Some are very sophisticated, and a great number are computer-based. Examples include games involving simulations of whole economic units, such as SimCity^{tm} . These can be used in the classroom to show the complexity of economics. Role playing games are popular in the social sciences.

One of most powerful ways games can enhance learning is by offering a problem-solving approach. To succeed in a game usually involves good decision-making. In order to make decisions, students must interpret and synthesize information. Hence games can force students to integrate their knowledge, and use it to solve a problem or achieve a goal. This is good training for real-life tasks as that is what will be expected of them, once they start working in their profession. Furthermore, these exercises allow students to make mistakes without grave consequences, and learn from them. In the words of Alfred, Lord Tennyson, 'And others' follies teach us not, Nor much their wisdom teaches. And most, of sterling, is what Our own experience preaches.' From an educational point of view, this is one of the most useful aspects of using game-type exercises in teaching. However, there is, I believe, another quality.

The *Concise Oxford Dictionary* defines a game as 'A contest played according to rules and decided by strength, skill or luck.' Within this definition are the two elements which help give games their value as a learning tool. They are 'contest' and 'rules'. These two factors can increase that important component necessary for learning success at all levels. I am referring to motivation.

There is little argument that, when it comes to advanced learning, motivation is a major determinant as to whether a student will succeed or not. At Massey University, we have a well established and comprehensive distance learning programme. Many courses are taught both internally and at a distance (extramurally). Both groups of students often undergo the same final examination. Initially, there is a high drop-out rate with extramural students as the pressures of jobs and families take their toll. However, students who complete the course tend to obtain higher grades than the internal ones, when the two groups are compared. This is despite the fact that the former have little person-to-person contact with the teacher, learning mainly from study guides, readings and assignments! They also have less free time to devote to studies due to their other commitments mentioned above. The reason for this grade imbalance seems to be attributed largely to motivation. Extramural students as a group are older, and are often actively seeking a qualification to improve their job prospects or promotion. Many are highly motivated and will often learn 'in spite of' rather than 'because of'

For the average college or university student, there are many motivating forces which come into play with regards to performance (Morgan, 1993). Regardless of which of these is most important to the individual student, they are more likely to work at a particular exercise if the activity appears relevant and meaningful to their life situation. In the case of plant pathology games with plant pathology students, they often see the task as a simulation of their future vocation. As such, it has intrinsic, vocational and personal value for them. Also, for many students, the teacher can play a major role in motivating them to learn. Albert Einstein once said 'It is the supreme art of the teacher to awaken the joy in creative expression and knowledge'. Teaching is indeed an art!

Having discussed the importance of motivation, let us return to games, and how 'contest' and 'rules' can increase the desire to learn. First, let us reflect on the competitive element found in games. The desire to do well, to succeed at a task, is a common human trait. Games are about a contest, not necessarily just against other people but also against oneself. This competitive element adds the fun, and it can also be enhanced by the teacher. In one of the author's own classes for example, students are divided into groups when they play the pest management game BEET (see below). A small prize (a chocolate bar each) is awarded to members of the winning group. The prize is trivial, but students work hard to win against the other groups and are genuinely disappointed when they do not! The point is, motivation is increased. In order to do well, the players must learn the principles embodied in the game and make decisions based upon them. I stay with them as they play, working as a facilitator, offering encouragement and keeping the competitive element alive. Research suggests that not every student responds to competition in a positive way (Garner, 1990). However, the students under discussion appear to respond well to the exercise described, possibly because they are undertaking advanced level studies in a subject area of their own choice, and hence have a keen interest in it. This would be true of most plant pathology students.

The second aspect of games that can increase motivation is that good games normally have clearly defined 'rules'. The slice of reality represented is finite and goal driven, and this helps focus the student's attention on the main principles being taught. Sometimes there can be more than one objective in the game, occasionally even conflicting ones. The point is, they are there to guide the player to work towards his or her objective. Competition and rules are inherent in a game. If a game is to be used as a learning tool, however, more than that is required. Along with the specific goals and objectives that are necessary for a 'good result', exercises should also have clearly defined learning goals and objectives. While the specific task in the game may be obvious (e.g. to make maximum profit from managing a crop) the learning goal should also be clearly stated. It is these learning goals and objectives that should be emphasized and discussed both prior to the exercise and at the debriefing afterwards.

Although the benefit of games in courses has been discussed, using too many games in a course can be counter-productive. They can be time-consuming and exhausting, for teachers and students alike. Their use is best limited to a few, well thought out areas (Mumford, 1993). Used properly, they can be very effective.

Computer-based Games for Teaching Plant Pathology: Some Examples

The usefulness of games as teaching tools has been discussed in the section above. Now let us look at some of the teaching packages related to plant pathology available at the time of writing. This is not an exhaustive list as many programs are written for in-house use and never find their way into the literature. I have restricted the list to games only. This implies students are given a distinct goal, and there is some skill or decision-making required in order to arrive at a satisfactory outcome.

APPLESCAB

APPLESCAB (Blaise *et al.*, 1987) simulates the management of apple scab (*Venturia inaequalis*) in an apple orchard. During the game, players can attempt to control scab over a number of years using a variety of fungicide strategies under many different conditions. The game uses a graphical interface showing daily information by way of charts and figures. At the end of each year, players receive a report summarizing the fungicides used, their costs, overall yields and profit and weather conditions throughout the season. Many parameters can be adjusted and output can include crop stage, lesion formation, weather, ascospore release and fungicide residue.

BEET

This decision-making management game incorporates not only plant diseases, but insect pests as well. Players are responsible for managing a 40 ha beet crop, rotated on a farm over 15 years (Mumford and Miller, 1995). During each season, decisions have to be made on control of soil and seedling pests, aphids (which can transmit beet yellows virus) and powdery mildew (Fig. 33.1). The game was written to teach students about pest management strategy, and to give them practice in decision-making in a stochastic environment. Like most simulations cited here, players receive a summary of costs and income at the end of each season. The game is best played with teams of students, competing against each other.

LATEBLIGHT

LATEBLIGHT (Fry *et al.*, 1991) is similar in many ways to APPLESCAB. This is a simulation in which students are required to minimize the damage caused to their potato crop by the fungus *Phytophthora infestans*. Players have control over many parameters, including potato variety, environmental factors, inoculum, management schedule, costs and market selling price. The effect of any decision

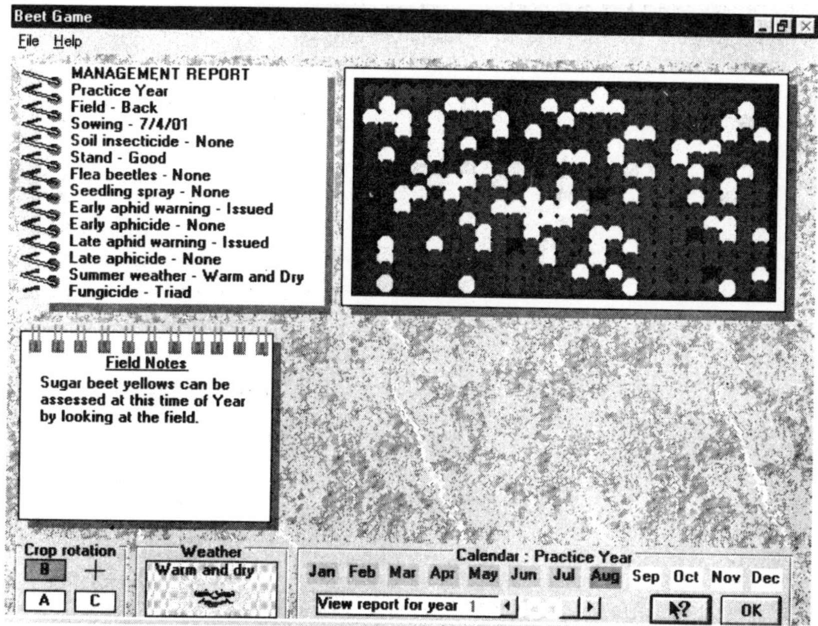

Fig. 33.1. A typical screen from BEET showing field with virus and mildew infection.

is displayed graphically as the season progresses in terms of percentage blight, defoliation, new infections, numbers of spores and fungicidal residues (Fig. 33.2). At the end of the season, a financial report can be generated.

TURFBLIGHT

This package places the student in charge of 5.1 ha of turf (Danneberger, 1988). The students attempt to control anthracnose (*Colletotrichum graminicola*) and pythium (*Pythium* spp.) through correct fertilizer and fungicide application. They must also consider drought stress and irrigate accordingly. The program is run over 153 days, and allows students to make daily decisions on the parameters above. Initial inputs include soil texture, budgetary limits and weather patterns. Daily readouts showing summaries of disease, drought, fungicide residues and weather are graphically displayed. At the conclusion of the game, turfgrass losses due to disease and drought are summarized, along with costs. Students can then compare proposed and actual budgets. Possible fungicide resistance and damage from excess fertilizer are also features of the game.

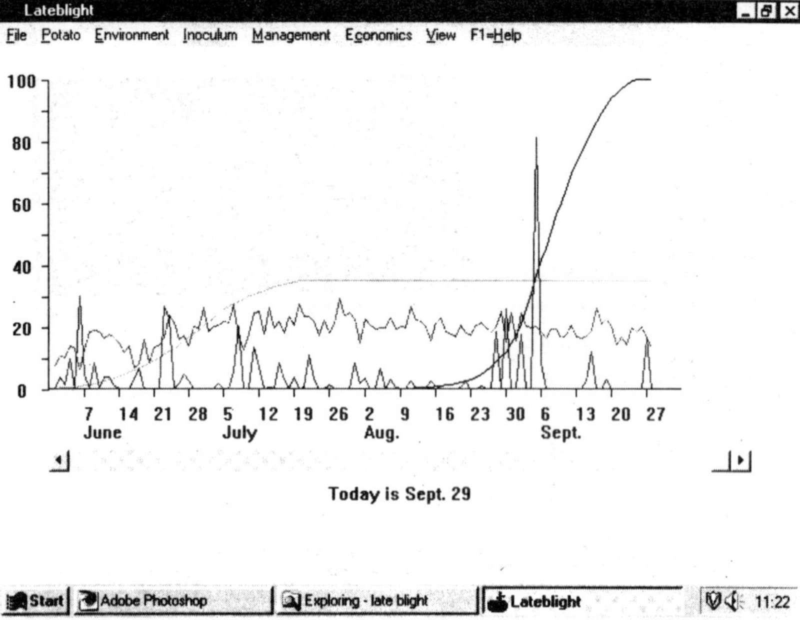

Fig. 33.2. A screen from LATEBLIGHT at the end of a season, showing leaf development, disease, and weather parameters.

DISEASEPRO

This program had its origins in DISTRAIN (Tomerlin and Howell, 1988), a program designed to improve disease assessment by presenting images of diseased leaves, and requiring students to estimate the disease severity. Once this is done, the actual severity is shown and the student starts again. After a number of iterations, the student's assessment becomes 'calibrated', and hence can provide accurate severity estimates.

DISEASEPRO was designed for peanut diseases but also provided similar spin-off packages like BARLEYPRO and ALFALFAPRO (Nutter and Schultz, 1995). It gives the student far more control over feedback and presentation parameters than DISTRAIN. Players can switch off the instant feedback to allow both a pre- and post-test when using the program. This enables a base-line of student accuracy to be set before using the software, so that improvements can be quantified afterwards. Other features include the ability to select the number of leaves, the size of the lesions, and charts showing various statistics between estimated and actual disease severity.

DIAGNOSIS

The DIAGNOSIS package has been described in Stewart *et al.* (1995) and was reviewed by Knight (1995). Essentially, the package helps teach the 'art' of diagnosing crop problems (Grogan, 1981). The user is presented with a problem scenario and must interrogate the program to diagnose the situation. Information is presented in the form of a text window, with optional graphics, sound and video (Fig. 33.3). By looking at different plant parts, the growing environment, and by asking a variety of questions, the student gradually accumulates knowledge about the problem and makes a diagnosis. Feedback consisting of the correct answer, the key points that should have been observed, and the points that an experienced extension agent or crop consultant would typically make to diagnose the problem are given once a final diagnosis is entered by the student (Fig. 33.4). A points score and accumulated costs are also shown. This feedback can be saved to disk for later assessment by teachers.

One point to note is that the package is not an expert system as the name might suggest. In the game, the student is assumed to be the expert and it is the student that is expected to provide the diagnosis, not the computer. What the computer does, is to provide a 'virtual reality' for them to practise their diagnostic skills. Observations are presented, and it is up to the player (i.e. the student) to determine the significance or otherwise of these.

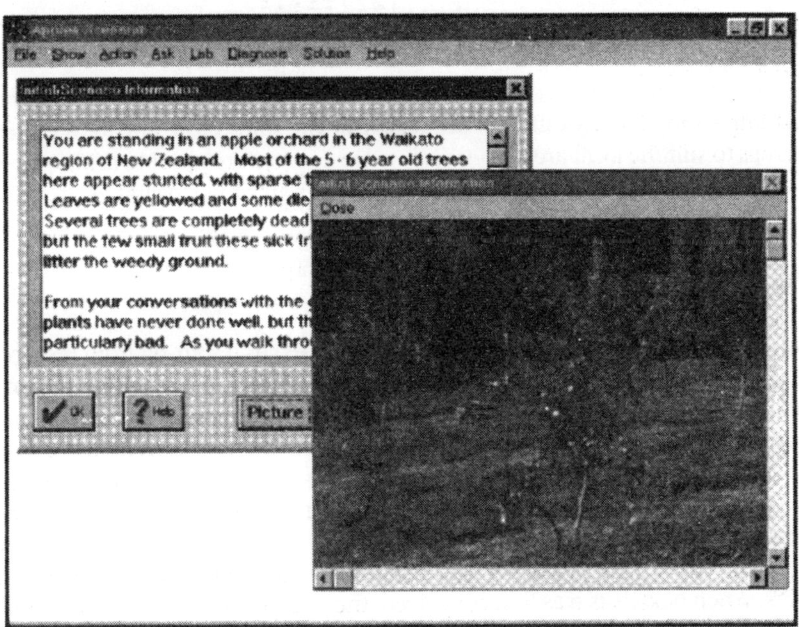

Fig. 33.3. A typical scenario introduction in DIAGNOSIS.

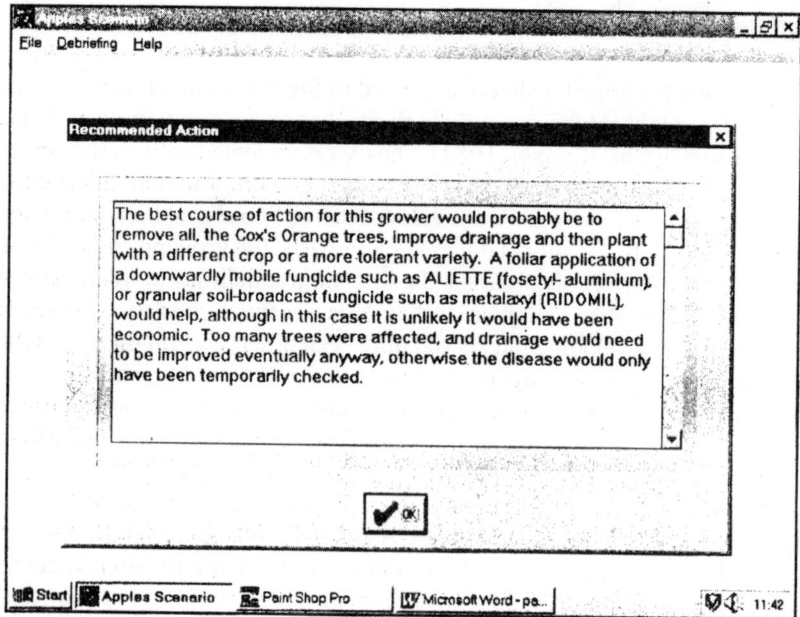

Fig. 33.4. Recommended action for a particular scenario from a DIAGNOSIS de-briefing.

A Scenario Builder complements DIAGNOSIS, to facilitate the creation or customisation of many different scenarios for the same crop, and for a variety of crops to suit the local area. The teacher decides on a crop problem, and enters textual information and selects relevant images and/or video clips.

The Genesis of an Educational Game

Many games, such as LATEBLIGHT, have been derived as a spin-off from research programmes. Converting a model from a research tool to a teaching tool can mean a considerable amount of reworking and time, in order to embed it in a course with the correct educational paradigms. Others have been designed from the start, as training aids. Having commenced from an educational viewpoint, the education paradigm can be an inherent and integral part of the programme, rather than an add-on to an existing model.

From the outset, DIAGNOSIS was designed to solve a teaching problem. In 1989, when DIAGNOSIS was first conceived, the author was taking plant protection students for a series of disease diagnosis practicals. In these classes, the emphasis was on giving students practical diagnostic skills in order to solve plant problems. The students would collect plants with an unknown (to them) condition and bring the specimens back to the laboratory, where they could examine them closely under supervision, and attempt to isolate and identify any pathogens or pests present. Appropriate literature was available for reference, and the students would use this material to gather information on signs and symptoms, or to work their way through a series of keys.

Since many of these students would be moving into careers as consultants or managers of various agricultural or horticultural enterprises, the emphasis of these series of laboratory-based exercises was on practical diagnosis, with the aim of providing a recommendation on the best management strategies for the problem. This approach worked well for many obvious pests and some of the more obvious diseases such as fungal leafspots/blights, mildews and rusts. However, the exercise was limited by the laboratory resources, specimens available and the difficulty of monitoring the student's approach to the problem in the field. Furthermore, it could only take place in the first semester (i.e. autumn) when specimens were freely available.

Field observations are critical for the correct diagnosis of many conditions, particularly where the cause is abiotic, such as frost or herbicide damage. Similarly, predisposing factors for biotic diseases can often be gleaned from the history of the crops and from discussions with personnel involved in their management. The spatial distribution pattern of symptoms in the field can also yield clues as to the cause of the problem. Many conditions are difficult, if not impossible, to diagnose without this information. Root problems can pose particular difficulties with the exercise described above, in that students may have collected material away from the site of infection, i.e. leaves rather than roots. In advance and during the exercises, the nature of this 'field detective work' was emphasized. Students were encouraged to carry with them a notebook, to record all these observations, including the past history of the crop, grower comments and any other useful information.

For a number of years previously, the author had been an ardent player of computer-based adventure games. Some readers may recognize vintage text-based games such as ZORK and Adventureland. In these 'adventures', the aim is to find the treasure, kill the vampire or solve the mystery. Many of the students were familiar with these programs and during the laboratory sessions the author would often compare the process of deduction used in these with that of diagnosing pest and disease conditions. In many ways they were similar. In a computer-based adventure game, the player had to explore the 'environment', uncover clues and interpret observations in order to come to the correct conclusions and so progress to the next stage.

One day the author was struck with the thought that, if the processes were similar, why not write an adventure type game where the mystery was a plant disease? Students could be put into a kind of 'virtual reality' where they would be required to examine plants in a field situation, converse with the grower, collect specimens to examine in the laboratory and, at the end, come to a reasonable conclusion. When students had finished, they could type their diagnosis and justification into the computer, which could then record it on disk for a tutor to mark later. Furthermore, the computer could record each step of the process, thereby monitoring the students' approach. It could then present the player with an appropriate debriefing, compiled by the tutor, but tailored to what the student did or did not do during the diagnostic procedure. This seemed an exciting concept! So the program was written, initially for an MS-DOS platform (Stewart, 1992), then as a Hypercard stack for the Macintosh and finally for Microsoft Windows 3.1 and Windows 95.

It is worth noting that DIAGNOSIS has added to rather than done away with the traditional laboratory classes in the author's course. Students still undergo hands-on practicals to learn techniques such as mounting, isolating and identifying plant pathogens.

Embedding Software into a Course

Like all teaching software, just how it is used in a course can determine its value. It is one thing to develop such packages within your own courses. As teachers, we know inherently what we are trying to achieve with the packages, and because they are developed within the framework of the course, they integrate well into them. It is another thing, however, to expect someone else to pick up the game and use it in their own course, even though they are teaching the same discipline. Exactly where and how should it be slotted into the course? Should it be played as a group exercise or individually, all at the same time or over a period of days? These are important questions which need to be addressed, if training packages like DIAGNOSIS are to be used more widely. These issues are equally valid for other computer-based exercises, not just for games.

Aims and objectives

Initially, the aims and objectives should be clearly spelt out. For DIAGNOSIS these were as follows.

Educational aims

These were to:

- re-enforce the processes undertaken in diagnosing a plant problem.
- experience and practise diagnostic situations, similar to those that a crop protection specialist may be faced with.
- gain a greater understanding of the importance of pathogens and pests in crop management.
- appreciate the impact of past management on present problems.

Learning objectives
The learning objectives will vary from scenario to scenario. In most, however, the student will:

- interpret clues correctly, both from their own observations and from conversations with personnel involved in management of the afflicted crop.
- take account of the predisposing factors which may have led to a specific disease or pest problem.
- take the logical steps required to gather information necessary for a successful diagnosis.

Prerequisites

With all exercises, whether computer-based or not, a base-knowledge is assumed. In DIAGNOSIS, students are required to have the following minimum knowledge:

- Some basic understanding of crop management and plant growth, and ability to identify plant parts for the crop under consideration.
- A knowledge of common diagnostic techniques such as isolation of fungi from plant tissue, and nematode extraction.
- A basic knowledge (biology, typical symptoms caused, and method of spread) of the causal agents of biotic plant problems, i.e. pests and pathogens.
- A knowledge of abiotic factors such as pesticide damage, frost, nutrient deficiencies, etc., and how these may affect plants.

Use at different academic levels

Often games can be tailored to the academic level required. In DIAGNOSIS, teachers can give the students assistance relative to their academic level by constructing/amending scenarios. Assume for example, leaves were sent for nutrient analysis in a scenario. In an advanced 300-level university course, the response from the testing laboratory may be simply to provide a list of nutrients and the percentages of each one. It is up to the student to decide if any of the figures appear abnormal. In a 100-level course, or a high school course, the reply might be 'The nutrients appear OK', or 'Nitrogen appears a little low'. During construction or amendment, teachers can also switch off or on some of the more complex diagnostic techniques to take into account the background knowledge and ability of their students.

Playing the games

When using teaching games with students, thought must be given to which external references will be allowed. If the game is simulating a real-life experience, then the resources that are available to a real life expert should also be available to the student. Some games have these built into them in the form of comprehensive Help Files. DIAGNOSIS is designed to be used with other resources, such as reference books on the crop, and its associated pests and diseases.

The teacher can elect to have the students work individually, or in groups. The exercise can be done in one session, or over a period of time. The availability of computer equipment will have some bearing on these decisions, as will the aims and objectives of the exercise. Playing in groups can help facilitate learning, as the players discuss possibilities between themselves. Playing individually allows the tutor to more accurately gauge the individual student's grasp of the processes.

In computer simulations, as in the real world, decisions can have financial implications. In crop management games, students often receive a balance sheet at the end of the season. In DIAGNOSIS, costs are built into some activities, such as sending a specimen to a taxonomist for identification, or using laboratory equipment. One option is to give students a budget within which they have to work, for the particular exercise. This has merit as it prevents wasteful extravagance and encourages students to think the problem through and focus on the most relevant material.

After completing the exercise, it is important that students receive a debriefing, to reinforce what has been learned. This can be a tutorial or a lecture preferably carried out immediately after the event. In the debriefing, the educational aims and objectives should be discussed as well as the technical aspects of the diagnosis process.

The debriefing is an important component in using DIAGNOSIS. The author usually discusses the debriefing with students as a group. In addition, the debriefing is appended to their disk file after they complete the game. This disk file can be printed out, their diagnosis marked, and the script (and the debriefing) returned to them at a later date. In the debriefing, the major clues the students needed to find to come to the right conclusion are outlined. Equally important is to draw attention to any 'red-herrings' that may have led them astray. The emphasis should be on the holistic nature of the diagnostic process and how the first stage is to draw up a list of possibilities, the second is to rank them in order of probability and the third is to look for further confirmatory evidence starting with the one highest on the list. If the evidence is not strong, then other alternatives on the list can be looked at.

Conclusions

In this paper, I have outlined some of the benefits of using computer-based, decision-making exercises and listed some well known ones. I have also given a personal view, using DIAGNOSIS as an example, of how they may originate and be used in a training course.

Computer-based games can aid in teaching plant pathology by forcing students to adopt a problem-solving approach, and by increasing the motivation to learn. By themselves, however, they are no panacea. They can be expensive to produce and utilize and, most importantly, they must be embedded correctly into the overall framework of a course. This means having clear educational aims and objectives for their use. If this is done, and adequate attention is given to the role of the teacher both prior to and after the exercise, they can be a powerful tool, giving the students a learning experience they could not obtain in any other way.

Acknowledgements

The author would like to thank Claire McLachlan-Smith, Teaching Consultant, of the Centre for University Extramural Studies, Massey University, for her advice and comments on this paper.

References

Blaise, P., Arneson, P.A. and Gessler, C. (1987) APPLESCAB: A teaching aid on microcomputers. *Plant Disease* 71, 574–578.

Danneberger, T.K. (1988) TURFBLIGHT: An interactive turfgrass management model for teaching. *Journal of Agronomic Education* 17, 73–76.

Fry, W.E., Milgroom, M.G., Doster, M.A., Bruhn, J.A. and Bruck, R.I. (1991) *LATEBLIGHT v 3.1: A Disease Management Game*. Computer program and manual, Department of Plant Pathology. Cornell University, Ithaca, New York.

Garner, R. (1990) When children and adults do not use learning strategies: towards a theory of settings. *Review of Educational Research* 60, 517–529.

Grogan, R.G. (1981) The science and art of plant disease diagnosis. *Annual Review of Phytopathology* 19, 333–351.

Knight, J.D. (1995) DIAGNOSIS for crop protection – a computer based teaching tool. *Integrated Pest Management Reviews* 1, 75–77.

Morgan, A. (1993) How do students come to be studying? In: *Improving your Students' Learning: Reflections on the Experience of Study*. Kogan Page, London, pp. 27–49.

Mumford, J.D. (1993) Pest management games. In: Norton, G.A. and Mumford, J.D. (eds) *Decision Tools for Pest Management*. CAB International, Wallingford, UK, pp 181–186.

Mumford, J. and Miller, S.H. (1995) *BEET v1.0. MERTaL*[tm] Courseware, (ed.) Health S.B, University of Aberdeen, Aberdeen, UK.

Nutter, F.W., Jr and Schultz, P.M. (1995) Improving the accuracy and precision of disease assessments: selection of methods and use of computer-aided training programs. *Canadian Journal of Plant Pathology* 17, 174–184.

Stewart, T.M. (1992) DIAGNOSIS, a microcomputer-based teaching aid. *Plant Disease* 76, 644–647.

Stewart, T.M., Blackshaw, B.P., Duncan, S., Dale, M.L., Zalucki, M.P. and Norton, G.A. (1995) DIAGNOSIS: a novel, multimedia, computer-based approach to training crop protection practitioners. *Crop Protection* 14, 241–246.

Tomerlin, J.R. and Howell, A. (1988) DISTRAIN: A computer program for training people to estimate disease severity on cereal diseases. *Plant Disease* 72, 455–459.

34

The Need to Rebuild our University Education Systems on an Information Technology Basis

P.H. Schalk[1] and W.H. Los[2]

[1]ETI, University of Amsterdam, Mauritskade 61, 1092 AD Amsterdam, The Netherlands
[2]Zoological Museum Amsterdam, University of Amsterdam Mauritskade 61, 1092 AD Amsterdam, The Netherlands
Fax: +31 20 525 7238/E-mail: peter@eti.bio.uva.nl

Introduction

In the last decade, governments in nearly all European countries have put severe financial constraints on educational institutions. Nowadays more universities in the 'western' countries are wrestling with an insufficient number of staff to fulfil their traditional teaching duties. As a consequence, the number of courses has decreased and education has become 'narrower'. In addition, curricula have been reduced, resulting in an alarming decrease in the knowledge level of the students, which in turn leads to a serious concern about the future academic capacity of the countries. This downward spiral has been recognized by the politicians and the time is right for counter measures.

New electronic media, such as CD-ROMs, the Internet, and special IT tools geared to teaching and self-learning, could provide an answer to the shortcomings of present-day higher education. In fact, like it or not, the wide scale application of electronic aids is inevitable and is penetrating deep into our societies. With Internet connections growing exponentially and more and more information becoming available on-line, a new information facility is becoming alive, or is partly already there. These electronic highways erase traditional borders and boundaries. Educational institutions should make use of these new possibilities and information, by building a new education system on top of these services. The wide scale application of IT to education will allow students to develop a self-study behaviour to use basic reference works at their own pace. With electronic teaching tools in place, teachers can concentrate on one-to-one education, if needed via teleconferencing, to solve the specific problems of individual students and in practical teaching.

Step by Step and Bit by Bit

It is clear that the step from present day 'traditional' education to an IT based approach in the future cannot be realized at once. Reference works, student readers, instructive videos, etc. will have to find a new life as computerized instruction tools and be made available electronically. Even more important is that the teachers will have to revise completely their thinking about education in terms of these new facilities and about new applications to convey concepts and transfer of knowledge. This is a major operation as most students know more about computers and software than their teachers.

It would be a mistake to try to make computer programmers out of the teachers. Instead they should be assisted in learning about IT and its application. When given knowledge about new opportunities, they should concentrate on the contents and didactics to create the new educational IT tools we need.

An operation like this will cost hundreds of millions of ECUs and therefore needs to be implemented step by step. However, it is important to realize that quite a few tools are already in existence. The BSPP/SA Conference on which this volume is based has demonstrated the wealth of tools that are available in the biological sciences. In addition to these tools, there are numerous scientific databases available on-line. What is needed is a master plan to connect these sources and forge them into an educational facility. Gaps should be identified and filled in by developing new tools. As the electronic networks allow for collaboration without physically being in the same place, the development of a new electronic basis for higher education may be easier to implement than one would say at first sight.

Biology as a Starting Point

The field of biology might provide a good starting point for a broad scale application of IT in higher education. Biology with its combination of different concepts, different kinds of information and 'attractive' pictorial information, provides an ideal experimental platform to develop new learning tools. Biology is already ahead of other disciplines, and biodiversity is an interesting starting point. The international driving force behind the Convention on Biological Diversity triggered many initiatives to pool resources and set up species databases and information systems that will form the basis for policy making. As traditional methods of sharing information simply cannot cope with the enormous amount of information that is involved, IT must be used to provide a solution.

Initiatives such as Species 2000, Systematics Agenda 2000, Diversitas and many others have shown the needs and benefits of computerized information, but they have been struggling to pave the way to a new information era. Many knowledge centres, such as natural history museums and universities, have developed specimen databases, species databases, computer-assisted identification tools, geographic information systems and systems that integrate several of these functions.

The databases of the 21 initial members of the Species 2000 Federation cover about 20% of the world's 1.7 million species and work is being carried out to connect these to form one powerful information tool on the Internet. ETI's 'World Biodiversity Database' CD-ROM Series, currently produces 15 titles of biodiversity information a year, and they are an example of 'stand alone' IT tools that can be applied everywhere where a computer is at hand (Schalk, 1992). Moreover, well over a thousand specialists are working with ETI's LINNAEUS II software (Schalk and Los, 1995) digitizing the earth's biosphere literature on a worldwide basis. LINNAEUS II is also being used to teach students how to classify organisms, build interactive identification keys and create geographic information systems to document and analyse biological diversity. The DELTA system (Dallwitz and Paine, 1986; Dallwitz *et al.*, Chapter 19, this volume) developed by CSIRO also has many users around the world who have created vast databases which are waiting to be used.

These are only a few examples, the field of biology is full of them, and many others are discussed elsewhere in this volume.

Changes for Developers

The European Commission has expressed its concern about the fact that Europe is lagging behind the USA and Asia in terms of information technology. If the European countries do not catch up soon, we will not only miss an enormous economic chance but also become seriously underdeveloped in this respect. It was decided to provide vast amounts of subsidy through special EU stimulation programmes to develop IT applications in various areas such as business and education. Many of the governments of the European member states are following this at the national level with additional programmes to stimulate IT developments.

It is the right time to join forces, to combine our existing IT knowledge and tools and to apply for an EU-funded and internationally coordinated development of the education system for the next millennium. Perhaps the field of biology would be a good starting place, and the BSPP/SA symposium has provided a first step in this direction. Whereas the developers currently work independently, and in most cases with insufficient financial support, a joint application may offer an answer.

Let's make a start!

References

Dallwitz, M.J. and Paine, T.A. (1986) *User's Guide to the DELTA System; a General System for Processing Taxonomic Descriptions.* 3rd edn. Report, Division of Entomology, CSIRO, 13, 1–106.

Schalk, P.H. (1992) Computer aided taxonomy. *Binary* 4, 23–26.

Schalk, P.H. and Los, W. (1995) Applications of interactive multimedia software in taxonomy and biodiversity studies. *Global Biodiversity* 7, 12–18.

35

CD-ROM as a Dissemination Medium in Practice: Crop Protection Case Studies in Africa

S.S. M'Boob*

Food and Agriculture Organization of the United Nations, Regional Office for Africa, PO Box 1628, Accra, Ghana

Introduction

Plant protection capabilities in the majority of African countries are inadequate to deal effectively with the various crop protection problems. FAO has conducted comprehensive surveys on plant protection in over 25 African countries, to identify and assess the major problems, constraints and needs (FAO, 1988). These surveys have shown that, although most countries have some form of infrastructure, many are lacking the most basic elements. In almost all cases, the delivery of advisory services to farmers and regulatory activities for plant import and export are limited by poor access to current scientific information and databases on crop pests in particular and crop protection in general, in addition to several other factors.

National Plant Protection Infrastructures

The general lack of plant protection infrastructures means that countries cannot deal adequately with the large variety of crop protection problems they face. The establishment or strengthening of national plant protection structures must be seen as the fundamental first step towards the improvement of crop protection capabilities in the region. In the short term, priority should be given to training of plant protection personnel and the building of national capacities to access current scientific information and databases on crop protection. In the medium and long term, emphasis should be given to the development and

*See Contributors List for present address. E-mail: fao-tza@fao.field.org

implementation of farmer participatory training in integrated pest management (IPM), pesticide registration and control, and plant quarantine and plant health certification schemes.

Research in Crop Protection

The need for research to develop sustainable, efficient and environmentally friendly means of controlling agricultural pests is as great as it ever was. Although there is a good amount of research activity throughout the region, the general situation is one of poorly focused, non-farmer-participative and isolated research programmes. The situation is compounded by declining resources, which means poor access to current scientific information as subscriptions to scientific journals are suspended.

The FAO bibliography of plant protection research (FAO, 1988) shows that there is a rich knowledge base and a modest resource of research expertise and experience in the region to be exploited. The capabilities of each individual country are weak and would need a modest degree of strengthening of the structures, programmes and resources. However, given the similarities of crop protection problems between countries sharing similar agro-ecosystems, one effective way of mobilizing the disparate expertise into a more effective whole would be through a system of networking between the various national plant protection research institutions. This would allow scientists from different countries to share information and technology and to coordinate their research efforts to solve identified problems of common concern. Considerable opportunities exist in the region to develop and strengthen research capabilities in plant protection in Africa within the framework of Technical Cooperation Networks (FAO, 1991a,b).

Extension and Training in Crop Protection

The FAO surveys on plant protection have revealed several weaknesses in plant protection extension and training in most countries (e.g. FAO, 1992). The shortcomings include, foremost, poor access to information on crop pests and recommended control strategies, weak linkages with research, poorly trained and unmotivated front-line staff and poor resources. The lack of sufficiently trained personnel, and the absence of regular staff training to update knowledge and skills in most countries, is considered the major constraint to the effective functioning of national plant protection services.

Pesticide Management

Pesticides at present constitute the main means of pest control in Africa. It is estimated that, at present, African countries import 500 million US dollars worth of pesticides annually, with a growth potential of 10% per annum.

Most African countries lack pesticide legislative control measures to ensure rational, safe and efficient use. This has led to the widespread occurrence of unacceptable practices with regard to distribution and use, posing a serious threat to human health and the environment. Difficult access to information and the lack of trained personnel are major constraints in this regard.

In recognition of this problem, FAO has, in cooperation with other UN agencies and the pesticide industry, developed the International Code of Conduct on the Distribution and Use of Pesticides, which was unanimously adopted by the FAO Conference in 1985 (FAO, 1986). The objectives of the Code are to identify potential hazards in the distribution and use of pesticides, to establish standards of conduct and to define responsibilities, mainly for governments, the pesticide industry, salesmen and users. The Code is accompanied by a set of technical guidelines covering the main issues of pesticide management.

The Code has been actively promoted in Africa through a series of sub-regional workshops, and follow-up activities to these workshops have included, direct assistance to individual countries in the drawing up and implementation of legislation and control schemes, the establishment of pesticide quality control laboratories and training. A major constraint in almost all countries is the lack of access to pesticide databases.

Parallel to activities related to the Code, FAO is actively promoting the development and implementation of Integrated Pest Management as a more sustainable method of pest control in Africa.

Plant Quarantine

Exotic pests continue to pose one of the most serious threats to agriculture in Africa. Preventing their entry, as well as containing those which have gained entry, and screening them out of imports and exports, are among the most onerous and important functions of national plant protection services. Yet, the majority of Plant Quarantine Services are uninformed, underresourced, understaffed and lack experience. The need for a realistic approach to plant quarantine in Africa has long been recognized by FAO. Consequently, feasibility surveys were conducted in 1989 on the development and implementation of sub-regional plans for improving plant import and export phytosanitary control and plant quarantine. The plan is predicated on the principle of concerted action to prevent the introduction and contain the spread of plant pests and contemplates the active participation of national plant protection services within the framework of sub-regional plant quarantine Networks. FAO's objective is to promote the establishment of a Network (FAO, 1993) for each of the 4 main sub-regions (SADC, East, West and Central Africa). These Networks will then provide the institutional framework for joint collaborative programmes and actions by national services and their partners for tackling and resolving common plant quarantine problems (Fig. 35.1). Among the priorities identified were: provision of prompt topical current awareness and information services, sharing information on pest distributions, establishing common pest lists, developing common inspection, certification and treatment standards, harmonizing phytosanitary standards for exports and imports, etc.

Information Management

The key element to the success of a Technical Cooperation Network is a strong information base to support the activities of the national institutions collaborating in the Network and the coordinating activities of the Secretariat. Since lack of information has been one of the major constraints to the development of

Fig. 35.1. Participants at the FAO/SADC Workshop on the establishment of a PQ Network for SADC (1993) being introduced to CD-ROM technology.

plant quarantine in the region, it became apparent to FAO that a strong information base on crop protection was needed at both the country and subregional levels, to support the activities of the Plant Quarantine Networks. Since most modern methods of information exchange can be very expensive to maintain, CD-ROM technology offered the best approach to providing rapid access to a very large volume of data with very minimal maintenance costs. The use of CD-ROM technology was made attractive by the ready availability of plant protection information databases from CAB INTERNATIONAL and the support and training services it provided. The CAB ABSTRACTS database is available on CD-ROM. CABPESTCD contains bibliographic citations and abstracts of the world literature of crop protection science and pest management since 1972. Subjects covered include entomology, plant pathology, nematology, weed science and vertebrate pests.

System requirements

Each participating country[1] was provided with a CD-ROM system, consisting of an IBM compatible personal computer with built-in CD-ROM drive, a printer and software including CABPESTCD.

Training in computer-based information systems

A one day hands-on training in basic computer operation and retrieval of information from CABPESTCD was provided to plant quarantine officials from 15 countries in southern and eastern Africa (Fig. 35.2). The training was part of a series of sub-regional workshops for the establishment of Technical Cooperation Networks for Plant Quarantine in southern and eastern Africa. The training consisted of a short introductory presentation on the work of CABI, CD-ROM technology and the development and coverage of CABPESTCD, followed by an intensive practical session. Familiarization with the work station and basic searching techniques were covered. By the end of the session, all participants were able to conduct effective, simple searches to extract relevant information from the disc. In addition, care of the disc and system management issues were addressed.

Impact of the CD-ROM system

The consensus of the participants at the end of the training was that the CD-ROM system would enable the national plant protection services to be better informed and to make hitherto unprecedented progress. The CD-ROM system was recognized as user friendly, invaluable and a very powerful tool for the

[1] SADC: Botswana, Lesotho, Malawi, Mozambique, Namibia, Swaziland, Zambia, Zimbabwe
Eastern Africa: Ethiopia, Kenya, Tanzania, Uganda, Zanzibar

Fig. 35.2. Participants practising retrieval of information from CABPESTCD.

successful operation of the plant quarantine networks. By providing ready access to a vast amount of information on crop pests, the CD-ROM system is proving a very useful tool in decision making in pest management, plant import/export certification, training and extension.

Conclusions

This brief overview of the major constraints to plant protection in Africa has shown that the lack of ready access by national plant protection services to the vast amount of available scientific information and data on crop protection is proving to be a limiting factor in the effectiveness of these organizations. This is especially true in the areas of pest diagnosis, research, pest management decisions, plant import/export certification, training and extension. The primary function of plant protection services in Africa, apart from carrying out emergency operations against migratory pests when these occur, is advisory and regulatory. For these functions to be carried out effectively, these services have to rely on data and information generated from within but mostly from outside the country or region. This means having ready access to relevant and needed published scientific data. Since, for budgetary reasons, most national plant protection services cannot and will not be able to afford for the foreseeable future to maintain modern methods of information management such as on-line access, the introduction of CD-ROM technology has responded to the immediate information needs of the national plant protection services.

References

FAO (1986) *International Code of Conduct on the Distribution and Use of Pesticides.* Adopted by the 23rd Session of the Food and Agriculture Organization Conference, November 1985.

FAO (1988) *Directory of Current Research on Plant Protection in Africa.* Food and Agriculture Organization, Rome. 145 pp.

FAO (1991a) *Plant Quarantine in Southern Africa: Feasibility of Establishing a Sub-regional Technical Cooperation Network.* Food and Agriculture Organization. Rome. 22 pp.

FAO (1991b) *Plant Quarantine in Eastern Africa: Feasibility of Establishing a Sub-regional Technical Cooperation Network.* Food and Agriculture Organization. Rome. 23 pp.

FAO (1992) *Bibliography of Extension Publications on Plant Protection.* Food and Agriculture Organization. Rome. 119 pp.

FAO (1993) *Report of the Sub-regional Workshop for the Establishment of a Technical Cooperation Network for Plant Quarantine in SADC Member States.* Food and Agriculture Organization. Accra. 43 pp.

Networked Communications in Extension

36

R. Ausher

Ministry of Agriculture and Rural Development, Extension Service, Department of Crop Protection, PO Box 7054, Tel Aviv 61070, Israel
Fax: +972 3697 1664/E-mail: ausher@agri.huji.ac.il

Introduction

Being part of the Agricultural Knowledge and Information System (Röling, 1986), agricultural extension organizations shape the right interfaces within the technology generation, diffusion and adoption continuum. Working directly with both research and growers, agricultural extension identifies bottlenecks in crop production, protection and marketing, develops technologies, feeds knowledge with data, and formulates recommendations. The role of extension advisers or specialists is to assist farmers in their decision making. Decision making is considered the core of the management process. In this context, the grower should be perceived as an entrepreneur who also has the role of a producer and manager. He or she has to decide how best to use the available resources to achieve goals using various technologies. The conditions for decision making which farmers face are typified as multiple personal goals in an unstable environment. Farmers have to cope both with structured decisions, mainly in their production-based routine, and with unstructured decisions in their planning, business orientation and management routine. They invest inputs in crops in order to reduce the risks that might endanger their investment (Ausher, 1995). In the particular area of protecting crops and ornamental plants from losses inflicted by noxious organisms (insects, nematodes, mites, pathogens and weeds) despite intensive pest management efforts, about 50% of the world's crops are lost to these organisms, at an estimated annual cost of about $400 billion (DeBach and Rosen, 1991). The magnitude of these losses urges us to bring information and knowledge closest to field-level decision making. Computer-communicated data, information and decision support systems are most appropriate aids to assist farmers in their decision making. Agricultural research and extension are expected to generate and diffuse crop production and protection technologies which, when adopted by farmers, could improve their production, working and living conditions. Given more timely information and knowledge, subsistence farmers in the developing world could turn gradually into commercial producers and next into agricultural entrepreneurs (Ausher, 1995).

Objectives

The premise of this review is a survey of the present situation and performance of networked communications in extension. It is based on the experience gathered by the agricultural extension service in Israel in several crop and animal husbandry areas. The review attempts to identify success and constraining indicators in order to enrich the current knowledge base and to develop guidelines for the establishment and improvement of networked communications in extension with focus on the crop protection domain. The suggested approach addresses both technical and managerial issues pertinent to the incorporation of networked communications in extension crop protection, and particularly in extension plant pathology.

Agricultural Extension in Israel

The extension system

As part of the Ministry of Agriculture and Rural Development, the Agricultural Extension Service employs 260 professionals, assisted by 70 support and clerical staff. The organization provides advisory services to a clientele of 30,000 growers. This farming population consists of private family farms (20%), and two types of cooperative farms (80%): Kibbutzim (collective farms), and Moshavim (family cooperatives). The extension system covers all crop and animal husbandry production branches. Four service disciplines, namely crop protection, irrigation and fertilization, farm machinery and farm management support the crop production units. The service is a two tier organization. The technical departments are located at headquarters in Tel Aviv and they accommodate the state extension specialists. Front line activity is carried out by regional extension specialists stationed at nine regional extension offices. The service reaches out to farmers through one-on-one field visits, and group activities: group visits, pre-seasonal and season-end sum-ups, seminars, courses, as well as publications, videotapes and computer programs. All professionals are involved in both field experimentation and technology dissemination. The extension system requests and encourages close cooperation between crop husbandry specialists and their counterparts in crop protection, irrigation, farm machinery and farm management at both state and regional levels. Farmers are not charged for field visits but their regional and state associations contribute annually around 10% of the service's budget and support the field trial programme. Extension specialists collaborate with researchers in the framework of joint field experimentation programmes defined by joint crop teams. In the last couple of years, the Ministry of Agriculture and Rural Development attempted to privatize the extension services. This move was met with strong resistance by employees and it was unsuccessful. The service intends to adopt a market-oriented, customer-sensitive approach, vying for customers, and signing branch-wide service agreements with clientele, with individual growers, and with their associations. This approach has to be staged in spite of severe and continuing personnel cuts which have affected the extension service during the last decade. During this period, its strength has been reduced from 550 to 330 employees.

Promotion of computer-supported extension

The service adopted the development of a computer-supported extension approach as a systematic policy since the late 1970s. The promotion of computer-supported applications in the extension system was geared to support the professional abilities of the individual extension specialist with personalized tools, and to support the extension system with general and standard tools. The main thrust is definitely software-oriented, involving the entire organization moving step by step from routine office use toward the sophistication of the message created by the specialists and its diffusion. Early adopters among both professionals and policy makers have been identified, usually among those who deal with problem-solving and handle quantitative data in their day-to-day activities (Ausher, 1995). Emphasis was put on the identification of problems at the field level with focus on the farmer:extension interface rather than defining more complex topics at the extension:research interface. So far, no in-house communication network was established. The introduction of computer applications and their conversion into extension methodologies has been closely monitored by the system. A large training effort was developed, courses being offered to both clerical staff and professionals.

Department of Crop Protection

This is one of the technical units of the extension service. It comprises 30 professionals. Seven state extension crop protection specialists are located at headquarters. They are in charge of the following crop areas: field and industrial crops, vegetable crops, herbs and spices, flower crops, deciduous fruit crops and grapes, subtropical fruit crops, and citrus. Twenty two regional extension crop protection specialists are located at nine regional extension offices and their range of specialization follows the major crops of their areas. Crop protection specialists manage all three crop protection disciplines, namely plant pathology, entomology and weed control. This set-up enables a smooth interaction with growers and crop specialists enhancing the capability to lead integrated pest management programmes, and handle with competence field diagnostic and pest control problems. A government-mandated reduction in pesticide use is the main objective of the group and it is achieved through the implementation of supervised control, fully-fledged IPM and biological control practices on most crops grown in the country. Crop protection extension specialists spend about 30% of their time on one-on-one field visits, 20% on joint visits with crop production extension specialists, and 40% on field trials and surveys. The department is responsible for the formulation of pest control guides and is involved in the pesticide registration process.

Promotion of computer-supported crop protection extension

Extension crop protection and plant pathology rely on collection of data and handling of information. These are information-intensive activities. The handling of disease monitoring data provides extension plant pathology programmes with a unique qualitative edge. By utilizing quantitative pest scouting data instead of random observations, the dialogue between growers, scouts and extension evolves into a more professional interaction. The very existence of series of field data makes extension plant pathology suitable for computer support, promising high rates of return to any computerization effort. There are many new information technologies under development that can greatly benefit extension plant pathology. These include spreadsheets to handle field records, weather predictions, degree-day calculations, databases comprising pesticide registration data, crop and pest models, and statistical analyses. Networked communications could turn these tools into easily accessible ones for growers, pest scouts and extension specialists. The following listing has been compiled of applications which are accessible and utilized by extension crop protection specialists in the framework of their advisory activities (Y. Melamed, 1996, personal communication).

- Computerized database of registered pesticides in Israel.
- Sunflower Disease Control Advisory.
- Wheat Disease Control Advisory.
- Computerized recording of pest monitoring, for crop protection extension specialists.
- Computerized database on selectivity of pesticides to beneficial insects.
- Whitefly – University of Florida, multimedia program.
- European corn borer, decision support system.
- Apple pest management – University of Pennsylvania.

- DISTRAIN – training program for pest scouts.
- Fruit flies in Africa – taxonomy program.

Networked Communications and the Pilot Project

Extension-related networks established in Israel could be divided into three categories: (i) farmer-operated networks; (ii) in-house extension-operated systems (Intranet); and (iii) access to Internet services. The second of these has led to the development of a Pilot Project in communication.

Farmer-operated networks

Regional networks

Farmers' regional enterprises were established by the kibbutz system in the 1960s. They provide central services such as storage, sorting, packing, ginning, marketing and computerized management, book-keeping and financial services. The computer units of these enterprises have established communication networks to the member farms for the interactive utilization of farm and branch management programmes. However, most of these systems do not carry technical information on cropping or production-level decision making. In the past, several regional networks contained technical information especially on cotton growing. A partial and empirical cotton production model containing irrigation, fertilization, pest control and defoliation components was in wide use. Actual crop development was monitored on sample plots and compared to normative, multi-annual development curves. The demise of this project was caused by its heavy development and maintenance costs and a sudden shift from 'Acala' to 'Pima' cotton varieties.

Extension-coordinated regional networks

A new regional network initiated by extension and regional growers' association has been launched in the Yezreel Valley. Forty collective and cooperative farms involved in the production of field crops are members of the TVTEL network. They are charged for software adjustment and system maintenance, and for actual access time. Basically the program acts as a Bulletin Board system providing e-mail services and access to Internet. Its technical component is based on normative crop costing, price lists of inputs and crop-related technical messages. Regional field crop, irrigation and fertilization, farm management, and crop protection extension specialists provide the technical information and messages. Regional growers' association and the state Cotton Production and Marketing Board are also suppliers of information. Crop protection routine is based on information gathered at the weekly meetings of regional cotton pest scouts and extension personnel, and at the weekly state-level meetings of Cotton Production and Marketing Board and state-level extension staff. The main findings and the formulated recommendations that ensue are circulated by e-mail to all member farms. The program is technically coordinated by one of the regional irrigation extension specialists and it is run on an experimental basis. It is followed with great interest by both growers and extension (A. Asencot, 1996, personal communication).

State-level networks

Dairy cattle. 'Doors' is a closed Bulletin Board system. It is serviced by the software house of Kibbutz Brur-Hail, and operated by the Milk Production and Marketing Board. Subscribers are charged US$ 100 per month. The system connects the grower with three main information sources which bear on day-to-day decision making: (i) the food centre, providing custom-tailored foodstuffs; (ii) the herd book, consulted mostly on a daily basis; and (iii) the milk quality control laboratory and quantitative marketing report service. Extension specialists cannot access hidden farm-level files but they do access the e-mail system. The technical intervention of extension specialists did not turn into a success story. Although there is room for updated messages on nutrition, fattening, and general information on production problems, extension specialists were slow in putting out updated messages due to an over-hierarchical clearance system. Soon obsolete information was met with indifference by growers and the section was closed (M. Rosen, 1996, personal communication).

Beef cattle. The attempt to establish a similar network for the beef cattle industry did not thrive. Since this industry is not organized around an active database like the herd book, there is no daily two-way flow of data, and subsequently the demand for information relevant to decision making is limited and curtailed to operate under low frequency.

Poultry. The network established for the poultry industry was not viable. The cost/benefit ratio of the system turned out to be negative. The network linked the food centres, extension and producers. However, two main sources of data, the veterinary laboratories and the slaughterhouses were not included. The extension service contributed weather information and normative costing but the latter was concomitantly available free of charge in a competing manual version. Users, charged US$ 30 per month for sign-up and for connection charges, found the system much too expensive and the information only partial. Actually, information from the food centres reaches growers more efficiently through facsimiles. One of the attractive attributes of the dairy cattle network, the brokerage of livestock and equipment was not tried out (Y. Aharoni, 1996, personal communication).

Flower crops. Almost ten years ago, a private software house, ELNET, established a videotex network for the flower production industry. The network contained daily marketing information for most growers who export their produce to the auctions in the Netherlands, Belgium and Germany. The system was used, sometimes daily, by growers to access the system and take decisions with respect to agrotechnical practices and harvesting according to the market situation. The extension service accessed the system and put out messages on a regional basis. This caused redundancies which could have been easily resolved by contributing extension-generated information on a crop basis. First, regional extension specialists sent in their observations to the state level, and subsequently state extension specialists formulated and put messages into the network. The information was sought by the growers but they complained about cumbersome access, poor communication lines and the vagaries of printing out information. Finally, due to this pressure, the Flower Board downgraded the technology level from the videotex network to a facsimile system. Although, the information from the marketplace reaches growers earlier (at around 1400 h instead of 1800 h as before), the information cannot be processed since it is not moved through file transfer. With the adoption of facsimile transfer, extension has ceased to send information and messages to growers. Growers are content with the straightforward performance of facsimile connection but they are increasingly aware of the fact that the system precludes data storage, processing and analysis. Since persisting growers cultivate on increasingly large holdings, their need for better management tools rather than for mere information becomes a major thrust. It is expected, therefore, that in the long run, a literate network containing extension-contributed information, sustaining file transfer and enabling current analyses of individualized information would replace the existing communication system (D. Wolfson, 1996, personal communication).

Lessons Learned

The experience accumulated with the establishment and operation of both farmer and extension-operated networks highlights generic success and failure indicators. These could be formulated as requirements for successful wiring up of prospective extension-related networks. They also bear on extension crop protection and plant pathology-centred programmes. The following are the main requirements to be met by networked communications in extension.

- Running in local language.
- Itemized billings, no sign-up, line or monthly fees.
- Easy login and logout.
- Closed network protecting users and their files but open to external networks such as Internet.
- Flow of comprehensive and updated information addressing farmer's real decision making needs.

- Clear added value.
- Extension specialists regarding the update of information flowing in the network as a day-to-day responsibility.
- Extension service regarding the network as a marketing channel, displaying information on activities and staff availability to promote off-network contacts;
- Simple clearance of messages contributed by extension specialists.
- File transfer of quantitative data to enhance its processing and analysis.
- Reliable server.

In-house extension-operated systems (Intranet)

The extension service has launched a new Pilot Project aimed at the establishment of an in-house communication network relying on e-mail technology and stimulating links with the external technical establishment through the Internet. Two technical divisions at headquarters have been earmarked for joining the pilot phase: the Floriculture department (Wolfson and Gelb, 1996) and the Farm Management division (Marom, 1996). Both units are reasonably endowed with extension specialists well-versed in microcomputer technologies. The experience of these two units will be closely followed up in order to stretch the lines of the network to the whole organization. The Floriculture department accommodates a group of 35 specialists who are backed up by three collaborating state extension specialists in crop protection, irrigation and fertilization, and farm machinery who all started tinkering with computers many years ago. The flower crop industry has developed a highly dynamic and aggressive research and development programme based on the massive annual introduction of new crops in order to present a rich and competitive bouquet on the marketplace. The following parameters in the area of knowledge handling are characteristic of the activity of this group of professionals. They render the group to be well-adapted for an easy assimilation of an in-house communication network.

- The need to find quick solutions to new and perplexing field situations and to share them with peers.
- Frequent consultation among specialists.
- State extension specialists coordinate a widespread field experimentation program.
- The need to organize a massive flow of information – the outcome of steady introduction of new crops and varieties.
- Most members of the group have access to modem-equipped microcomputers.
- Growers and their representatives are aware of the need for efficient knowledge management.
- The establishment of a database is underway, containing information and know-how on flower cropping.

- Since most flower crop production extension specialists are equipped with cellular phones, a comparative study of the communicative efficiency of mobile phones and the network poses a methodological challenge.

The second group of professionals earmarked for the evaluation of an in-house network is that of the farm management extension specialists. As members of the Farm Management division, they are:

- large-scale developers of software, and spreadsheet-based applications, graphs and textual material.
- extensive users of quantitative data and models, and networks could enhance both their input and output.
- all well-equipped with computing facilities and are experienced users (Y. Levi and D. Wolfson, 1996, personal communication).

The Pilot Project is aimed at the integral assimilation of networking by two technical units. The monitoring of this process will focus on: (i) the system-wide payoffs of e-mail; and (ii) exploring for knowledge sources offered on the Internet.

The follow-up will pay special attention to bottlenecks in the assimilation of these technologies by extension specialists, training needs, implications on the organization and management of the extension system, changes in the time allocation of the involved extension specialists and in the efficiency of their advisory performance. A small group of extension executives had been made responsible for the close follow-up of the Pilot Project. They will listen to frontline specialists who are usually the first to spot shifts. Unfortunately, the pilot scheme comes at a time of financial constraints in which the extension service faces difficulties in obtaining the necessary resources to launch this pilot activity smoothly.

Access to Internet services

The Pilot Project launched by the extension service attempts to examine the effectiveness of three main Internet-offered services: e-mail, discussion groups, and search of databases (Waksman and Harkin, 1996). Further, the service intends to establish a site on Internet (S. Segal, 1996, personal communication). At present, around 15 extension specialists are already accessing the Internet on an individual basis. The pilot project will encourage all extension specialists of the Floriculture department and Farm Management division to access the system. State extension specialists are expected to lead the incorporation of Internet-related activities into the routine of the crop teams chaired by them. They are counted upon to act in the forefront as both users and suppliers of information.

Conclusions

The lessons learned from the already operating networks are of a generic nature, and will be used in the design and operation of any prospective networks that bring together growers and extension workers. In practice so far, research has not yet adhered to the currently operating systems, and most networks focus on the extension:grower interface. Due to funding constraints, the extension system is embarking on the information highway on a carefully designed pilot project basis, which will be closely observed by extension management. Network-derived benefits will be converted system-wide into extension methodologies (Varner, 1996). These will be gradually shared with additional units within the service. Several integrative crop-oriented networks already contain crop protection components and could contribute to the gradual evolution of fully-fledged crop protection and plant pathology systems. In the light of continuing staff reductions, the extension service expects that new opportunities will abound from networked communications. Hopefully, communication networks could make up for reduction in staff, and increased travel costs, and help to set a bolder course, instilling a spirit of service industry and providing the right tools for its support. When the dust settles, with good luck and good work, the extension system will have a clear picture of how to systematically support both internal and external communication networks, the final aim being to professionalize personnel activities and to improve their advisory capabilities.

Acknowledgements

The author is grateful to Y. Aharoni, A. Asencot, E. Gelb, Y. Levi, D. Marom, Y. Melamed, M. Rosen, S. Segal, D. Wolfson, staff members of the Extension Service, Ministry of Agriculture and Rural Development, Israel for the valuable information they have shared with him and for challenging discussions.

References

Ausher, R. (1995) Microcomputers in support of the generation and diffusion of agricultural technologies in developing countries. *Outlook on Agriculture* 24, 147–150.

DeBach, P. and Rosen, D. (1991) *Biological Control by Natural Enemies*. University Press, Cambridge. 440 pp.

Harsh, S.B. (1996) Farms information systems: current trends and future perspectives. *Proceedings ICCTA'96/VIAS/NNAA Congress on ICT Applications in Agriculture*, Wageningen, pp. 3–17.

Marom, D. (1996) Evaluation of computer-use benefits for farm management extension advisors. *Proceedings ICCTA'96/VIAS/NNAA Congress on ICT Applications in Agriculture*, Wageningen, pp. 573–575.

Röling, N. (1986) Advisory work in the context of the future evolution of agriculture: conceptual aspects – the European viewpoint. *Proceedings Seminar on Perspectives for Advisory Work in the Context of Future Evolution of Agriculture. EUR Report 11267.* Louvain, pp. 101–116.

Varner, M. (1996) Trends in electronic communication methods for cooperative work groups. *Proceedings ICCTA'96/VIAS/NNAA Congress on ICT Applications in Agriculture*, Wageningen, pp. 177–181.

Waksman, G. and Harkin, M. (1996) Green telematics: is Internet a breakthrough for farmers and extension workers? *Proceedings ICCTA'96/VIAS/NNAA Congress on ICT Applications in Agriculture*, Wageningen, pp. 182–188.

Wolfson, D. and Gelb, E.M. (1996) The role of extension in developing information systems for floriculture producers. *Proceedings ICCTA '96/VIAS/NNAA Congress on ICT Applications in Agriculture*, Wageningen, pp. 119–122.

37

Modern Information and Communication Needs in Agriculture for Developing Countries

S. Nagarajan

Directorate of Wheat Research, Indian Council of Agricultural Research, Post Box 158, Kunjpura Road, Karnal 132001, Haryana, India
Fax: +91(0)184 251390/E-mail: dowr@x400.nicgw.nic.in

Introduction

The developing countries spill over several continents and are diverse in their agricultural development, standard of living, technical development and endowment of natural resources. These botanically rich nations are economically poor but are being increasingly viewed by the developed nations as their future markets. The market force that operates in these nations and creates demand for information based products is very diverse, whilst the cultural, social and linguistic needs complicate the product development, as information needs to be made available in several understandable forms. In this chapter the discussions will be confined to the situation in South Asia, but the concepts are applicable to many other similar situations.

Agricultural Renaissance in South Asia

Agricultural production in British India (1900–1947) increased at a very insignificant rate. Agriculture to a large extent remained rainfed and scientific advances were limited. Only by extending the cultivated area and marginal access to irrigation was total food grain production kept ahead of population growth. The real great leap forward in the production of wheat, rice, potatoes, eggs and milk, occurred after 1965 and production is still increasing at a rate far above that of population growth (Fig. 37.1). This was primarily a consequence of an increase in productivity, better fertilizer and other input usage, and access to high quality seeds of improved varieties, crossbreed animals, improved poultry strains, etc. The rapid agricultural growth generated income, promoted manufacture, transport, sale and trade. Over the last thirty years the managerial skills of the farmer have undergone a change. Farmers view agriculture as an industry and several agricultural advisory services, sale points, opportunities for consultancies and information transfer have opened up. We are at the beginning of the development of a highly technological agricultural industry as far as India and, to a large extent, Pakistan are concerned.

Communication Needs in Graduate Education

In India there are 29 Agricultural Universities with affiliate colleges in Agriculture, Horticulture, Forestry, Agricultural Engineering, Animal Sciences, Dairy Sciences, Fisheries, Home Science and Business Management, etc. Nearly 10,000 students complete their basic degree in agriculture annually and many of them proceed to advanced degrees. There is a greater demand for new educational communication tools, audiovisual cassettes, especially for topics such as advanced virology and molecular biology, conservation of ecological resources and others. The Discovery channel of the TV is rated by many youngsters as a good source of new knowledge. Previously, journals that used to arrive by mail were thc only source of scientific information. Reading and assimilat-

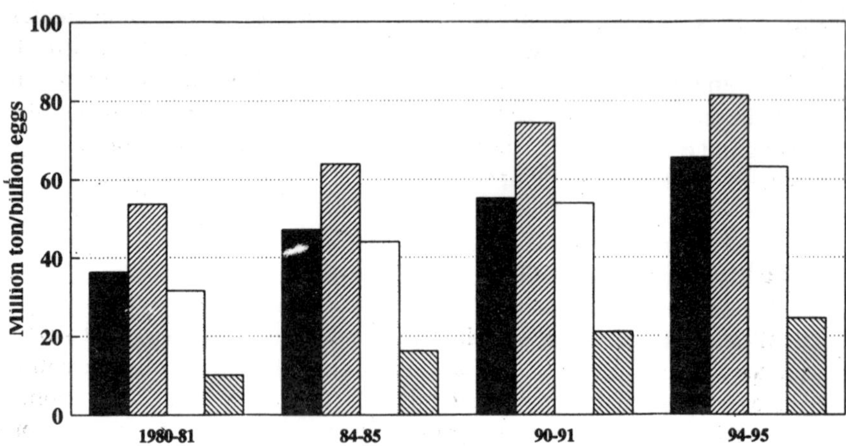

Fig. 37.1. Growth in production of rice, wheat, milk and eggs as an indicator of overall agricultural growth in India. (From Anonymous, 1996.) ■ Wheat; □ milk; ▨ rice; and ▧ eggs.

ing this information for project planning and appraisal was quite time-consuming. Nowadays this process is faster and literature scanning using CD-ROM systems of the major abstracts and access to several other information retrieval systems are being used in a routine manner by those working in our premier institutions. Some institutions such as the Indian Agricultural Research Institute, New Delhi, extend the facility to literature scan to those who do not have access to it at a reasonable cost. Also, through the AGRIS computer network, nearly 75 research institutes under the Indian Council of Agricultural Research (ICAR) and the State Agricultural Universities (SAUs) are connected through e-mail, which permits students and researchers to scan the literature available in one library whilst sitting in their own institutions. In my opinion, this is just at the beginning and will expand in the future.

Information on weather is an important input in deciding several agricultural operations. Weather parameters have been recorded daily at several places for the last hundred years by the Indian Meteorological Department and these data form a huge information base. Through the Indian satellite programme the Department of Science and Technology, Government of India has linked several automatic weather recording stations with the mainframe supercomputer at New Delhi and has provided on-line connection to several users to tap the latest weather information. Despite all this, the user list of this information is short as there is a lack of formal training in the usage of modern systems of information flow.

Geographic Information System for Sustainable Agriculture

The Geographic Information System (GIS) is a powerful tool in measuring ground surface features such as the type of crop cover, area cultivated, land classification and in measuring the productivity of the system based on greenness index. In India, satellite images are used for several agriculture related issues (Navalgund *et al.*, 1991). For example, the Landsat scans covering South Asia have been successfully used to separate out the wheat fields severely damaged by *Puccinia striiformis* f.sp. *tritici* in the Punjab of Pakistan from the healthy wheats of the Indian Punjab (Nagarajan *et al.*, 1984).

Cotton is one of the most important commercial crops grown in India where the area cultivated is over 7.5 million ha. Obtaining an accurate estimate of this area was attempted by the Space Applications Center, of the Department of Space, Ahmedabad, and it was concluded that satellite data acquisition can be used for assessing the area under cotton with a high degree of confidence and the accuracy was greater when the scan is made at the critical crop growth stage (Ajai, 1992). Similarly, satellite images have been used very successfully to assess the area under citrus and those orchards suffering from decline disease. Citrus growing is a big business around Nagpur in central India and has been suffering for the last two decades from the complex decline disease, which makes the leaves pale green and, under extreme conditions, the twigs, branches and even the trees die. Citrus decline is one main reason for the drop in fruit production in this savanna area. Indian Remote Sensing data were used to differentiate the less green decline affected orchards from the healthy by performing a supervised classification with LISS-II data (Sinha and Karale, 1992). It was noted that diseased orchards accounted for 85 ha or 5% of the total area under citrus in that belt. Such diagnostic uses for loss assessments, and for identifying areas that are disease free for fresh planting, are going to be a new area of remote sensing consultancy. This technology can be applied for several situations in the tropics where commercial plantations are declining and need rejuvenation.

Extension and Education Oriented Information Flow

Communicating good agronomy practices, varietal details, etc. to farmers occurred mainly through radio in the 1950s and by film and publications in the 1960s. When the green revolution was sweeping India, the need was felt for a more convincing and efficient medium of communication. Through a joint project between ICAR and ISRO an attempt was made for satellite TV linkage with villages to telecast agriculture-related programmes. It was an instant success. This paved the way for video filming of agri-related information, skills and knowledge such that the presence of private entrepreneurs in this industry is now substantial. For high-tech aspects of agriculture such as drip irrigation, plastic culture techniques, prawn culture, or mushroom growing, brief audio documents are available. There is a growing demand from agro-consultancies and the agri-related industries, as these films are used for educating their clients and users.

In the 1960s, audiovisual films were used to support printed information, pictures and flow diagrams. In their keeness to see the 'cinema' many villagers were drawn into the exercise of information transfer and understanding. The main means of information transfer was at farmers fairs, shows, and field discussions, where this see, feel, get convinced and adopt attitude was fostered. Nowadays, with surplus agriculture, intensive farming and the need to rationalize the input management, a different mode of information transfer is needed. Short term courses offered to elite farmers have been made attractive and many of them are interested in learning the finer aspects of commercial agriculture. Multimedia and menu-driven software is emerging as a new and powerful tool in information transfer. A multimedia document of twenty minutes duration would cost around Rs.100,000 and can be used for several aspects of advanced learning. Apart from being a novelty to farmers and county agri-specialists, this medium also permits them to rate and upgrade their own knowledge and skills. The DWR (ICAR) is in the process of making such an information base for widespread use. New creations have already been made for integrated pest management (IPM), rice and several other issues. Once a good multimedia CD is made, it can be given an audio text in several languages and therefore can be made at a competitive and affordable price. Interactive systems such as menu-driven programmes which use pop-up questions on the screen can lead to a pictorial or written information base. For example, many of the pest identification and taxonomic CD systems developed by CABI are ideal. However, in developing countries where the cost is higher, many do not have access to computer facilities with CD drives and those who do have this facility do not necessarily use these information systems for agricultural purposes. It may be necessary to educate the clients and their masters in their own country by organizing an area programme.

Efforts of the Indian Wheat Programme

The International Maize and Wheat Improvement Center (CIMMYT), Mexico and the National Agricultural Research Systems (NARS) interacted to undertake an analysis of multilocational data on pedigree details, yield information for wheat, etc., using several statistical packages. This information on the details of crosses/ varieties/genotypes was used in developing the Genetic Resources Information Package GRIP – I version 1, and can be periodically updated by adding new information. The GRIP for wheat has a main Menu with two systems of access to data base either by query or by other options (Fig. 37.2). The information regarding an accession can be obtained by feeding the cross

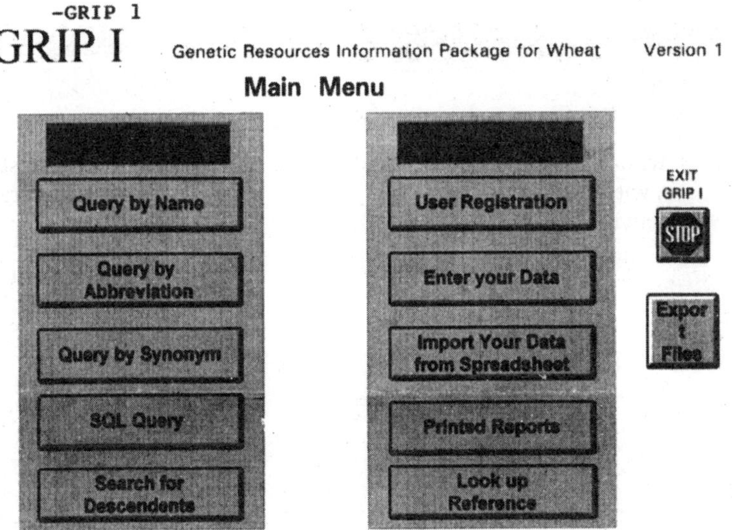

Fig. 37.2. Genetic resources information package for wheat – GRIP I.

identity (CID) or the selection identity (SID). In situations where this detail is not available, it can be obtained by giving the genotype name or the accepted abbreviation. Once the CID number is given, the genotype profile appears on the monitor (Fig. 37.3). In order to get the complete cross lineage details, the dendrogram of the Wheat Pedigree Management System (WPMS) file can be opened. For the same example, i.e. wheat variety Veéry (CID 7691), that was developed by CIMMYT at Mexico in 1977, six ancestral parental lineage details can be retrieved in the form of a dendrogram (Fig. 37.4). The genetic coefficient available in the information base indicates that Veery is derived from a gene pool involving at least twelve different wheats (Table 37.1). Using this GRIP software, similar information for a number of genotypes/varieties can be obtained on a worldwide basis. The package is open ended, and can be loaded with additional CID accessions and thus serves as a useful information base for plant breeders and those involved in crop improvement activities.

To What Use Can this Information be Put?

With the growing concern for ecological and environmental issues, we feel the need to cut down on pesticide usage by breeding disease resistant varieties. The information provided by WHIP (Wheat Improvement Programme), developed by us, permits:

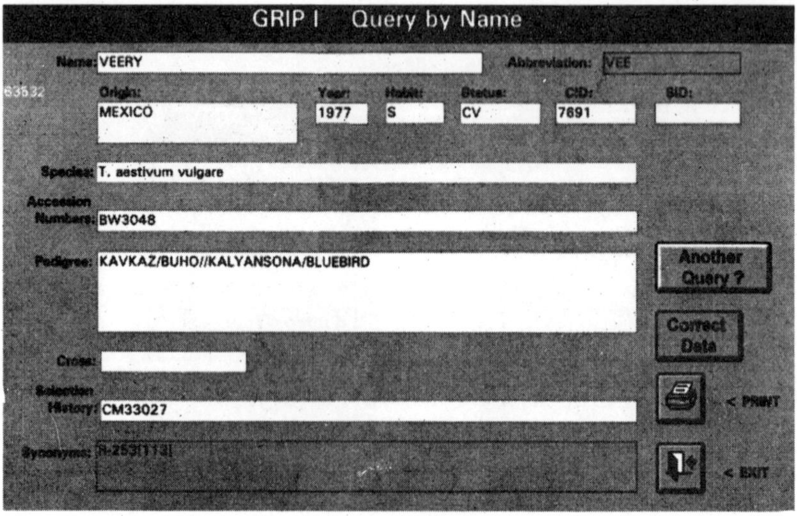

Fig. 37.3. Genotype profile of the genotype Veery.

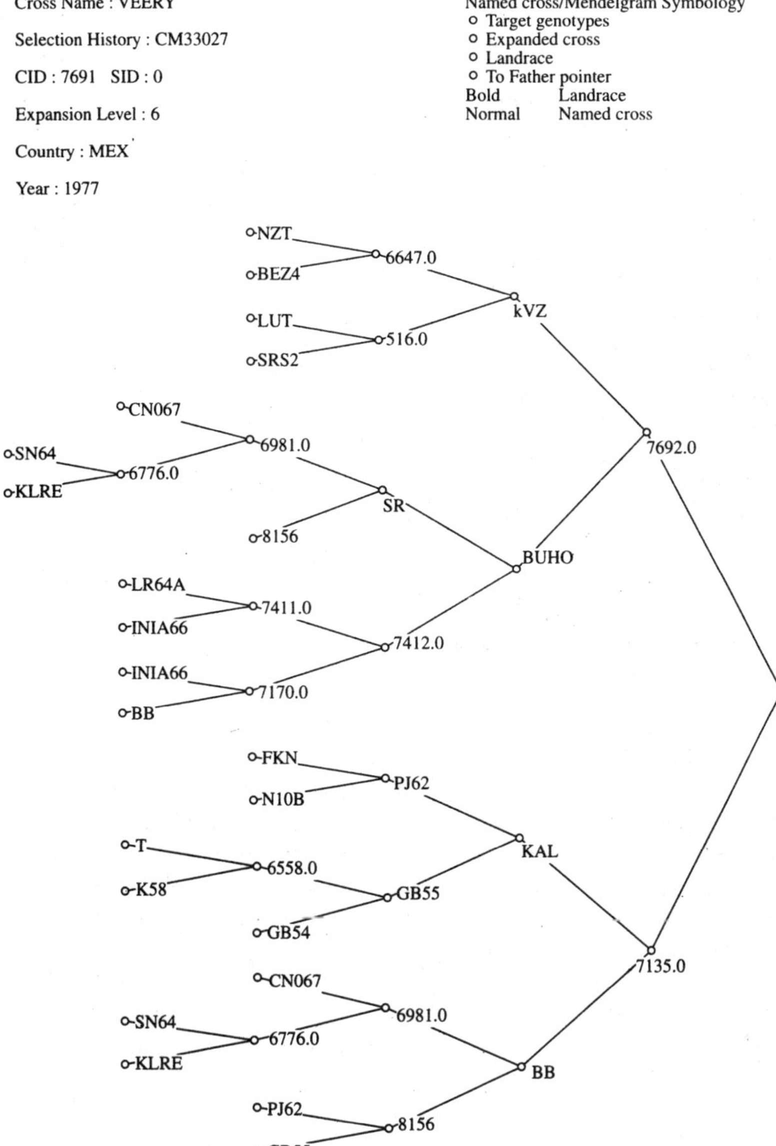

Fig. 37.4. WPMS Dendrogram for Veery.

1. Once a wheat variety becomes susceptible to a rust disease or other pathogen, it needs to be replaced. This boom and bust cycle (Vanderplank, 1963) repeats every 5 to 7 years in India demanding varietal change. Prior to

Table 37.1. Coefficient of parentage of accessions involved in the Veery development

Name	Abbreviation	Country	Contribution*
Ciano F 67	CNO67	Argentina	9.375000
Neuzucht	NZT	Germany	6.250000
Bezostaja 4	BEZ4	Soviet Union	6.250000
Lutescens 17	LUT	Soviet Union	6.250000
Skorospelka 2	SRS2	Soviet Union	6.250000
8156	8156	Mexico	6.250000
Inia F 66	INIA66	Mexico	6.250000
FKN	FKN	USA	6.250000
Norin 10 Brevor	N10B	USA	6.250000
Gabo 54	GB54	Mexico	6.250000
Penjamo T62	PJ62	Mexico	6.250000
Gabo 55	GB55	Mexico	6.250000
Sonora 64	SN64	NGA	4.687500
Klein Rendidor	KLRE	Argentina	4.687500
Lerma Rojo 64A	LR64A	Mexico	3.125000
Bluebird	BB	Mexico	3.125000
Timstein	T	Aus	3.125000
Kenya 58	K58	Kenya	3.125000

*Contribution (%) of various genotypes to Veery genome.

recommending replacement of a variety, it is necessary to determine if the new replacement genotype is genetically different from the existing one, and the extent of genetic variation that has been created. Erecting a spatial and temporal diversity is one of the important disease management strategies adopted by developing nations in mitigating the loss caused by plant pathogens. By quantifying the pedigree variation, an educated guess can be made on the choice of the correct substitute genotype.

2. At the moment the WHIP does not provide the information on the resistant genes (*Yr*, *Lr* and *Sr* etc.) that a given variety possesses. The scope of WHIP can be widened to infer probable rust resistance genes in these lines by creating appropriate software to compare the seedling reaction matrix of the various resistance genes (isogenic lines) with the genotypes with CID. Following Browder and Eversmeyer (1980), a comparison of matrixes can infer the probable resistant genes in a given genotype. These postulations will gain strength if supportive inferences are added by scanning the parental lineage of the genotype and checking the source of various resistance genes postulated from the already available information bases (McIntosh *et al.*, 1990, 1995).

3. With the PVR and IPR regimes becoming a reality, WHIP will permit the plant breeder to check the parental lineage, and verify if any registered materials are involved in the material he has developed.

Teleconferencing

Teleconferencing on important and urgent issues is a quick method of information gathering. Recently, for example the American Phytopathological Society had a teleconference on Karnal bunt (*Tilletia indica*), which has unfortunately spread to the US from the southern border. Since much of the information on this pathogen and the disease is available in India, the DWR participated in the e-mail conference. Access to global information through the government-run e-mail national network (nicnet), proved to be a major impediment in the efficient participation of Indian scientists in the conference. With time, this limitation can be overcome as many new companies are extending access to other networks and this will permit rapid shuttling of the information back and forth.

Identification Manuals

There is a growing need for accurate identification of insects, plant diseases, weeds and other pests in order to execute a correct management strategy. Broad-spectrum chemicals have been replaced by systemic and selective molecules. Correct diagnosis of the causal organism is a prerequisite to recommendations for control. The user friendly identification manuals developed by CABI have proved to be both timely and useful. If it was not for the high cost, these menu-driven identification and advisory systems would be of great use. The Indian Institute of Horticultural Research at Bangalore has also developed similar packages for several horticultural crops. CABI does not cover all the various tropical crops and their diseases in an appropriate form. Possibly there is scope for joint programmes with CABI, so that the products developed could address the regional market effectively. Plantation crops with their highly efficient production and marketing systems might be a first target for such a joint venture.

Simulations

Crop growth models are efficient tools based on scientific knowledge. They are primarily information based and are very useful in estimating production potentials, optimising crop production, input usage, predicting disease pest levels for prophylactic action, and assessing the environmental effects on the production system. In India, a wheat growth model WTGROWS (Aggarwal *et al.*, 1994) was developed to analyse the effect of climatic variables on the productivity of wheat in the tropical and subtropical environments. These models have been validated for making yield assessments and seem to be useful. These simulators are new information tools which can be made user friendly to meet the needs of agri-consultants who assist the farmers in several of their decision making processes.

References

Anonymous (1996) *Agricultural Statistics at a Glance.* 11th Brochure, Directorate of Economics and Statistics, Department of Agriculture and Co-operation. Ministry of Agriculture, Government of India, 140 pp.

Aggarwal, P.K., Kalra, N., Singh, A.K. and Sinha, S.K. (1994) Analyzing the limitations set by climatic factors, genotype, water and nitrogen availability on productivity of wheat. I. The model description, parametrization and validation. *Field Crops Research* 38, 73–91.

Ajai (1992) Cotton acreage estimation and condition assessment using IRS-data In: Karale, R.L. (ed.) *Natural Resources Management – a New Perspective.* NNRMS, Department of Space, Government of India, Bangalore, pp. 266–273.

Browder, L.E. and Eversmeyer, M.G. (1980) Sorting of *Puccinia recondita*: *Triticum* infection-type data sets toward the gene-for-gene model. *Phytopathology* 70, 666–670.

McIntosh, R.A., Hart, G.E. and Gale, M.D. (1990) Catalogue of gene symbols for wheat (1990 supplement). *Wheat Information Service* 70, 29–50.

McIntosh, R.A., Wellings, C.R. and Park, R.F. (1995) *Wheat Rusts: an Atlas of Resistance Genes.* CSIRO Publications, Melbourne, Australia, 200 pp.

Nagarajan, S., Seibold, G., Kranz, J., Saari, E.E. and Joshi, L .M. (1984) Monitoring wheat rust epidemics with landsat-2 satellite. *Phytopathology* 74, 585–587.

Navalgund, R.R., Parihar, J.S., Ajai and Nageshwara Rao, P.P. (1991) Crop inventory using remotely sensed data. *Current Science* 61, 162–171.

Sinha, A.K. and Karale, R.L. (1992) Preliminary investigation on area and health assessment of orange plantations using IRS-data. In: *Natural Resources Management – a New Perspective.* NNRMS, Department of Space, Government of India, Bangalore. pp. 274–279.

Vanderplank, J.E. (1963) *Plant Diseases: Epidemics and Control.* Academic Press, New York, London, 349 pp.

38

Electronic Publishing in Plant Pathology: Predicting the Unpredictable

R. Campbell and A. McLean-Inglis

Blackwell Science Ltd, Osney Mead, Oxford OX2 0EL, UK
Fax: +44(0) 1865 721205/E-mail: robert.campbell@blacksci.co.uk

Introduction

One of us was once assured by a French professor of business studies that the present is determined by the future, and that to make God laugh we should show him our 5-year plan. There are certainly plenty of examples of hopelessly wrong predictions in information technology to amuse us. For example, 'I think there is a world market for about five computers' (T.J. Watson, Chairman of IBM, 1943) and 'Computers in the future may weigh no more than 1½ tons' (Popular Mechanics, 1949).

To help us formulate more realistic predictions, the current fashion in management schools is to focus on purpose, people and process, rather than proffer detailed forecasts and meaningless mission statements. Those controlling the tattered remains of research funding should perhaps concentrate on general objectives linked to the purpose of their funding, and search out the best people to determine the exact paths and achieve those objectives.

The objectives in electronic publishing have been clear enough: to make information more accessible more quickly. To provide greater accessibility we must bear in mind price as well as technology. The market is demanding more value through improved production and delivery efficiency.

The latest forecasts from the European Commission on electronic publishing are safely vague, warning against the problem of information overload and the resulting disorientation. They urge publishers to develop systems which will enable users to access with relative ease the huge knowledge resources becoming available, enabling them to 'stand high on the mountain of knowledge'. They also predict rapid growth of the Internet from the current 50 million users, pointing out that even now a million new users per month are coming online (Baubin, 1996).

The Future

In the standard futuristic scenario of the 1980s, a plant pathologist would arrive at his laboratory in the morning, announce his arrival to his computer, and ask for his messages and the day's schedule. The machine's response would be 'spoken' or displayed on a flat screen, much like print on paper. The researcher might then ask if any comments had arrived from his co-authors on the first draft of a paper, transmitted to them earlier. Comments from co-authors and suggestions for additional text would be displayed, with the new material clearly differentiated on the screen from the original. Should the researcher ask the computer if anything relevant had been published in the last two days, references, complete with abstracts, would be shown on the screen. The researcher could request the full text of these and, if one referred to a paper of interest, he could ask to see its abstract and then the full text. At the end of the search, this last paper could be printed out for further perusal.

Such an information system would have kept the researcher completely up-to-date, with no need to visit the library. On transmission of the final draft of the paper to the preferred journal, the authors could be confident that their survey of the literature was complete. The journal's editor would then select referees from a database of appropriate candidates, showing how often and when they had last been used and their typical speed of response. The editor would easily be able to check whether the work submitted was innovative and original. If accepted, the paper would be edited electronically and speedily made available to the research community.

Such scenarios ignored the conclusion from John Senders after the first formal trial of the online scientific journal: 'We have visited the future and it doesn't work' (Senders, 1980). By 1978 the potential economies in computerizing journal publishing were being analysed (King and Roderer, 1978; King, 1980) and Singleton (1981) predicted many of the issues now facing us. Discussions then focused on the 'cross-over point', when electronic systems would become less expensive than print-on-paper, but we now no longer see the switch to electronic publishing as so clear cut. Publishers have, for example, incorporated digitization in their production systems for several years, to lower the cost of origination and thus maintain the viability of journals despite falling circulations (Bodinham and Campbell, 1992).

The Present

This vision or promise of the future could be said to have largely determined the present. As expected, information technology and telecommunications systems have relentlessly become more powerful and cheaper. The Internet has brought the world together, fulfilling the promise of the 'global village' more quickly than we might have imagined in the early 1980s.

Speech recognition and computerized translation systems have progressed relatively slowly. Although a range of prototype screens have been produced, the researcher's physical relationship with his machine has not developed as quickly as predicted. Perhaps we shall adapt; who knows how children brought up with the screen will relate to it? We do know, for example, that the process of reading text has developed through greater availability of and access to text. Before the seventh century, few could read silently; some centuries earlier Saint Augustine professed astonishment that Saint Ambrose could read without moving his lips. From the Middle Ages until the late 1700s we read 'intensively'. The few books in existence were read over and over again, usually aloud and in groups. Reading 'extensively', rushing from one item to the next and rarely rereading, is a relatively modern process (Birkerts, 1996).

Let us consider the three elements of the information system: discovery, delivery and publication itself.

Discovery

The development of searching systems and the interlinking of the research literature has taken place rapidly. The process of discovery, of finding the required item, has been the first to benefit from online access to databases. The World Wide Web (WWW) has encouraged the development of natural language search engines and more recently 'intelligent' agents, software that will look for the required information on behalf of the user.

The suppliers (publishers) are following what might be termed a 'neutral database strategy' (Page *et al.*, 1997). Journal articles ready for publication can be set up as PostScript files, able to generate both hard copy and Portable Document Format (PDF) files for electronic document delivery. At a minimum, article headers (title, authors, abstract, key words) might also be coded in Standard General Mark-up Language (SGML) and supplied in this form to current awareness/secondary databases.

The next step is for publishers to work together to establish a system for hypertext linking between article files. Very soon, some publishers will have linked their databases to the major secondary databases so, for example, a scientist searching Medline and needing more than an abstract will be able to click through to a file of the full article. When coming across a reference of particular interest, another click can take the user back to Medline to look at the abstract of that reference.

Secondary databases in other subjects, including plant pathology, should follow when the market demands. Such systems would offer added scientific value; a plant pathologist could use hypertext links to move from reading a paper to genome databases, an image library of diseases or an Internet newsletter.

The development of properly trained agents may make search engines and linking technologies obsolete (Hemingway, 1997). The traditional engines build indexes of the words in the text database. When a user searches for a word, or a Boolean expression, the engine consults its index in order to locate all occurrences of the word(s) in the database and builds a results list. When used by a skilled information scientist, these traditional techniques yield result sets which are small and highly relevant. In unskilled hands, such tools are cumbersome, and the user risks retrieving an unmanageably large results set or missing important results.

The more modern engines apply some form of ranking to the results set, so that even if an unskilled searcher has retrieved a huge results set, he will only need to inspect the first few hits to find his most meaningful results. The simpler ranking engines count how often the search term appears in each hit article, whereas the more modern engines use probabilistic techniques to compute relevance, and may go further to suggest related articles not containing the search term. These engines judge the context when matching the search term to the words in the database, hugely benefiting unskilled searchers.

Intelligent agent technology is being tested by publishers. Intelligent agents also have a fuzzy nature to their searching. Agents are 'trained', rather than given short discrete searches to perform. Although still not as definitive as a Boolean search, such training allows the user to be more expressive in composing the initial search requirement than in a probabilistic search. Agents' training is continuous. The agent should detect which of its search results are pursued by the user, and adjust its parameters so as to perform better next time the search is run. This feature should prove valuable for information scientists who are conducting repeat searches over time, looking for new articles. Agents may also be programmed to alert the user when a new article appears. Alerting services are not new, but information needs change over time, and the hit lists returned by an alerting service which acts on a old search expression will gradually diminish in usefulness. Agents, with their continuous revision, should give good results for as long as the user needs information on the agent's topic.

Delivery

Speed of delivery is certainly a problem at present. Many publishers hold a database of PDF article files on a server and have a reasonably secure system of charging (by electronic subscription, account or credit card). If a user in Europe accesses a long article with half-tones it can take several minutes to transmit; if the user is in the United States at a busy time, or is using a modem, the transmission time can exceed one hour. The most popular PDF software, Acrobat, has been improved to allow page by page transmission, enabling a user to read the first page whilst waiting for more. Fibre-optic cabling and new transmission protocols will also reduce transmission times.

Although the CD-ROM approach has also been followed, with publishers offering 'archival' CD-ROMS of past issues, and secondary providers delivering collections of articles to libraries, the general view is that the market will prefer online access. The signs are that users are prepared to put up with delays on the Internet when accessing a journal, for the sake of access to the very latest articles.

The much discussed improvements to the Internet will doubtless improve matters. Publishers will also be setting up mirror site servers around the world and using third parties such as OCLC, the North American library service, to provide Intranet facilities. Large consortia of libraries may take article files and supply their Local Area Network (LAN) from their own servers. The British Higher Education Institutes (HEIs) can expect considerable improvement through a new five year funding programme recently announced by the Joint Information Systems Committee (JISC) (Brindley, 1996).

Publication

As outlined in the 1980s scenario of the future, improved telecommunications are now enabling authors to write papers more quickly. Certainly, journal publishers are being offered more papers than ever before (Campbell and Halliday, 1985). Editorial offices use special management software to help them process papers more swiftly (Page *et al.*, 1997). The current widespread belief that electronic publication can achieve faster refereeing of papers is sometimes used to argue that the electronic journal is superior to the traditional journal. This is part of the artificial debate on the merits of the electronic journal versus the conventional print on paper journal. Established journals are being produced digitally and offered in electronic formats as well as hard copy. Editorial office software and e-mail can enable the editor to run the refereeing process more efficiently but this will apply to any type of journal, paper or electronic. The time needed by the referee is independent of the medium of publication.

A new primary publishing system in high-energy physics has achieved rapid, low-cost publication (Ginsparg, 1996), but by abandoning refereeing and relying on a large grant to run the 'e-print archives' at Los Alamos National Laboratory. The system is entirely automated, including submission and indexing, and allows access via e-mail, anonymous ftp and the WWW. An important feature is the formal mode of communication in which each entry is archived and indexed for retrieval. Conversely, Usenet newsgroups and bulletin boards represent an informal mode of communication with unindexed entries that typically disappear after a short time. There is probably a degree of quality control, as the cost of most experiments in high-energy physics requires a fairly rigorous approval system before any experiment leading to an 'e-print' abstract can be started. Although the Ginsparg model has not been followed in subjects where peer review is seen as the basis of primary publication, it will undoubtedly stimulate change.

Most publishers are currently preparing for the unpredictable, if that is possible. A recent report from the European Commission (Baubin, 1996) forecasts that although electronic publishing currently makes up around 1.5% of the total science, technology and medicine (STM) market, by 2000 it will reach 15%. We are told that 'electronic publishing can be profitable, but few publishers will reach a breakthrough in profitability before 2000'. There is no significant market for electronic subscriptions, yet most major STM publishers have expensively streamlined their production process to generate PDF files. These are stored on servers and accessed, using new customer management software, over the Internet, either direct or via intermediaries offering electronic subscription facilities. These systems can also handle electronic document delivery, allowing a publisher to sell a document identified by a user through one of the discovery services improved by publisher cooperation.

What Can Be Assumed?

Publishers are assuming that the basic product – the refereed research paper – will remain the same, though coded for ordering and tracking usage, and that publication through the most appropriate current technology will continue. The popularity of Paul Ginsparg's pre-print server has worried other physics publishers and there has been a proliferation of electronic journals (Okerson, 1996), but as Tenopir and King (1996) have pointed out, some of the decisions concerning electronic publishing may be based on old myths. For example, the universal cost per reading has remained relatively constant over the last 20 years (with inflation considered). 'Costs are merely shifting from one participant to another – individual to library or institution – or from one kind of cost to another kind – subscription to interlibrary loan or document delivery'.

Journals have been dismissed as an inefficient and expensive means of transmitting information about research, with citation data taken literally to mean that most papers are read little if at all. Yet Tenopir and King over 20 years of research have shown that journal articles average nearly 1000 readings, and the average amount of reading by university scientists has apparently increased from 150 articles per year in 1977 to 190 today. The educational role of journals is also too easily forgotten by academic administrators who tend to relate the provision of journals only to the requirements of their research community, as demonstrated in the influential Follett review of academic libraries in the UK (Follett, 1993). Trials have shown that students, including undergraduates, are a significant component of overall usage. The Follett report failed to pick up the possibility that the successful researcher who can attract large grants will have more influence than the librarian over the nature and purchasing of the new information systems. Pearce (1996) has pointed out this opportunity for publishers who can think themselves out of simply publishing more pages and upping the price to libraries.

Even if the continuing role of the journal, offered in a range of formats, is accepted there is also the issue of ownership. Librarians who have long complained about the rising price of journals have led the debate over whether publishers are entitled to claim ownership of material supplied by the research community to them at no charge and sold back at great cost. Some librarians have suggested that the academic community, e.g. the librarians and perhaps university presses, should take over primary publishing (Bennett, 1996; Carrigan, 1996). Almost universally they disregard the vital role of the academic societies in primary publishing and are understandably obsessed by the high-priced journals put out by a handful of international publishers (Campbell, 1992). The librarian at Louisiana State University, Chuck Hamaker, has maintained a high profile through his attacks on journal pricing and his recent policy of cancelling expensive journals and using the funds saved to make the cancelled titles 'available' through document delivery. An account of this in *Forbes* (Hayes, 1995) led to a drop in the share price of Reed Elsevier (publishers of some 1200 journals) the next day.

Many academic societies publish themselves and many others use commercial publishers and university presses (for example, 180 societies publish with Blackwell Science Ltd). There is no sign of them wanting to publish with an academic library organization.

Part of the problem is that librarians do not fully understand the role of the publisher, standing slightly outside the academic community and therefore able to impart objectively some level of quality control and status. It is clear that authors choose to submit papers to a particular journal primarily on the basis of its perceived status: perceived readership is of secondary importance and, surprisingly, speed of publication is relatively insignificant (Coles, 1993: Schauder, 1994). The lack of status and, therefore, value on a publications list in a curriculum vitae has held back the development of electronic journals which have no hard copy edition.

The librarians may well complain about journal prices, but they too have contributed to the problem. It is often commented that librarians have not fought sufficiently hard to preserve their dwindling budgets. When they have cancelled journals and used instead inter-library loan, the latter has made no contribution to the costs of the original publication. Consequently, reduced circulation has forced publishers and societies to charge the remaining subscribers more. Some publishers have responded by experimenting with new pricing models, supplying a range of publications under a licence with added services, and suggesting that the price of the licence be related to usage, rather than the amount of material made available. This would enable smaller universities to match larger, richer universities in providing facilities for their students (Campbell *et al.*, 1996), but the initiative has not been well received by the more influential librarians who are, of course, all from larger libraries (Okerson, 1996). It is difficult to see how they can defy the logic of charging by use, which can be monitored with online systems.

As journals cannot survive indefinitely on ever declining circulations, we can safely assume that new publishing systems will be developed to fulfil the requirements of the research community, with the librarian's role changing to that of a broker, negotiating rights and licences to services and lists of publications to give their users the benefits of economies of scale. The 'free at the point of use' principle will probably be maintained to enable students to have equal access to the research literature. Publishers have the expertise and resources to develop new services, but those with a tendency towards complacency could lose their position to new players, if they do not introduce significant value innovation. Societies will readily drop their current publishers if more attractive options emerge and without the vital association with societies many publishers will perish. With new systems of discovery and delivery in place, the international marketplace is accessible to all.

Change, however, is lagging behind the potential of the new technology as the conventionally published article remains the basis of research, teaching and the development of a career in both. We should also bear in mind the warning from DG XIII of the European Commission that users are having difficulty coping with the growth and lack of organization of content available online.

The Electronic Textbook – a Synthesis

Only primary publishing has been discussed up to this point, yet new technology is also changing our approach to textbook publishing. Grycz (1996) has recently outlined a definition of the textbook of the future, which is near to the model that we are likely to work towards over the next five years. It will fulfil the following criteria: it will be a multi-author work, customizable by teacher and student; will use online databases to remain current; will be delivered inexpensively via campus or regional replicating services; will include dynamic interactive segments via cable, TV or telecommunications service providers; will offer a suite of helper applications; and will offer the basis for continuing education after completion of formal coursework.

Popular features of current electronic medical textbooks, such as self-assessment, quizbanks and tutor assessment features could be added to this definition. The investment in such a system would demand a large potential market to give any chance of a profitable return. Consequently, the current view amongst publishers and librarians is that the electronic journal is likely to succeed before the electronic textbook.

Plant pathology, with only several thousand undergraduates each year, could only support a less ambitious scheme, but nevertheless it is interesting to speculate. The core text could be an extensive set of tutorials, perhaps based on an existing textbook, with optional units covering diseases specific to each geographical region. The text would be supported with animations showing, for example, the process of infection, the usual graphical presentation of data, colour images of diseased plants and pathogens, quizbanks (perhaps using some of the colour images), epidemiological and disease management models, a glossary, and classic papers able to be printed out as facsimiles of the originals. This material could be delivered on a CD-ROM, allowing for cheap reproduction of the colour.

Inside the CD-ROM case the user would find a form, to be returned to the publisher if access to a supporting online service, holding titles and abstracts of recent papers, were desired. The British Society for Plant Pathology could consider including access for one year to the journal *Plant Pathology*, to introduce students to their journal and thus, indirectly, to the Society. The online service could also include a quizbank, with an annual prize sponsored by the Society. Students could interact with other keen users of the textbook through a list server. The product should enable users to 'stand high on the mountain of knowledge' in plant pathology.

Even this model would be relatively expensive to implement, but it is likely that academic administrators will be looking to self-tuition systems to enable universities to put through more students at lower cost. Funds may well be found to provide access to electronic textbooks in the new learning centres.

Back to the Future

Although the WWW and low cost microcomputers were not envisaged in 1980, much of the way primary publishing has evolved was predicted then. Now we have the technology we need to think how to make more imaginative use of it. Predictions of how the technology will develop can help. For example, DG XIII of the European Commission predicts that most people will soon be carrying their computers, or wearing them, and by the year 2005 there will be more wireless telephones than wired telephones. The plant pathologist can expect, therefore, to be able to consult a database while in the field and feed in data for analysis, although advances in battery technology will be required.

In the next five to ten years the way the products of primary publishing are packaged and sold must change, and new technology gives more scope here. For the first time, usage of copyright material can be monitored; this should enable publishers, subscription agents and/or other new players and librarians to negotiate fairer means of charging. At present, the physical aspects of packaging and delivering information, e.g. a print-on-paper journal issue, dictate a basic minimum charge for supplying a journal to, say, members of a society. If members are comfortable with accessing a server for the same information using a password that comes with membership, then the cost comes down and there is less constraint on the amount of material published. This could also lead to a fresh approach to the format of journals.

At present, abstracts are used by secondary services without payment to the original publisher, yet a good abstract, and especially a structured abstract, will be as much as many will see of the paper. As mentioned above, publishers are working together to enable users to access abstracts more easily online. This could undercut the vital subscription revenue. One way out of this might be for publishers to provide only a very short abstract, or 'taster', free of charge aimed at encouraging the user to want to see more of the paper. The actual abstract could be a little more comprehensive and structured to become in effect an 'executive summary' of the paper and as such an integral part of the paper which is not given away without charge.

Further changes will depend on the status of peer review and other measures of 'quality' such as the citation index. Ginsparg's pre-print server has enabled high energy physicists to record quickly their findings on a reasonably formal basis, but physicists are still sending more papers each year to the print-on-paper peer-reviewed journals. The Institute for Scientific Information (ISI) has only just started to cover a handful of electronic journals. Research assessment exercises tend to use Impact Factor in evaluating the published work of researchers, and no electronic journal has an Impact Factor. The established journals which are now produced and delivered electronically, although virtually all their revenue comes from subscriptions to the print-on-paper edition, seem to offer the best of both worlds.

Electronic document delivery should show how many times a paper has been ordered; such data could be used to assess the value of a paper. Publishers may wish to make available such listings. With cooperation they should also be able to indicate 'forward citations', i.e. when a paper is cited in a later paper the reference to that paper can be appended to the bottom of the original paper. Over time, an indication of the value of the paper can be built up (usage and forward citations) which might be more useful than the presence of the paper in a low or high Impact Factor journal.

Although the time taken by referees is likely to be a constant, articles should be published more quickly as they are handled and transmitted in machine readable form. This might give scope for discriminatory pricing; a commercial research organization might be prepared to pay more for earlier access than a largely teaching organization. A benefit to members of the Society might be access to articles as soon as they enter the database, before paper publication; they might even have access to the production database indicating papers about to be published.

In the immediate future, therefore, we expect the market to catch up on the current technology and we shall be using this period to implement new, fairer methods of charging, to improve the discovery process through cooperation, and perhaps establish new ways of evaluating what is published. We may also see the textbook, which has been a relatively stable concept, benefit from the potential of new technology and help establish a link between teaching and research. In the longer term there may be some reaction against the relentless pace of change, and publishers will need to take account of this.

Why, for example, do vineyards have to use corks when there are better and cheaper methods of sealing the bottle?

Acknowledgements

We should like to thank Richard Lucier (UCSF) and our colleagues Jim Krosschell, Clive Hemingway and Anne Stanford for their advice on an early draft. The views expressed, however, are ours and do not necessarily represent those of Blackwell Science Ltd.

References

Baubin, T. (ed.) (1996) *Strategic Developments for the European Publishing Industry towards the Year 2000*. European Commission, DG XIII/E, Brussels, 46 pp.

Bennett, S. (1996) Re-engineering scholarly communication: thoughts addressed to authors. *Scholarly Publishing* 27, 185–196.

Birkerts, S. (1996) Reading in the electronic era – 1. *Logos* 7, 211–114.

Bodinham, M. and Campbell, R.M. (1992) The survival of the journal: a simple spreadsheet model for predicting the outcome. *Learned Publishing* 5, 153–159.

Brindley, L. (1996) Subscript for a free and easy system. *The Times Higher Education Supplement*, October 11, iii.

Campbell, R.M. (1992) The commercial role in journal publishing: past, present and future. *Logos* 3, 27–33.

Campbell, R.M. and Halliday, T. (1985) Why so many papers? *Scholarly Publishing* 16, 313–316.

Campbell, R.M., Bannerman, I. and Stevens, A. (1996) Cheaper by the case. *Interlending and Document Supply* 24 (3), 4–5.

Carrigan, D.P. (1996) Commercial journal publishers and university libraries: retrospect and prospect. *Scholarly Publishing* 27, 208–221.

Coles, B. (ed.) (1993) *The Scientific, Technical and Medical Information System in the UK*. A report of the Royal Society, British Library and the Association of Learned and Professional Society Publishers. British Library R&D Report 6123, the Royal Society, London.

Follett, B. (ed.) (1993) *Joint Funding Councils' Libraries Review Group: Report*. HEFCE, Bristol.

Ginsparg, P. (1996) Winners and losers in the global research village. In: Shaw, D. and Moore, H. (eds) *Electronic Publishing in Science, Proceedings of the Joint ICSU Press/UNESCO Expert Conference. February 1996*, ICSU Press, UNESCO, Paris, pp. 83–88.

Grycz, C.J. (1996) The textbook of the future. *Logos* 7, 204–208.

Hayes, J.R. (1995) The Internet's first victim? *Forbes*, December 18, pp. 200–201.

Hemingway, C.H (1997) Successful journal publishing on the Internet: hit or myth? *Learned Publishing* 10, 127–131.

King, D.W. (1980) Electronic alternatives to paper-based publishing in science and technology. In: Hills, P.J. (ed.) *The Future of the Printed Word*. Francis Pinter, London, pp. 99–110.

King, D.W. and Roderer, N.K. (1978) *Systems Analysis of Scientific and Technical Communication in the United States: The Electronic Alternative to Communication through Paper-Based Journals* (King Research, Inc., NTIS PB 281 847-PB 281 851).

Okerson, A. (1996) A librarian's view of some economic issues in electronic scientific publishing. In: Shaw, D. and Moore, H. (eds) *Electronic Publishing in Science, Proceedings of the Joint ICSU Press/UNESCO Expert Conference February 1996*. ICSU Press, UNESCO, Paris, pp. 143–150.

Page, G., Campbell, R.M. and Meadows, J. (1997) *Journal Publishing*. Cambridge University Press, Cambridge, 420 pp.

Pearce, A. (1996) Publisher, know your niche! *Learned Publishing* 9, 203–205.

Schauder, D. (1994) Electronic publishing of professional articles: attitudes of academics and implications for the scholarly communication industry. *Journal of the American Society for Information Science* 45(2), 73–100.

Senders, J. (1980) The electronic journal. In: Anthony, L.J. (ed.) *EURIM 4: Proceedings of a European Conference in Primary Publication: Impact on Producers and Users, 23–26 March 1980*. Aslib, London, pp. 14–16.

Shaw, D. and Moore, H. (1996) *Electronic Publishing in Science, Proceedings of the Joint ICSU Press/UNESCO Expert Conference February 1996*. ICSU Press, UNESCO, Paris, 198 pp.

Singleton, A. (1981) The electronic journal and its relatives. *Scholarly Publishing* 13, 3–18.

Tenopir, C. and King, D.W. (1996) Setting the record straight on journal publishing: myth vs. reality. *The Library Journal* March, 32–35.

39

The Life Sciences and the Information Revolution

S. Blackmore

The Natural History Museum, Cromwell Road, London SW7 5BD, UK
Fax: +44 171 938 9232/E-mail: sb@nhm.ac.uk

Introduction

This short chapter presents a personal perspective on the impact of the information revolution on the biological sciences. It comes from someone who is a plant systematist by training and has no specific qualifications concerning information technology. However, one way or another, I have recently been much involved in discussions about biodiversity and information systems. In the last 12 months these have included workshops and conferences convened by the Bioinformatics Subcommittee of the Biology and Biotechnology Research Council (BBSRC) and the Engineering and Physical Sciences Research Council (EPSRC), the American Association for the Advancement of Science, the Megascience Forum of the Organization for Economic Cooperation and Development (OECD) and the UK Systematics Forum. The views presented here are derived from some of the recurrent themes at these workshops.

First, has there been an information revolution as a result of computerized information technology? Certainly, revolutionary developments are taking place, but in biology they have a long way to go before an older order be swept aside to establish a new information age. Many of the necessary technologies exist but most of the knowledge gained concerning biology is not in computerized form. The rapidly expanding information superhighway, for example, can provide a means of interconnection between disparate sources of information but in biology some of the most obvious kinds of information are not currently accessible.

One of my main concerns is that current computerized representations of the real world, with its complex biological, chemical and physical interactions, are a poor proxy for reality. Focusing on the life sciences, it is clear that the information currently available in computerized information systems and models does little justice to the complexity and diversity of biological organisms in the real world. One might argue that this is inevitable and, of course, sometimes the simplicity of computer models is quite deliberate. We deliberately model dynamic processes and systems by building them up from the simplest set of rules that appears to behave as we think nature should. The earliest predictive models of global warming were based on an Earth that lacked vegetation, an essentially physical view of the world. More recent models are sufficiently refined to include a layer of green vegetation over the land and are beginning to include biological processes in the model. This is the normal pattern of progress in the development of global models and no doubt we can look forward to increasing levels of sophistication.

What alarms me is that we, in the international scientific community, are clearly not doing enough to create the most basic representations of the real biological world within our computer systems. Foremost among the basic details needed in biological information systems are the names and the systematic relationships of all known species. I consider this lack to be a cause for considerable concern because it can be shown that it has severe financial consequences for the global economy (McNeely, 1988; Systematics Agenda 2000, 1994; Heywood and Watson, 1995). To elaborate upon this problem I will first explore the idea that diversity and complexity are inherent properties in biology and then discuss the key elements of the biological knowledge base that need to be brought within the reach of information technology. I will consider some of the reasons why this has not already happened and end by outlining some possible solutions.

Biodiversity and Biological Complexity

The concept of biodiversity is still fairly recent but it captures and expresses the complexity of life on Earth very effectively. Since the 1992 Earth Summit and the Rio Convention on Biological Diversity, biodiversity has generally been defined as having three conceptual levels. In the words of the Convention, '*Biological diversity* means the variability among living organisms from all sources including, *inter alia*, terrestrial, marine and other aquatic ecosystems and all ecological complexes of which they are part; this includes diversity within species, between species and of ecosystems.'

The conceptual levels of biodiversity are easily and intuitively grasped. After all, the diversity of ecosystems and landscape are a familiar part of human experience, as is the diversity of species with which we share the planet. Knowledge of species, and to some extent of ecosystems, is essential for human survival (Blackmore, 1996) since we cannot escape our dependence on biodiversity for food, medicines and materials (Wilson, 1992) or for the ecological services provided by healthy, diverse ecosystems (Mooney *et al.*, 1995). Even today we begin training in the basic knowledge needed to survive at an early age. The first books we buy for our children are usually simple, illustrated taxonomic accounts of the mammals and birds of our agricultural and domestic environments! Almost before we attempt to impart any other knowledge to them we teach them the names and the sounds of familiar animals. Genetic diversity seems a more abstract notion although most people are aware of the inheritance of traits within their own family. The point I want to emphasize is that biodiversity is very familiar to us, a grasp of it is built into all human cultures because our cultural knowledge and experience of the world are constructed around our understanding of its diversity. This applies both to the formal body of scientific knowledge we have built up about the biological world and to the great diversity of informal systems in different cultures.

However, it is important to emphasize that the inherent complexity and diversity of biology make it a science that cannot be reduced to a few simple laws or underlying principles. Interestingly, it is this inherent complexity that makes information technology so desirable for processing and analysing biological information and at the same time makes the task of computerization far from trivial.

Biodiversity and Bioinformatics

One of the most pressing challenges for the application of information technology is the creation of computerized information systems that can handle and integrate the large pool of knowledge relating to biodiversity that we have built up throughout human history.

The present situation is that, in addition to the wealth of uncomputerized information about biology, there are many large computerized databases ranging from those containing gene sequences, to taxonomic databases and information systems that provide abstracts of the scientific literature. The recent *Global Biodiversity Assessment* provided a concise survey of such resources (Heywood and Watson, 1995). These diverse systems are progressively being linked together. The World Wide Web, in particular, is rapidly enabling greater levels of interconnection between information systems. *Species 2000* provides a good example of the potential for a global system of databases of taxonomic names that can be accessed over the World Wide Web (Bisby, 1994). At the same time, ever more powerful bioinformatic tools for analysis and modelling are being developed, usually within specialist disciplines.

Unfortunately, at present there is nothing that integrates the computerized biological knowledge base. There is no skeleton on which the information can hang. This is true in other areas of knowledge, of course, but the concern of this article lies with biology. The wider issue is, how will we move from a written and oral culture to a computerized culture? I mean this only in terms of the systems used to store, retrieve and analyse knowledge; I have no more desire to be 'silicomorphized' than Peter Cochrane and Chris Walker (Chapter 40, this volume).

The biological knowledge base in both written and oral forms is vast, so much so that keeping track with past knowledge is increasingly challenging. In many areas of biology 'old' literature, from 20 years ago, is often simply not being cited! Some scientists claim that their fields advance so rapidly that literature more than a year old is no longer relevant. Such an attitude is ill-informed. It is true that within a narrowly defined scientific context progress can be made without knowing or fully understanding the historical background of the science. But this is not to appreciate that civilization rests on the knowledge and understanding built up by successive generations. We cannot afford to lose the most basic knowledge of the biosphere that sustains us.

So, what is lacking in current information systems? The most significant problem is that the connections between observations derived from individual biological entities are missing. We need to create information systems that can link together different classes of information. In order to do this we need to connect biological observations through the names of the organisms that were studied. This is equally true both for formal scientific knowledge systems and for informal or ethnobiological systems. It is important to recognize that neither system of names is absolutely fixed and unchanging because diversity is an inherent property of living things; the boundaries between taxa are not absolutes. Understanding of the scope and definition of a biological entity can change as we develop more sophisticated scientific hypotheses about the entity itself and its relationships to others.

But if names are so important why have the names of 1.7 million species so far known not been computerized and then used to link together diverse classes of information? This and related questions have been discussed many times (see for example, May, 1988, 1990; Gaston and May, 1992; Blackmore, 1996) and has been a central theme in several recent workshops in the UK and the USA. Robert May has argued that the need for quick access to existing knowledge, and for rapid additions to it, is hampered by traditions that have perhaps been too slow to embrace new information technologies. This is undoubtedly a correct diagnosis of part of the problem, but there are a variety of other barriers to progress including technological difficulties, questions of design and the sheer cost of creating information systems.

I am not qualified to delve into the technicalities and details of the information technology required to establish global biodiversity information systems, a subject that others have addressed in this volume. Instead I will briefly argue in favour of the planned development of such systems rather than their *ad hoc* emergence. To illustrate the role of specimen-based information I will use the example of Mexican birds and I will conclude by discussing the funding of biodiversity information systems.

I consider that, despite many successful projects and initiatives, there is an immediate need to design the central connecting elements of a dispersed global biodiversity information system. Design is needed for several reasons:

1. To ensure that the system can deliver the information that is needed by users in the form in which they require it.
2. The nomenclatural and taxonomic complexity of biological names require sophisticated systems that can interconnect related observations via common or scientific names, and independently of whether the latter are currently applied or not. The issue of names, and whether it is possible or desirable to achieve stability of names, has been a contentious subject in recent years (see for example, Hawksworth, 1991). In my opinion there is a need within bioinformatics for the capability of handling all names that have been or are applied to organisms, and this transcends the debates about nomenclatural procedures and practices.
3. Information systems need to be able to express information about current and past concepts of the phylogenetic relationships between taxa.
4. The emergent processes by which the World Wide Web grows and extends have yet to give rise to the required level of interoperability and interconnection.

There is no reason to believe that a system with these properties, that meets the needs of users, will ever emerge without a phase of active market research, international consensus building and bioinformatic design.

The current Bioinformatic Initiative of the OECD Megascience Forum is ideally placed to discuss and resolve these issues. The Biodiversity Subgroup of the initiative is currently developing and refining a model for the core of a global biodiversity information system that draws on the best experiences gained around the world.

The Role of Biological Specimens

Although I have emphasized the central role of names in connecting together biological facts, it is evident that names alone are not sufficient to meet many practical needs. It is the links between information, facilitated by names, that are important because they enable computerized models to draw more completely on our understanding of biodiversity and biological interactions.

I would like to outline the particular importance of the links between names and preserved specimens in museum collections. Although these collections are often appreciated primarily for their heritage value alone, they are in fact a highly utilitarian resource. By answering questions about the identity and distribution of disease vector organisms, wild relatives of crops, parasites, invasive weeds and a host of other species that impact on our lives, these collections contribute millions of pounds annually to the global economy.

Increasingly, the answers are expected in computerized form, on-line and immediately accessible from anywhere in the world. Since the Rio Convention on Biological Diversity, an increasingly common question is, what information can the collection provide about which species of organisms are present in a particular country? Although most of the institutions that hold such information resources would be more than willing to help by answering such questions, few of the major reference collections in museums are computerized.

The Natural History Museum in London has global collections of over 68 million specimens collected over a period of 400 years and is committed to making computerized information about these collections available. However, at the current rate of progress this Herculean task will take 450 years to complete. Clearly, the middle of the next millennium is too late for the delivery of information that is needed today. Naturally, many institutions are researching ways of accelerating the processes of data entry by human or automated means but this still remains a major challenge (Blackmore, 1996).

Given the scale of the task of creating specimen information databases, some people have expressed doubts about whether the effort is worthwhile. Some question the value of information derived from museum specimens arguing that it might have few practical applications. One frequently cited drawback to specimen-based information is that specimens themselves are rarely collected according to any valid sampling procedure and merely reflect the itinerary of the collector. This may be so, but in contrast to observational records, only museum specimens can provide a permanent archive that is amenable to re-evaluation and further research.

The interest of several countries in compiling information systems based on museum specimens in collections around the world demonstrates that there is a growing need for such information. One example, kindly provided by Jorgé Soberon of the Comisíon Nacional para Conocimiento y de la Biodiversidad (CONABIO) in Mexico relates to specimen information on birds. Birds are widely used as bioindicators of environmental change and, with a vast following of amateur enthusiasts, observational records for birds are abundant (Bibby *et al.*, 1992). CONABIO decided that information on which species of birds live in Mexico and where they occur was a valuable basis for decision making that could be derived from museum specimens. CONABIO drew up formal data-sharing agreements with collection-holding institutes in Mexico, the USA and Europe and sent teams to visit the collections and transfer specimen information into computerized databases and geographical information systems. In the UK, these included the two largest collections, at The Natural History Museum and the National Museums and Galleries on Merseyside.

Figure 39.1 shows the distribution of bird specimens based on the collections held in Mexico and those held in Mexico and the USA. The European specimen information that has been gathered is not included. The Figure shows that, on the basis of specimens held within Mexico, specimen-based bird records are lacking for many regions of the country. On this basis alone, specimens do

Fig. 39.1. Distribution of birds in Mexico, based on specimens preserved in museum collections. Above. point distribution data derived from collections held in Mexico. Below, point distributions derived from collections in Mexico and the United States of America. (Source: Jorgé Soberon, Comisíon Nacional para Conocimiento y de la Biodiversidad (CONABIO), Mexico.)

indeed provide a very skewed impression of the Mexican avifauna. The addition of information from outside Mexico provides a much more complete coverage of the country. Once all major collections of Mexican birds have been included in the study an impressively detailed geographical information system will be available. Viewed at a global level, the many millions of specimens in major collections around the world collectively represent an information resource of vast potential. As a result of the Mexican bird project, it is already possible to use this system to answer questions such as which species of birds show changes in their distribution through time? These changes can be compared and related to changes in vegetation, based both on plant specimens and vegetation maps.

As more information is transferred into such information systems, they will progressively become more accurate as a model of biological complexity and more valuable as a tool for modelling and prediction. It will be possible to model biological interactions and to ask questions such as, how are changes in bird distribution likely to affect their role in pollination and seed dispersal and what will the consequences be for the future of the vegetation? The more our detailed knowledge of biodiversity migrates into computerized information systems, the more we can begin to address complex questions. And perhaps one day the models used to predict global warming will be sufficiently sophisticated to differentiate between vegetation types in Mexico and those in other countries.

The crucial next steps in the development of global biological information systems are the establishment of a system that can link observations by means of names and the transfer of the existing knowledge base into computerized form.

Conclusions

The crunch is, who will pay for the next steps? The design of an internationally operable biodiversity information system that can create links between disparate biological information sources needs input from both information technology and biology. This is properly within the remit of national and international research funding agencies. They should support it, perhaps by funding the proposals that form the eventual outcome of the current initiative of the OECD Megascience Forum.

Transferring knowledge from manual retrieval systems to computerized form in order to create the databases is too vast a task to be funded in the same way and will undoubtedly require funding from a wide diversity of sources. These ought to include international development agencies, and especially the Global Environment Facility, working to help countries meet their obligations under the Convention on Biological Diversity. Article 7 of the convention, in particular, emphasizes the need to identify and monitor the components of biodiversity, both processes that are dependent upon ready access to taxonomic information, not least for the purposes of identification. As I have argued elsewhere, the benefits that would accrue from this process would be to both developing and developed countries, even though most of the information resources are located in developed nations (Blackmore, 1996). Some projects involving the conversion of information systems will be appropriate for research council funding especially where the contribution the information systems can make to wealth creation and the quality of life can readily be identified. Institutions that have important information resources will undoubtedly contribute to the creation of databases through their own initiative but are likely to need substantial new sources of funding to accelerate the process. It is perhaps surprising that very few commercial companies in the business of publishing and providing information have actively sought involvement in the creation of biodiversity information systems.

Whether there is an immediate commercial potential or not, it is clear from the recent political focus on biodiversity that this enterprise is urgent and needs a massive increase in the level of activity (Wilson, 1992; Heywood and Watson, 1995). Immediate actions need to focus on key groups of organisms for which there is the greatest current informational need. In my opinion, these include plants, because detailed information on vegetation is required for so many aspects of the management and utilization of natural resources. Birds and mammals are also a high priority, as key indicator organisms for which substantial observational records exist, and so are other indicator groups. Other organisms, less readily categorized by taxonomic group, that are a high priority for bioinformatics include pests and beneficial organisms that impact most directly on humanity. I want to emphasize that these are groups for which the mobilization of information resources is the highest priority. In *Systematics Agenda 2000*, this activity corresponds to the third of three important but distinct contributions of systematics. It is obvious that the priorities for the other key missions would differ. Mission one, for example, concerns the inventorying of as yet undescribed species and few would disagree that the highest priorities include invertebrates and microorganisms, and organisms in relatively poorly explored environments including the marine world.

For information technology to achieve its full potential in biology, the emphasis in bioinformatics needs to remain on biological diversity until a workable informatic core has been established. Achieving this will be an interdisciplinary endeavour, requiring substantial intellectual and financial resources.

Acknowledgements

I am grateful to Jorgé Soberon for permission to use the diagram in Fig. 39.1 and to A. Navarro and H. Benitez of the Museum of Zoology of the School of Sciences, National University of Mexico who assembled and maintain the database from which the maps were generated.

References

Bibby, C.J., Collar, N.J., Crosby, M.J., Heath, M.F., Imboden, Ch., Johnson, T.H., Long, A.J., Stattersfield, A.J. and Thirgood, S.J. (1992) *Putting Biodiversity on the Map: Priority Areas for Global Conservation*. International Council for Bird Protection (ICBP), Cambridge.

Bisby, F.A. (1994) Global master species databases and biodiversity. *Biology International* 29, 33–40.

Blackmore, S. (1996) Knowing the Earth's biodiversity: challenges for the infrastructure of systematic biology. *Science* 274, 63–64.

Gaston, K.J. and May, R.M. (1992) Taxonomy of taxonomists. *Nature* 356, 281–282.

Hawksworth, D.L. (ed.) (1991) *Improving the Stability of Names: Needs and Options. Regnum Vegetabile No 123*. Koeltz Scientific Books, Königstein.

Heywood, V.H. and Watson, R.T. (eds) (1995) *Global Biodiversity Assessment*. Cambridge University Press, Cambridge.

May, R.M. (1988) How many species are there on Earth. *Science* 241, 1441–1449.

May, R.M. (1990) Taxonomy as destiny. *Nature* 347, 129–130.

McNeely, J.A. (1988) *Economics and Biological Diversity: Developing and Using Economic Incentives to Conserve Biological Diversity*. IUCN, Gland.

Mooney, H.A., Lubchenco, J., Dirzo, R. and Sala, O.E. (eds) (1995) Biodiversity and ecosystem functioning: basic principles. In: Heywood, V.H. and Watson, R.T. (eds) *Global Biodiversity Assessment*. Cambridge University Press, Cambridge, pp. 281–325.

Wilson, E.O. (1992) *The Diversity of Life*. Harvard University Press, Cambridge, Massachusetts.

Systematics Agenda 2000 (1994) *Systematics Agenda 2000: Charting the Biosphere*. Produced by Systematics Agenda 2000, a Consortium of the American Society of Plant Taxonomists, the Society of Systematic Biologists, the Willi Hennig Society in cooperation with the Association of Systematics Collections, New York.

Biology, Computers, Sex and Sorting?

40

P. Cochrane and C. Winter

BT Laboratories, Martlesham Heath, Ipswich, Suffolk IP5 7RE, UK
E-mail: pcochrane@bt-sys.bt.co.uk/wintercs@boat.bt.com

Introduction

Considering the obvious synergy of biology and computing, it is remarkable that practitioners seldom communicate thoughts and experiences across the discipline divide. The absence of algorithms and a history of computer modelling in biology; the language used to describe events in each domain; the precision required in computer simulations and their brittleness; and the complex and interlinked behaviours of biological organisms have combined to minimize the impact of each field on the other.

For over 50 years, computer science has stared longingly at biology for inspiration. Von Neumann spent years trying to build self-replicating machines, believing this to be the key to life. The cellular automata that arose from this work are still key to alternative parallel computing architectures. Turing considered morphogenesis one of the most interesting phenomena to study and wrote seminal papers on the subject. Artificial Intelligence (AI) and Artificial Life (AL), two major disciplines in computer science, draw their philosophy from biology. So why isn't there more interchange between the disciplines?

Here we describe areas of biology and computer science now creating cross discipline developments of extreme importance to both camps. In addition, we outline the way these new developments and state-of-the-art correlation, visualization and modelling techniques could be utilized by biologists to aid and advance the understanding of the complex systems found in nature.

It is useful to start by listing those techniques already adopted by computer science that have biological roots:

 Information Technology, Plant Pathology and Biodiversity (eds P. Bridge, P. Jeffries, D.R. Morse and P.R. Scott)

- Psychology
  - expert rule systems
  - case-base reasoning
  - fuzzy logic
- Neuroscience
  - neural networks
  - cellular automata
  - noisy neurons
- Evolution
  - evolutionary computing
  - genetic algorithms
  - genetic programs

However, there are a wide range of phenomena still remaining to be examined:

- Bacterial mobility – nanotechnology and microscopic robots using protein motors.
- Behavioural sciences – predictable rather than deterministic computers.
- Immune systems – computers with learning defence mechanisms.
- Metabolic pathway control – as an example of asynchronous parallel computing.
- Morphogenesis – programs that grow and develop.
- Neuro-science – dynamic neural networks that need no external teachers.
- Neural science – interfaces direct from silicon to carbon nervous systems.
- Sensor technology – probably essential for true machine intelligence.
- Many more.

The attraction of biological systems is that they are error and noise tolerant, robust, asynchronous, chaotic, and non-deterministic. They represent a computing domain far from Von Neumann machines and may be absolutely necessary to cope with the emerging and chaotic world of networked machines.

Although raw computer power has increased the ability to use modelling to explain the results and discoveries in engineering physics, economics, and many other fields, their application to biology seems to have been relatively neglected. Biologists should be at the forefront of system modelling, because they are trying to decode the most complex systems in nature. Curiously, biology has been a source of inspiration for some of the most important concepts and tools in computing. Today, there is a fast growing use of neural networks, genetic algorithms, and attempts at sexual reproduction in software.

Learning from Biology

Evolutionary software

Designing systems using biological tools not only frees us from constraints placed by human understanding, it also brings other benefits. Evolutionary systems, by their very nature, produce robust, adaptable systems rather than solutions that are highly optimized and brittle. In a fast changing and chaotic world, adaptability, rather than optimality, is the key to long term survival.

In developing evolutionary systems to run as computer models we naturally start with biological analogies. Evolution is driven by a cycle of reproduction (which must produce a diversity of offspring) and selection (which weeds out, statistically, the strong from the weak). Selection is relatively straightforward with a test environment which compares the performance of different programs. A technique called co-evolution ensures that the programs are always tested against the hardest examples in the population. This speeds up the process over a constant set of tests or testing against every possible example, which is a very lengthy and time consuming process. The key design techniques lie in the nature of the reproductive process, and turn out to have profound implications for the selection of adaptable systems.

Currently, the most widely used systems are Genetic Algorithms (GA) and Genetic Programs (GP). In the former case, a problem is represented as a set of symbols in a linear string for which parameters or values must be found. The symbols form an effective program that is calculated by the test harness and used to generate the individuals fitness. Sex (in the form of cross-over) consists of swapping parts of two parental strings to produce an offspring. Mutation is a random change of one of the symbols. In Genetic Programs the program is represented as a branching decision tree where sub-branches are swapped over. Both techniques have proved very powerful for finding solutions to small, but complex problems.

To date GAs and GPs employing these techniques appear to be limited to small programs of about 100 symbols. In contrast, DNA and biological evolution appears to involve millions of symbols – providing an existence theorem for process improvement. One major limitation of both techniques is that the creatures compete indirectly through an external fitness function which never changes – their world is static. This is a common mistake of drawing direct parallels between the disciplines and an incomplete extrapolation from, or modelling, biology. The real world of biological organisms consists of dynamic entities, and so, static systems modelling can lead to major errors and spurious artefacts.

When we move to a silicon world, we are no longer bound by the limits imposed on carbon-based biological organisms. The first of these concerns children. In biological systems, there is no mechanism by which parents can learn from their offspring's success in producing new and improved children. Improvement is a slow, generation by generation process. This need not be the case for silicon. For instance, we have designed an evolutionary system where a parent produces two offspring and uses their success or failure to judge how to produce successively better children. These virtual children are then a partial learning process. On a classical set of test problems, the virtual children system outperforms the best existing algorithm (DHC) on three out of five of the tests set, and is beaten on two out of five. However the virtual children algorithm makes no hidden assumptions, unlike the DHC which exploits the symmetry in two of the problems to solve them fast. This is a clear warning that when we learn from biology we should not be slavishly bound by the best biological outcomes. In silicon we can be both recursive and postscriptive at the same time, using the benefit of hindsight to advantage.

The N-sex problem

Sometimes the simplest questions can have much deeper implications than first apparent. One case we have recently examined is 'Why are there only two sexes in the smartest life forms?'. This question has been part of a wider programme to investigate where engineering principles can be used in biology, and where biological concepts can change the way we design silicon systems. The answer to this question has impact on a much wider audience than might be expected, embracing dynamic systems, learning theory, business and economic modelling as well as biology.

Sex and evolution are tightly bound; and from these roots also spring many other ideas related to the adaptability and optimization of biological organisms and engineering systems. These concepts can return to affect biology in many ways, in particular to equip biologists with powerful and unusual search/compare/find tools for handling large, complex and error-prone data sets.

To examine 'why two sexes?', a simulation was devised in a world of six species each containing 100 starting individuals. The environment also contained 10,000 individuals of a seventh species. We could consider the first group as flowers and the second group as insects (specifically bees), with the second group looking to gain energy and the first trying to get themselves pollinated at minimal cost. Critically the 'bees' were a diverse group wanting different features from the flowers in terms of colour, shape and scent to convince them the best pollen was there. In each generation, the bees moved around the flowers to achieve the highest pay back. The best flowers breed, the poorer ones die. Now in this world there is a diverse and complex task (how do I make the most 'profit', i.e. attract bees at minimum cost) which is made worse by the choices of competing species. Of course, this analogue is much wider than flowers and bees; it could represent the way companies attract customers, or adaptive algorithms learn to manage traffic flow across international networks.

A number of system parameters were systematically varied, and 2000 trials run for each combination of variables from randomly selected starting points. The six species each had a different number of parents for their offspring, either asexual or between two and six. The other variables included:

1. Are all parents of one sex or are there as many sexes as parents?
2. Do the bees select flowers on cost alone or by a wider range of features?
3. Do the bees change their preferences with time?

The assumption at the beginning of experimentation was that two sexes would win because it happened in nature, and we could think of good statistical reasons why three or more would be more disruptive. The results, however, contradicted our beliefs. We found that (1) had little effect – it simply led to high sex species dying out slightly more rapidly on occasion due to the disappearance of one or more sexes from the population. In a similar manner, (3) was of slight consequence – all the dynamics arose from interactions between companies rather than the small changes in the environment. The effects observed were dominated by (2) and what we decided constituted a success.

The results can be summarized as follows:

1. If total profit is the measure of success and bees use a general measure to select which flower *then six sexes is best.*
2. If total profit is the measure of success and the bee selects on minimum cost *then four sexes wins.*
3. If efficiency (profit per bee) *then four sexes wins every time.*

Why should this be? A first clue lies in early results as we were developing the system and had a less complex world. Then the battle was between three and four sexes for dominance; but as a more complex and dynamic environment was created, four prevailed. This implies that system dynamics determines the best learning algorithm. A little thought would suggest this is axiomatic. But, because most work on learning systems is inspired by chasing optimal solutions where time is immaterial, the answers normally found are very different. Basically two sexes leads to less disruption than multi-sex, and when a good solution is found the population is not dispersed so easily from it. However, the disruptive effect of multi-sex sees population searches much faster than two sexes. If there is sufficient time, two sexes is best, otherwise it is three, four or more that search progressively faster, but find poorer overall optima. More evidence for this lies in the time-series evolution of single example populations (Fig. 40.1).

The percentage swings in the six sex case is much larger than for two as evolution progresses. Thus the answer to the long held question 'why are there two sexes in biology?' can be found in the dynamics of the world. When we are trying to evolve adaptive computer systems, the optimum number of sexes may be very different.

Our conclusions are simple but profound: in biology two sexes may be best, but in a computer many-sexes may be better. It just depends on the type of problem to be solved. The models show another interesting phenomenon; if you start with a big enough advantage and a moderate amount of flexibility, your smaller, smarter competitor may be unable to win regardless. Many believe that if you are smarter and more adaptable you win in the long run. Evolutionary modelling shows this is not the case. We do not have to be as adaptable as smaller competitors, only sufficiently adaptable to beat them.

(a)

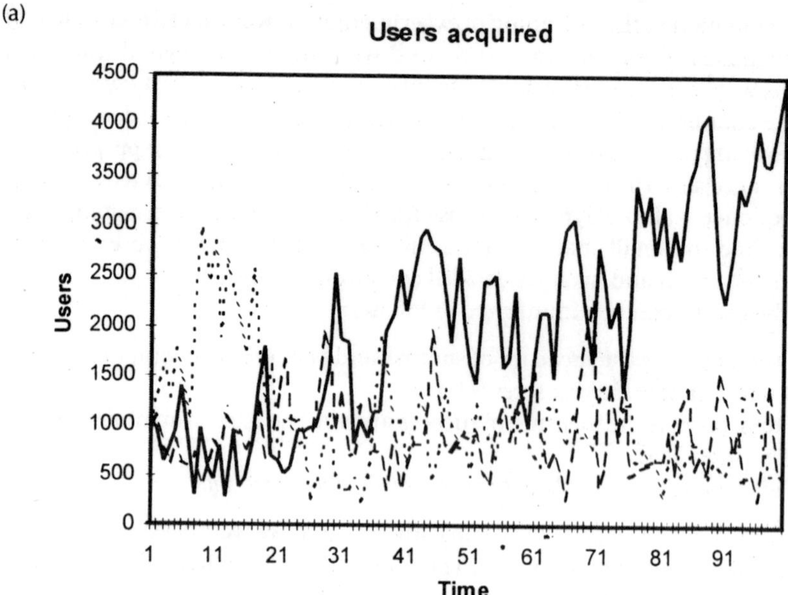

(b)

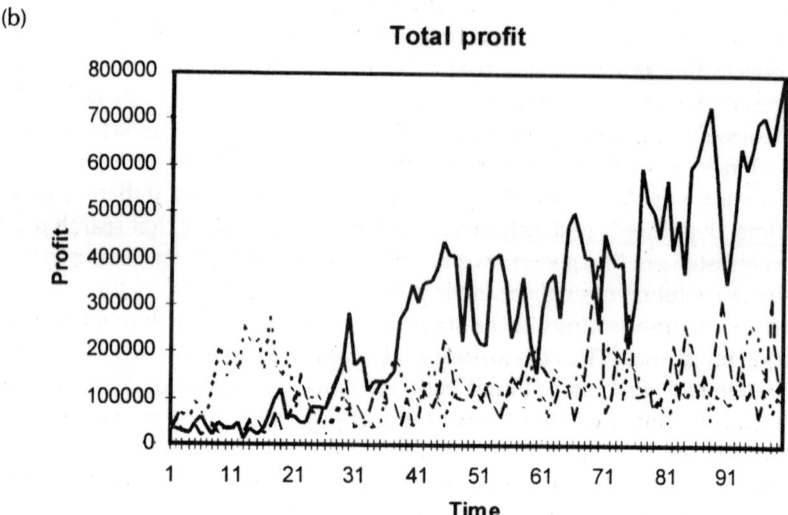

Fig. 40.1. (a) A typical plot showing the number of individuals in three species (solid = six sexes, dashes = four sexes, dots = two sexes) over 100 generations. Note the wild swings in the six sex population size. (b) Based on total profit, six sexes come out best due to the number of users.

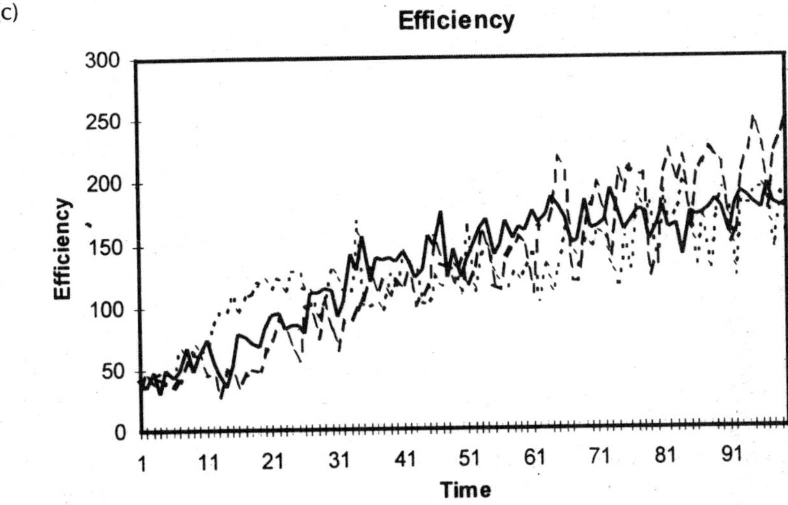

Fig. 40.1. (c) Based on efficiency, four sexes wins.

Information processors

The traditional Von Neumann architecture has proved extremely useful to the computing industry and is well suited for many processing applications, particularly those which map well to serial algorithms and arithmetic tasks. Processing proceeds in a logical, discrete, sequential, synchronous and deterministic manner. It is not well suited to implementation of algorithms which are inherently parallel and asynchronous (unclocked), or tasks which involve significant elements of pattern matching. Here we describe a model for information processing inspired by the processing of molecules within biological cells which overcome many of these problems.

Within cells, increasingly complex molecules are produced from simpler components in an intricate series of production lines. The resulting products, mostly proteins, maintain the cell's structure and operation. The reactions which transform molecules into the required products are controlled by enzymes which selectively accelerate the necessary chemical reactions by many orders of magnitude. Probably the two most important features of enzymes are that they exhibit specificity for particular molecules (substrates) and indirectly, via the reactions they catalyse, carry out particular functions which transform substrates into products. The specificity is associated with one or more enzyme binding sites to which substrate(s) can bind, whilst an enzyme's active site plays the key part in enabling, and accelerating, a particular reaction and thus provides the enzyme's function.

The specificity and function of a particular enzyme derive from the enzyme's shape and the three-dimensional positions of its constituent amino acids determined by a process known as protein folding. Whilst enzymes exhibit immense pattern specificity, they only carry out one of four chemical operations – break, make or rotate bonds, or transport a molecule. In biology, we frequently observe a duality of complexity and simplicity that is hard to explain.

The information processing is carried out despite the fact that information (substrates) moves around the system asynchronously at random. Overall control and integrity is maintained by manipulating concentrations (and thus the rate of processing). Robustness occurs through the massive parallelism. The system we have examined as an analogue of these processes consists of many processing elements. These are analogous to enzymes in that they have specificity, via pattern matching, for their operands (data as opposed to molecules) and cause the data to be transformed in some way. The interactions and dynamics of the system taken as a whole, operating in parallel and asynchronously, comprise the information processing system. Each enzyme performs an interim task in the context of the overall system in a similar manner to a particular machine working in a factory production line.

We are interested in looking at the processing capabilities of biological cells and what can be learnt and used in the design of information processing systems. By analogy between molecules and data, and enzymes and processing units, cells can be viewed as processors which take simple building blocks and transform them to build increasingly complex structures. As cells have solved many complex problems in the interests of survival and have clearly proved to be evolvable, we may learn important lessons for designing silicon systems which demonstrate similar attributes.

The attributes of the biological counterpart which are of particular interest include robustness against small fluctuations (homeostasis), processing dynamics (enzyme kinetics), pattern matching of enzymes leading to specificity for substrates, and evolvability. A model for information processing designed with such an architecture is very different from the traditional Von Neumann architecture, so we design systems allowing us to explore the impact of each of these issues.

The information processing architecture consists of an enclosure which we call a sac which includes processing enzymes, data items and, potentially, further constituent sacs corresponding to intracellular vesicles. Enzymes are capable of pattern matching data thus providing enzyme specificity and carrying out operations on the data which provide the enzyme functionality. The specificity provides a mechanism for consistent identification of appropriate classes of data to operate upon. The functionality typically transforms the data in some way, such as combining data into a more complex data structure or manipulating data field values. For real enzymes, the specificity and functionality result from the enzymes' shape; the enzyme attributes described here could be derived from shape via a simple analogue of protein folding.

A system based on the architecture outlined was built to investigate a particular information processing application, namely that of sorting a set of integers into an ordered list (Fig. 40.2). To carry out this task, a data representation was defined as a list comprising a low-end terminator, one or more integers, and a high-end terminator. Two types of processing enzyme were defined and named as join and break. A join enzyme simultaneously binds a high-end and low-end terminator of two separate data items. The operation it performs is to concatenate the data item lists into a single list by removal of the terminators, but only if the integer value adjacent to the high-end marker is less than the integer value adjacent to the low-end marker. This enzyme acting alone results in ordered lists of integers, but with typically large steps between successive integers. We also define a break enzyme which simultaneously binds the high-end and low-end markers of one data item in addition to another data item of unit length. The operation it performs is to break the first list into two by addition of terminators, but only if the value of the single integer is both greater than the integer adjacent to the low-end marker and less than the integer adjacent to the high-end marker.

Having defined the enzymes, it only remains to define the system kinetics which determine the rate at which data is processed, and the concentrations of data and enzymes. In this implementation, data items and enzymes were

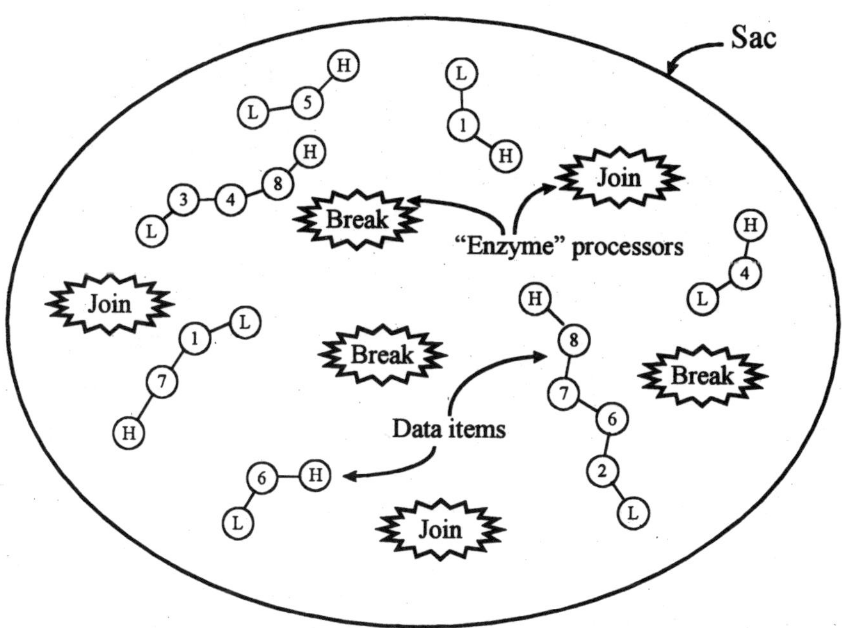

Fig. 40.2. A schematic of an information processor using 'enzymes'.

assigned random positions on a toroidal surface with random initial velocities. Binding takes place when data and enzyme approach within a fixed distance and the necessary conditions are met. Concentrations depend on the number of join and break enzymes and data items introduced into the system. Following the binding of data to an enzyme, the transformed data (the reaction products) are released with random velocities after the appropriate function has been applied to the original data items.

A single sac was created containing only data items of unit length (a single integer between low and high terminators) together with multiple join and break enzymes. Each integer to be sorted is converted into multiple instances of the corresponding unit length data list with the given integer value. The system was then iterated through time with the join and break enzymes manipulating the data lists. As the number of data items falls due to lists joining together, more unit length data lists are introduced to compensate. These are created using integers selected randomly from the set of integers we wish to sort.

Simulation results are presented in Fig. 40.3, in which sorted list frequency is plotted against time. It can be seen that sorted lists grow in length over time

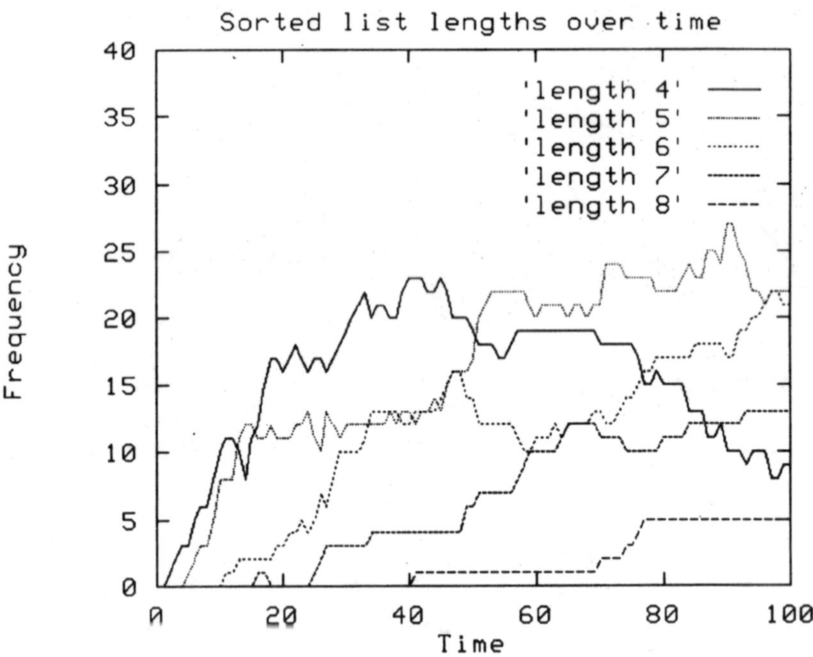

Fig. 40.3. Dynamics of list processing by a 'chemical' information processor. The rise and decay of intermediate products follow a similar pattern to that observed in chemical reactions.

as might be expected, with the longest lists representing the best solutions so far. Intermediate lengths of string initially rise in concentration and then fall, exactly as would be anticipated for intermediates in a chemical reaction with limited starting materials.

The simulation illustrates that an information processing architecture (Fig. 40.2) can successfully carry out specific application tasks. The processing happens in a highly parallel manner and its dynamics depend on the concentrations of both data items and enzymes present. Note for instance that the break enzymes can only perform their role while a supply of unit length data lists is present.

Whilst we would not propose that this architecture is efficient for number sorting in comparison with traditional algorithms, it does have interesting properties. Processing is highly parallel, depends on concentrations of data items and reaction dynamics, and we can evaluate the best solution available at any time by examining the longest list to date. In addition, the system should be relatively stable to small perturbations such as fluctuations in concentration – whether there is one more or one less instance of a particular data item present will simply alter the reaction rates slightly. This homeostasis may prove important to the system being evolvable.

Biology and Computers

Dynamic visualization and correlation techniques

Imagine a 3D virtual world with a flock of boids (bird like objects). These boids are programmed with simple rules – they turn away from objects, try to fly level and try to head towards the centre of mass of a group of boids. The net effect is simple flocking behaviour representing a dynamic, correlated system. It proves to be extremely resistant to perturbation and the flock behaviour is easily observed (but not easily deduced from the rules).

Now imagine each boid, rather than flying towards the centre of the flock, flies towards other boids with similar characteristics – being artificial, these could be anything from colour to a liking for architecture. If each boid has a group of characteristics and a group of interests what happens is that dynamic flocks appear, separating and merging with individuals moving from one to another. The flock rules make self-correlated behaviour appear, with boids of similar characteristics and interests flocking together. Our human eye is quick to spot groupings in this dynamic world. It should also be noted that these flocking systems are noise resistant.

So, one way to examine DNA sequences for homology, is to give each boid a DNA sequence so they attract with closeness of match. The resulting dynamic patterns of boid groups of similar sequences are visually separated in seconds. Curiously, self-correlating behaviour like this was first studied as a way of improving the appearance of sophisticated graphic displays by making them look more natural and less scripted; but it has yet to be re-used in its original environment – biology. Where data is complex or noisy, it is better to use the dynamic pattern recognizing ability of the human eye than the limited use of pre-programmed correlation analysis and static graphs.

If a graph is worth a thousand words, experience on data visualization of complex systems has shown a 3D movie is worth a thousand graphs (Peter Cochrane, 1996).

Here the movie further enhances understanding since the dynamic movement of boids between clusters emphasizes and shows low correlation links as well as the high correlation flocks.

Conclusions

We have briefly outlined some of the more important and exciting developments being brought about by the coalescing of computing and biological disciplines. So far, the intellectual richness of our world has been dominated by men and women from many different ethnic groups, scientific and social disciplines. Soon, we will have a new partner – the machine. To ignore the creativity of future computing power will be to deny ourselves progress, whilst continuing to isolate the scientific disciplines inside their boxes of interest will seriously denude our creative abilities.

Already, electronic engineers have discovered new aspects to sex, reproduction, genetics, sorting, and chaos in biology, that are hidden by the conventional thinking of a biological society. How rich a marriage it could be – biological computing, software that breeds, machines that think and reproduce might not be the dream they are currently perceived to be.

Further Reading

Adleman, L.M. (1994) Molecular computation of solutions to combinatorial problems. *Science* 226, 1021–1024.

Conrad, M. (1990) Molecular computing. In: Yovits, M.C. (ed.) *Advances in Computers*. Academic Press, New York, USA pp. 236–324.

Conrad, M. (1992) Guest editor's introduction: molecular computing paradigms. *IEEE Computer* 25.

Dawkins, R. (1990) *The Blind Watchmaker*. Penguin, London, UK.

Holland, J.H. (1975) *Adaptation in Natural and Artificial Systems*. University of Michigan Press. Michigan, USA.

Kauffman, S.A. (1993) *The Origins of Order: Self-organization and Selection in Evolution*. Oxford University Press. Oxford, UK.

Koza, J. (1993) *Genetic Programming*. MIT Press. Cambridge, USA.

Langton, C.G., Taylor, C., Farmer, J.D. and Rasmussen, S. (1992) *Artificial Life II.* Addison-Wesley, Redwood City, USA.

Langton, C.G. (1994) *Artificial Life III*, Addison-Wesley, Redwood City, USA.

Paton, R. (1994) *Computing with Biological Metaphors.* Chapman and Hall, London, UK.

Pohl, I. and Shaw, A. (1981) *The Nature of Computation: an Introduction to Computer Science.* Computer Science Press, London, UK.

Rummelhart, D.E. and McClelland, J.L. (1986) *Parallel Distributed Processing.* MIT Press, Cambridge, USA.

Wasserman, P.D. (1989) *Neural Computing: Theory and Practice.* Van Nostrand Reinhold, Amsterdam, The Netherlands.

Sites to Visit

http://www.labs.bt.com
http://www.labs.bt.com/people/cochrap/index.htm
http://www.labs.bt.com/people/wintercs/index.htm
http://alife.santafe.edu
http://www.cogs.susx.ac.uk

Index

Figures in **bold** indicate major references.
Figures in *italic* refer to diagrams, photographs and tables.